# Excited States
# of Biopolymers

# Excited States
# of Biopolymers

Edited by
## Robert F. Steiner
*University of Maryland*
*Catonsville, Maryland*

Plenum Press • New York and London

Library of Congress Cataloging in Publication Data

Main entry under title:

Excited states of biopolymers.

Bibliography: p.
Includes index.
1. Macromolecules — Analysis. 2. Fluorescence spectroscopy. I. Steiner, Robert F.,
1929–
QP801.P64E93 1983                574.19′125                83-11220
ISBN 0-306-41316-7

© 1983 Plenum Press, New York
A Division of Plenum Publishing Corporation
233 Spring Street, New York, N.Y. 10013

Printed in the United States of America

# Contributors

**Robert M. Dowben,** Department of Pathology, Baylor University Medical Center, Dallas, Texas 75246

**Richard P. Haugland,** Molecular Probes, Inc., Junction City, Oregon 97448

**R. M. Hochstrasser,** Department of Chemistry, University of Pennsylvania, Philadelphia, Pennsylvania 19104

**Yukio Kubota,** Department of Chemistry, Faculty of Science, Yamaguchi University, Yamaguchi 753, Japan

**Tsung-I Lin,** Department of Pathology, Baylor University Medical Center, Dallas, Texas 75246

**William W. Mantulin,** Department of Medicine, Baylor College of Medicine, and The Methodist Hospital, Houston, Texas 77030

**Henry J. Pownall,** Department of Medicine, Baylor College of Medicine, and The Methodist Hospital, Houston, Texas 77030

**Robert F. Steiner,** Department of Chemistry, University of Maryland Baltimore County, Catonsville, Maryland 21228

# Preface

During the past decade, fluorescence techniques have come to occupy a position of central importance in biochemistry. Such areas as laser techniques, radiation-less energy transfer, and nanosecond fluorometry have evolved from esoteric research specialties into standard procedures that are applied routinely to bio-chemical problems. Indeed, discussion of the above three areas occupies the greater part of this book. Its level and approach are appropriate for the biological or physical scientist who is interested in applying fluorescence techniques, but is not necessarily an expert in this area.

The coverage of the literature has, in general, been selective rather than exhaustive. It is likely that what is summarized here will prove resistant to the erosion of time and provide a basis for the future evolution of this rapidly developing area of science.

Robert F. Steiner

*Catonsville, Maryland*

# Contents

## Chapter 3

Nanosecond Pulse Fluorimetry of Proteins
*Tsung-I Lin and Robert M. Dowben*

Chapter 4

The Use of Fluorescence Anisotropy Decay in the Study of Biological Macromolecules
*Robert F. Steiner*

Chapter 5

Plasma Lipoproteins: Fluorescence as a Probe of Structure and Dynamics
*William W. Mantulin and Henry J. Pownall*

Chapter 6

Fluorescent Dye–Nucleic Acid Complexes
*Robert F. Steiner and Yukio Kubota*

# Some Principles Governing the Luminescence of Organic Molecules

R. M. HOCHSTRASSER

## 1. Introduction

During the 12 years that transpired since the original article under this title appeared, little has occurred that modifies the fundamental principles governing luminescence of complex systems. However, the emphasis in the field of luminescence has changed markedly because the laser has blossomed during these years into a readily accessible tool for research. Concepts that were highly specialized ten years ago are now commonplace, so in this sense the range of principles relevant to current and future luminescence research has greatly expanded. This article is therefore aimed at enlarging the scope of the material rather than simply updating the various topics that were treated previously. Of these topics only the general discussions of spontaneous emission and transition mechanisms will be repeated here. The subject of radiationless transitions has also developed considerably, but it has been reviewed frequently during the past years (see Section 1 of the Bibliography). One area that is of growing importance is resonance Raman spectroscopy, so the relationship of this to luminescence is described fully here. The fundamental differences between coherent and incoherent optical processes is also discussed in order to introduce the reader to concepts that properly describe molecular systems that are heavily pumped by laser light sources.

The emission spectra of molecules corresponding to electronic transitions can be the result of spontaneous or induced transitions from higher to lower energy

R. M. HOCHSTRASSER • Department of Chemistry, University of Pennsylvania, Philadelphia, Pennsylvania 19104.

states. The spontaneous emission rate obtained in an experiment corresponds to the rate constant for the decay of the radiant state into *all possible channels*. These channels include not only the generation of the photons that are detected but the creation of other excited states through relaxation or energy transfer, or of new species by virtue of chemical reactions. The inverse of this emission rate constant is the fluorescence lifetime $\tau$. The lifetime $\tau$ can be measured not only by fluorescence but by any technique that detects the decay of the initially populated state or the rate of creation of other states or products that derive from it in unimolecular or pseudounimolecular processes. Thus

$$1/\tau = \sum \text{(specific rate constants for all decay channels)}$$

$$\equiv \text{rate constant for decay of initial state}$$

$$= \text{exponential appearance constant of any one channel} \qquad (1)$$

The decay channel that corresponds to the emission of photons in the region of the electronic transition frequency is characterized by a rate constant $1/\tau_r$ where $\tau_r$ is the radiative lifetime.

# 2. Spontaneous Emission

## 2.1. General Considerations

The radiative lifetime $\tau_r$ is a property that characterizes the strength of the interaction of the molecular system with radiation in a particular frequency range. It is related to the transition moment dipole for the transition involved. The molecular absorption coefficient (or extinction coefficient) is the simplest experimental measure of the interaction between the molecule and radiation, so it is natural that absorption spectroscopy may be used to measure the radiative lifetime. The key fundamental facts needed to establish this relationship between radiative lifetime and absorption coefficient involve the concepts of induced and spontaneous emission. The total rate of transitions $W_{0i}(s^{-1})$ occurring between a ground state 0 to an upper state $i$ is proportional to the incident light intensity $I(\nu_E)$ at the absorption frequency $\nu_A$. The rate of downward transitions $W_{i0}$ is proportional to the incident intensity at the emission frequency:

$$W_{0i} = B_{0i}I(\nu_A)h\nu_A; \qquad W_{i0} = B_{i0}I(\nu_E)h\nu_E \qquad (2)$$

It is now easily shown that the radiative rate $1/\tau_r$ at frequency $\nu_E$ in a medium having index of refraction $n$ (at the emitted frequency) is related to $B_{i0}$ as follows:

$$1/\tau_{\rm r} = (8\pi\nu_{\rm E}^2 n^3/c^2)B_{i0} \tag{3}$$

For a simple two-state system $\nu_{\rm A} = \nu_{\rm E}$ and, of course, $B_{0i} = B_{i0}$. The calculation of $W_{0i}$ for a single frequency is done phenomenologically so that $B_{0i}$ is known in terms of the concentration and the absorption coefficient. Before outlining these arguments it will be useful to summarize the practical units commonly used in the optical field.

The power $P$ of a light source having essentially only one frequency is measured in watts; 1 W is equivalent to $1\,{\rm j\,s^{-1}}$ or $10^{-7}\,{\rm erg\,s^{-1}}$. A pulse that lasts for a time $\tau_{\rm p}$ will have a peak power of $P = E_{\rm p}/\tau_{\rm p}$, where $E_{\rm p}$ is the energy (joules) of the pulse such that it contains $N_{\rm p}$ photons with $N_{\rm p} = 10^7 E_{\rm p}/h\nu$, where $\nu$ is the frequency ($\rm s^{-1}$ or hertz) of the laser light. The average power $\bar{P}$ of a laser corresponds to the rate at which energy can be delivered continuously. For example, a 1-W cw laser obviously has an average power of 1 W. A pulsed laser producing one 10-ps pulse of 1 mJ in every second has an average power of only $10^{-3}$ W, but each pulse has a peak power of $(10^{-3}/10^{-11}) = 10^8\ {\rm W} \equiv$ 100 MW.

The rate of excitation of a molecule by light is controlled by the size of the quantity $|\mu E_0/\hbar|^2$, where $\mu$ is the transition moment dipole and $E_0$ is the peak electric field (statvolt $\rm cm^{-1}$) of the light wave $E_0 \cos \omega t$ that is incident on the sample. The power density, or flux, of radiation is $P/A = cE_0^2 8\pi$, where $A$ is the cross-sectional area of the beam. Thus the number of transitions per second occurring for a molecular system in contact with such radiation is proportional to $P/A$. In this article we will use the light intensity $I$ in units of photon $\rm s^{-1}\ cm^{-2}$, so that $I = 10^7 P/Ah\nu$, or, for a pulse of length $\tau_{\rm p}, I = E_{\rm p}/Ah\nu\tau_{\rm p}$.

Strickler and Berg (see Section 2 of the Bibliography) have given a method of obtaining $\tau_{\rm r}$ when the mean absorption and emission frequencies are not equal, as will generally be the case in polyatomic spectra. In that case, the induced absorption coefficient is $\Sigma_{\nu'} B_{\nu'} (i \leftarrow 0)$, where $\nu'$ represents excited-state vibrational quantum numbers, whereas the spontaneous emission rate is $\Sigma_{\nu''} A_{\nu''}$ $(i \rightarrow 0)$, where $\nu''$ are ground-state vibrational quantum numbers. The problem is readily solved for the case where all the emission originates on a single excited level $i$, and all the absorption originates from a single ground-state level 0. This is a useful approximation for molecules in condensed media (but not in gases) where rapid internal conversion to a rather narrow distribution of levels associated with the lowest excited state normally occurs prior to any significant emission. The lifetime of the emission is then given by

$$\frac{1}{\tau_{\rm r}} = \frac{2.88 \times 10^{-9} n^2}{\langle \nu_{\rm E}^{-3} \rangle} \int_{\rm Band} \epsilon(\bar{\nu})\, d(\ln \bar{\nu}) \tag{4}$$

To obtain (4) we have assumed no dispersion in the refractive index $n$. The integral is to be taken over the whole absorption band corresponding to the electronic state that will ultimately emit; $\langle \nu_E^{-3} \rangle$ is the mean value of $\nu_E^{-3}$, and $\bar{\nu}$ is measured in $cm^{-1}$ ($\bar{\nu} = \nu/c$). Other approximate expressions for the lifetime are

$$\frac{1}{\tau_r} \approx \frac{2.9 \times 10^{-9} n^2}{\langle \bar{\nu}_E^{-3} \rangle \bar{\nu}} \int \epsilon(\bar{\nu}_A) d\bar{\nu}_A \approx f \bar{\nu}_{PE}^2 \tag{5}$$

where $f$ is the dimensionless oscillator strength for the transition, and $\bar{\nu}_{PE}$ is the peak emission wave number ($cm^{-1}$). The quantity $\int \epsilon(\bar{\nu}_A) d\bar{\nu}_A$ is the area under the absorption band of the emitting state expressed as molar extinction versus wave number.

## 2.2. Luminescence from Nearby States

The decay of emission from a group of closely spaced states is a fairly common occurrence in condensed-phase spectra. Often the medium can be considered to provide a quasi-continuum of levels such as those derived from lattice vibrations of any type — in ordered or in random media — and we may assume that an equilibrium distribution over excited states can be maintained under conditions of steady excitation and at temperatures far from $0°K$. The realization of this equilibrium depends on the spontaneous emission rates being much slower than the rate of exchange of energy between excited molecules and the lattice. In these circumstances, the observed lifetime behavior will be always exponential, with $1/\tau$ dependent on temperature if more than one electronic state of the system is thermally accessible. This is a common situation in condensed phases since there are usually a number of optically accessible electronic states present in a relatively narrow energy region. Under conditions of low spectral resolution ($\sim kT$), the emission spectra at different temperatures would be quite similar. The same considerations apply in understanding the luminescence spectra of molecules having thermally accessible nonradiative pathways: The decay would be exponential with a half-life corresponding to a Boltzmann averaged mean of the rate constants. It follows that molecules having nearby states (a few $kT$) should show an exponential decay of emissions at all temperatures but with a lifetime that depends on the temperature.

## 2.3. Multiple State Decay

In condensed phases molecules are rapidly exchanging energy with the surrounding medium. The relaxation by the medium is extremely fast compared with $1/\tau_r$, the radiative rate, and for most molecules — certainly for all fluorescent systems — it is fast compared with $1/\tau$. This means that the interaction with the medium maintains the system of excited molecules in quasi-thermal equilibrium on the time scale of the luminescence. The observed lifetime of the fluorescence is therefore temperature dependent in principle. The molecular states between which energy can be exchanged sufficiently rapidly under the influence of the medium can be of many different types. The exchange of vibrational energy with the medium is always fast enough for this equilibrium to be established in the case of normally fluorescent molecules, and in certain cases the electronic relaxation processes induced by the medium are also sufficiently rapid. At normal temperatures we are usually concerned with states within a few $kT$ of the coolest excited molecules.

In a system having many excited levels $i$ with populations $n_i(t)$ at time $t$, the equilibrium assumption implies

$$n_i(t) = (1/Z) \exp(-E_i/kT) N(t) \tag{6}$$

where $N(t)$ is the total number of excited molecules in the system at time $t$, and $Z$ is the partition function. The observed lifetime (exponential decay) is therefore given by

$$1/\tau = (1/Z) \sum_i k_i \exp(-E_i/kT) \tag{7}$$

where $k_i^{-1}$ is the lifetime of level $i$. At sufficiently high temperatures the value of $1/\tau$ becomes simply the average of the emission rates for the individual levels. The lifetime $\tau$ is observed in experiments that probe any of the populations $n_i$, or all of them. It is apparent that it will be necessary to understand the variations in $k_i$ that occur in particular systems in order to be able to predict the temperature dependence of the lifetime. For many systems the variations of lifetime among vibrational levels with frequencies less than ca. 1000 cm$^{-1}$ are not large enough to cause significant temperature effects on the observed lifetime. On the other hand, there are many cases where the relevant information about individual vibronic levels of the molecule is just not available. When different electronic states are nearby and the system is able to undergo transitions between them under the influence of the medium, then the effects of temperature on the exponential decay time of luminescence are often more dramatic.

# 3. Molecular Luminescence Characteristics

A spectroscopist who deals frequently with the luminescence of organic molecules would seldom have difficulty estimating the expected natural lifetime of emission from almost any molecule provided he could guess the orbital characteristics of the emitting state. These order-of-magnitude estimates turn out to be very reliable, indicating that the qualitative theoretical ideas and the experimental experience on which they depend are reliably connected. The theoretical basis is concerned with three main points.

(a) The vast majority of organic molecule (and other) luminescences involve the spontaneous emission of dipole radiation. As such, the probability of emission is proportional to the electric dipole matrix element connecting the initial and final states.

(b) Certain properties of the low-energy states of organic (and other) molecules are apparently quite well described by rather simple molecular-orbital approximations. Thus it appears that the emissive states of most molecules can be considered, to a first approximation, to have specific atomic-orbital character-istics. Indeed, it turns out that two electron configurations are often quite suitable to obtain reliable estimates of certain properties, hence the wide usage of terms such as $\pi\pi^*$, $n\pi^*$, $\sigma\pi^*$, $\pi\sigma^*$, $\sigma\sigma^*$, charge transfer, and so on.

(c) The absorption or extinction coefficients of most transitions can be guessed within an order of magnitude or so from knowledge of the transition type. Simple MO calculations based on two electron pictures yield the extinction coefficients to within a factor of 5–10.

The effect of nuclear motions (molecular vibrations) on the simple configu-rations, and hence on the transition moment, must be taken into account. The spin–orbit interaction must be considered to account for transitions between states having different total electron spin. It is the essence of the perturbation approach that these interactions, all of which lead to mixtures of configurations for the states of interest, can be estimated qualitatively by employing relatively simple features of the orbitals in the two-electron configurations that were given as the model first approximations.

## 3.1. The Transition Dipole Moment

The transition dipole moment $\mathbf{m}_{0i}$ is given by

$$\mathbf{m}_{0i} = e \int \psi_i \psi_0^* \mathbf{r} \, d\mathbf{r} \tag{8}$$

$\mathbf{m}_{0i}$ is the dipole moment of the charge density distribution $e\psi_i\psi_0^*$ – known as the transition density – where $\psi_i$ and $\psi_0$ are the wave functions for the excited state and ground state, respectively. The position vectors of all the electrons are lumped together in the factor $\mathbf{r}$, so that $\mathbf{r} = \Sigma_n \mathbf{r}_n$ for an $n$-electron system, and $\mathbf{r}_n = x_n\hat{i} + y_n\hat{j} + z_n\hat{k}$. The substitution of two-electron configu-rations for $\psi_i$ and $\psi_0$ can give an approximate idea of the magnitude of $\mathbf{m}_{0i}$. A zero or nonzero judgment can be made by inspecting the multipole structure of the transition density, since unless the transition density has the structure of a dipole, the transition will be forbidden. Sometimes it is only necessary to examine the multipole character of $\psi_i\psi_0^*$ at a single atomic center in the molecules. Such is the case when the initial or the final state involves a mostly localized molecular orbital, for example, a nonbonding orbital. Some common transition-density multipoles are: dipole for $\psi_0 = s$, $\psi_i = p$ and for $\psi_0 = \pi$, $\psi_i = \pi^*$; quadrupole for $\omega_0 = p_x$, $\omega_i = p_y$ on the same center. If the molecule has no symmetry, then all multipole components of the transition density exist in principle. These local transition-density concepts may be used successfully with molecules that have no symmetry if the electronic transition is a localized one. In the spirit of the correspondence principle, the transition moment may be regarded as the oscillating dipole moment that gives rise to the emitted electromagnetic field.

The relationship between the transition dipole moment $\mathbf{m}_{0i}$ and experimental spectroscopic parameters is as follows:

$$|m_{0i}|^2 = \frac{9.2 \times 10^{-39}}{n} \int \epsilon \, d(\ln \lambda) \qquad (9)$$

The units of $\mathbf{m}_{0i}$ are esu cm. The relationship between $\mathbf{r}$ and $\mathbf{m}_{0i} = e\mathbf{d}_{0i}$ involves the frequency:

$$1/\tau \approx 7 \times 10^{-6} \langle \nu_E^{-3} \rangle^{-1} d_{0i}^2 \approx 10^{-5} \nu_E^3 d_{0i}^2 \qquad (10)$$

where $d_{0i}$, the dipole length, is expressed in Å and $\bar{\nu}_E$ is expressed in cm$^{-1}$. For fairly symmetric, narrow emission spectra, $\langle \bar{\nu}_E^{-3} \rangle^{-1}$ is roughly $\bar{\nu}_E^3$, the cube of the mean emission frequency (expressed in cm$^{-1}$). Thus, for usual organic solvents and near ultraviolet emission, the lifetime is given approximately by $1/\tau = 4 \times 10^8 d_{0i}^2$. Thus we see that $d_{0i}$ is approximately 1 Å for a $\tau$ of approximately 3 ns.

The transition-moment direction for a molecular system, once understood with reference to the molecular framework, can be used in structural determinations of many types. Unfortunately, unless a molecule has certain symmetry characteristics, the transition-moment direction is not readily obtained. When there are symmetry planes or axes, the molecular transition moments must lie parallel or perpendicular to them so that their directions are readily obtained experimentally. In the next section some of the principles governing the determination of transition moments in unsymmetric molecules using fluorescence methods are described. As it happens, many of the interesting chromophores that occur in biological systems have little or no intrinsic symmetry.

## 3.2. Determination of Transition-Moment Directions from Fluorescence

The fluorescence from molecules in condensed phases mainly results from transitions to the ground state from the lowest excited state. Thus the emitted light is always generated by a dipole characteristic of that lowest state. If more than one state of the system is involved in the absorption process, then the dipole emission intensity will be the appropriately weighted sum of the signal obtained from each absorbing state considered separately. For example, the state could be degenerate, or the molecule could have two quite different electronic states in the same energy region.

In order to properly connect experimental fluorescence information based on laboratory coordinates to the molecular axes in a system that is initially a random distribution, it is necessary to have prior knowledge of the direction of the transition moment in the molecular frame responsible for the initial light absorption. For low-symmetry molecules such information is seldom available. However, it is possible to accomplish fluorescence-polarization measurements with essentially fixed distributions of molecules by using rigid systems or through the excitation of fluid systems by means of picosecond or even better with subpicosecond pulses.

We now consider a typical problem in fluorescence polarization where the absorption occurs from a ground state (0) to two electronic states of the molecule, $a$ and $b$, but the emission occurs from a single state, $a$. The extinction coefficients for these two states at a particular wavelength are $\epsilon_a(\lambda)$ and $\epsilon_b(\lambda)$. The emission intensity is the product of the probabilities of absorption and reemission. We assume that the transition moment to the state $b$ is located $(\pi/2 - \alpha)$ from the $0 \rightarrow a$ transition moment. In a molecule having axes of symmetry, $\alpha$ would be zero or $\pi/2$ in the case where the moments for $0 \rightarrow a$ and $0 \rightarrow b$ transitions were perpendicular or parallel. The emitted signal for choice of polarization $\hat{i}$ for the incident light, and $\hat{e}$ for the emitted light, is $\{\epsilon_a|\hat{i} \cdot \hat{\mu}_a|^2 + \epsilon_b|\hat{i} \cdot \hat{\mu}_b|^2\}|\hat{\mu}_a \cdot \hat{e}|^2$. In this relation $\hat{\mu}_a$ and $\hat{\mu}_b$ are unit vectors defining the directions of the dipole moments for transitions from the ground state to states $a$ and $b$. The conventional experimental arrangement involves setting the polarizer and the analyzer for $90°$ scattered light at the laboratory direction $Z$ and comparing the signal with that obtained by rotating the analyzer through $\pi/2$. The ratio of the latter to the former is the depolarization ratio $\rho$, given by

$$\rho(\lambda) = (\beta + 2)/(3\beta + 1) \tag{11}$$

where $\beta = (1/\cos^2 \alpha)[\epsilon_a(\lambda)/\epsilon_b(\lambda) + \sin^2 \alpha]$. Since $\beta$ is measured in the experiment, the wavelength dependence of $\epsilon_a(\lambda)/\epsilon_b(\lambda)$ can be obtained (because $\alpha$ is considered a constant) by making measurements of $\rho$ for different excitation wavelengths. The result (11) yields the well-known values of $\rho = 1/3$ for $\epsilon_b = 0$ ($\beta \to \infty$) and $\rho = 2$ for the case $\epsilon_a = 0$, $\alpha = 0$.

A similar problem arises when a molecule has two states $a$ and $b$ which not only can both absorb the exciting light but also can both fluoresce. When the two emitting states are maintained in thermal equilibrium during the emission process, one cannot of course distinguish them by time gating the fluorescence (cf. Section 2.4). In this case the fraction of the emission that occurs as if from state $a$, namely $\alpha_a$, is determined by the temperature and the wavelength at which the fluorescence is detected. The observed signal is proportional to $\{\epsilon_a|\hat{\mu}_a \cdot \hat{i}|^2 + \epsilon_b|\hat{\mu}_b \cdot \hat{i}|^2\}\{\alpha_a|\hat{\mu}_a \cdot \hat{e}|^2 + (1 - \alpha_a)|\hat{\mu}_b \cdot \hat{e}|^2\}$ averaged over the random ensemble of emitters. The result for $\rho(\lambda)$ is:

$$\rho(\lambda) = \frac{3(1 - \beta)(1 - \alpha_a)\cos 2\alpha + \beta(3 - \alpha_a) + (3 + \alpha_a)}{4(1 - \beta)(1 - \alpha_a)\cos 2\alpha + 2\beta(2 + \alpha_a) + 2(2 - \alpha_a)} \tag{12}$$

Clearly equation (12) for the depolarization ratio transforms into (11) when there is only one emitting state (i.e., when $\alpha_a = 1$). The extreme values of $\rho$ are now simple functions of $\alpha$ and $\alpha_a$.

The results (11) and (12) arise from our assumption that the absorption spectrum consists of two independent transitions $0 \to a$ and $0 \to b$. Such would be the case if the levels of two electronic states overlapped but did not interact. In the case where the two sets of levels interact, a 'mixed state' is excited by the light, in which a superposition of levels from $a$ and $b$ are first created. If the initially excited state relaxes to a pure '$a$' state before emitting the depolarization ratio (say for $\alpha = \pi/2$) is the same as given in (11) except that $\epsilon_a/\epsilon_b$ is replaced by $(\epsilon_a/\epsilon_b)(C_a^2/C_b^2)$ where $C_a^2$ and $C_b^2$ measure the probability of occurrence of the $a$ and $b$ states in the superposition state. In the case of chromophores like tryptophan having overlapping and interacting $L_a$ and $L_b$ bands the factor $C_a^2/C_b^2$ may have to be incorporated. The effects of this subtlety will show up in the excitation wavelength dependence of the depolarization ratio.

## 3.3. Polarization of Fluorescence from Crystals

In crystals the molecules, and hence the transition-moment vectors, all have a definite angular relationship to the axes of the crystal. The anisotropy of the fluorescence of crystals therefore yields direct information on the direction of the transition moment of the emitting state. While the molecules are arranged in a known fashion in crystals, their proximity can introduce some complications in the interpretation of the fluorescence data.

The exchange of excitation energy between molecules in a crystal results in the formation of energy bands (exciton bands). Excitation of the crystal at normal temperatures usually leads to essentially localized excitations which then hop from molecule to molecule. This situation is always expected to prevail when $kT$ is comparable with the bandwidth of the excitons, and for many molecular systems this is likely to be an appropriate picture at $300°K$. It follows that the emission intensity from the crystal can be considered to be the average of the intensity of emission from each molecule in the system. To the extent that molecules are related by symmetry transformations of the crystal, they will each have the same probability of emitting light. If there is more than one molecule in the asymmetric unit of the crystal, then the luminescence of the crystal will originate from each of these sublattices in accordance with its thermal population. Generally, in a crystal-phase experiment the irradiated crystal is oriented in the laboratory such that light emitted perpendicular to a principal face is analyzed for its state of polarization referred to the crystal-axis system. The polarization ratio $\rho_{ij}$ is in this case just the ratio of the intensity of light emitted polarized parallel to any chosen pair of crystal axes $i$ and $j$, so that $\rho_{ij} = \mu_i^2/\mu_j^2$ ($\mu_i$ is the transition moment projected onto the $i$th crystal axis). If the sublattices are located at energy $E_n$ (this is the so-called asymmetric site splitting), and $\theta_{ni}^{\alpha}$ is the angle between the transition moment in the $\alpha$th molecule in the unit cell and the $i$th crystal axis for sublattice $n$, $\rho_{ij}$ would be

$$\rho_{ij} = \sum_{\alpha, h} \cos^2 \theta_{ni}^{\alpha} \exp\left(-E_n/kT\right) \bigg/ \sum_{\alpha, h} \cos^2 \theta_{nj}^{\alpha} \exp\left(-E_n/kT\right) \qquad (13)$$

For the common case of only one molecule in the asymmetric unit we can omit the index $n$, and (13) reduces to $\rho_{ij} = \Sigma_{\alpha} \cos^2 \theta_i^{\alpha}/\Sigma_{\alpha} \cos^2 \theta_j^{\alpha}$. The angles $\theta_i^{\alpha}$ and $\theta_j^{\alpha}$ are obtained from the crystal-structure data, or conversely $\rho_{ij}$ may be used to locate the direction of the transition moment. This latter process usually does not lead to a unique determination of the direction in the molecular frame.

The measurements actually provide the values of $\mu_i^2$:

$$\mu_i^2 = 1 \bigg/ \sum_j \rho_{ji} \qquad (14)$$

It follows that two possible values of $\mu_i$, namely $\pm |\mu_i|$, are found for each value of $i$, so that four distinctly different directions in the crystal give rise to the same set of values of $\rho_{ij}$. What is measured in the fluorescence experiment are $\mu_a^2, \mu_b^2$, and $\mu_c^2$* determined up to a constant value, so that $\mu_a^2 + \mu_b^2 = 1 - \mu_c^2$. The four

---

* The symbols $a$, $b$, and $c$ as used here are three mutually perpendicular principal directions of the crystals. In the case of monoclinic and triclinic crystals the crystallographic axes $a$, $b$ and $c$ are not mutually perpendicular and should not be confused with the designations used here. We also assume, for convenience, that the $b$ direction is a symmetry-determined principal axis so that $\cos \theta_{xb}^{\alpha} = \cos \theta_{xb}^{\beta}$ for molecules $\alpha$ and $\beta$ in the unit cell.

results for $\hat{\mu}$ in the crystal axis system are given by

$$\hat{\mu} = |\mu_a|\hat{a} \begin{pmatrix} + \\ + \\ - \\ - \end{pmatrix} |\mu_b|\hat{b} \begin{pmatrix} + \\ - \\ + \\ - \end{pmatrix} |\mu_c|\hat{c} \tag{15}$$

The transition moment to be determined, when written in the coordinate frame of the $\alpha$th molecule in the unit cell, is $\hat{\mu}_\alpha = \mu_x\hat{x}_\alpha + \mu_y\hat{y}_\alpha + \mu_z\hat{z}_\alpha$. A determination of $\hat{\mu}_\alpha$ would allow specification of the transition-moment vector, but it would not be known which of the systems of axes ($\alpha$) it referred to. The quantities $\mu_a^2, \mu_b^2$, and $\mu_c^2$ derive from all the differently oriented groups of molecules in the crystal. However, if the experiment is carried out with polarizations parallel to the principal axes (these are parallel and perpendicular to the symmetry axes and planes of the lattice), each group of molecules contributes to the same extent. In such an experiment $\mu_i = \mu_x \cos \epsilon_{xi} + \mu_y \cos \epsilon_{yi} + \mu_z \cos \epsilon_{zi}$. In many cases it is possible to assume that the transition moment is confined to the $xy$ plane of the molecule $\alpha$ (written as $x_\alpha y_\alpha$). The molecular transition moment then takes the form

$$\hat{\mu}_\alpha = \frac{(\mu_a \cos \theta_{yb} - \mu_b \cos \theta_{ya}^\alpha)x_\alpha + (\mu_b \cos \theta_{xa}^\alpha - \mu_a \cos \theta_{xb})y_\alpha}{(\cos \theta_{xa}^\alpha \cos \theta_{yb} - \cos \theta_{xb} \cos \theta_{ya}^\alpha)} \tag{16}$$

This defines the components $\mu_x$ and $\mu_y$: The angle that $\hat{\mu}$ makes with $y$, say, is $\tan^{-1}(\mu_y/\mu_x)$. It must be remembered that the plane $x_\alpha y_\alpha$ is not parallel to the plane $x_\beta y_\beta$, so that confining the transition moment to a plane such as $x_\alpha y_\alpha$ [as in equation (16)] provides a unique choice of $\hat{\mu}$ for given values of $\mu_a, \mu_b$, and $\mu_c$. The only ambiguities that remain concern the different sign possibilities for $\mu_a$ and $\mu_b$. In general there are *two* different directions of $\hat{\mu}$ in the $xy$ plane due to this ambiguity, as is easily seen by substituting the values of $|\mu_i|$ from (15) into (16).

The crystal method of obtaining transition-moment directions certainly appears to be the method of choice. The ambiguity described above can easily be resolved by studying the chromophore in a number of different crystal structures. Most amino acids and nucleotides can be found in many different structures corresponding to different derivatives in which the chromophore is not perturbed significantly. Since energy is transferred throughout crystals, the possibility of trapping by impurities must always be considered. Such effects are lessened by carrying out the experiments at the highest feasible temperature.

# 4. Principles of Luminescence Experiments Carried Out with Lasers

What are the physical phenomena that are important to consider when lasers are used for experiments? The unique characteristics that can be obtained with lasers are:

(1) Narrow spectral bandwidth and high average power per unit bandwidth (cw lasers)
(2) Narrow spectral bandwidth and very high peak powers (pulsed lasers)
(3) Extremely brief pulses of light of virtually any duration down to a few tenths of a picosecond.

## 4.1. Nonlinear Processes and Optical Pumping

In a linear process such as conventional light absorption, the same number of transitions will occur whether the laser is continuous (cw) and applied for a certain time, or whether it is pulsed with the same number of photons being absorbed. Of course the population of excited states that are produced after irradiation may not be the same in these two situations because the opportunities for decay are quite different in the two cases. In a nonlinear process the effects of slow and fast optical pumping are not so similar, and processes become dependent on the peak power of the light source. These points are made clear in the following example, common in luminescence experiments, of two states 1 and 2 coupled by radiation at frequency $\nu_{21} = (E_2 - E_1)/nh$. Suppose $n$ photons are required to effect the transition between the two states so that the upper-state population changes in time according to

$$\dot{n}_2 = P^n n_1 \tag{17}$$

where $n_1$ and $n_2$ are the populations. First we consider the case where $n_1$ is not changed much by the interaction with light so that $n_1(\tau) \simeq n_1(0)$. Thus, if there were no relaxation, the number of upward transitions would be given by the population $n_2$ after time $t$:

$$n_2(t) = n_1(0) \int_0^t P^n(t') dt' \tag{18}$$

In the case $n = 1$, the integral is simply the pulse energy $E_p \approx P\tau_p$. The effect of having $n > 1$ is readily adduced by supposing a pulse shape having a specific analytic form. For example, for a pulse $(E_p/\tau_p) \exp(-t/\tau_p)$ having a width $\tau_p$ and energy $E_p$ we find

$$n_2(t \gg \tau_p) = n_1(0) \cdot E_p P^{n-1}/n \tag{19}$$

so that the generated population at times long compared with the laser pulse depends sensitively on the *peak power* of the pumping laser.

The optical pumping of a two-level system is a problem of great interest in photophysics, and the results, necessary in order to deal with almost all laser experiments, are not always described simply as described above. One approach that is often used in laser-induced experiments is that of conventional rate equations which, for the case where the excited state can relax with lifetime $\tau$ and the absorption cross section is $\sigma(cm^2)$, are

$$\dot{n}_2 = I\sigma n_1 - (1/\tau + I\sigma)n_2 \qquad (20)$$

$$\dot{n}_1 = I\sigma n_1 + (1/\tau + I\sigma)n_2 \qquad (21)$$

The appropriateness of these particular rate equations depends on a number of factors. In the first place the decay of the excited state is considered to complement the repopulation of the ground state. This will be an excellent approximation when the bottlenecks in the relaxation $2 \rightarrow 1$ all have lifetimes much shorter than $\tau$. However, the excitation of molecules in solution seldom meets this criterion. The more common situation is that the pump radiation couples the ground state to a set of levels that have very short lifetimes determined by vibrational relaxation into a bottleneck state. Under these circumstances the stimulated emission terms, $I\sigma n_2$, can be dropped from each of the equations to yield

$$\dot{n}_2 = I\sigma n_1 - (1/\tau)n_2 \qquad (22)$$

$$\dot{n}_1 = I\sigma n_1 + (1/\tau)n_2 \qquad (23)$$

The populations $n_1$ and $n_2$ are now those of the *ground state* and the *bottleneck state*. This bottleneck state is often the lowest singlet or triplet excited state. The quantity $\sigma$ is the absorption coefficient at the laser frequency. When very short light pulses are used, these relations may be invalid because they depend on neglecting population in the initially pumped levels for which the inequality $I \ll 1/\tau_v\sigma$ must hold. The cross section $\sigma$ is defined for $N$ molecules per $cm^3$ by Beer's Law $I = I_0 \exp\{-\sigma Nx\}$, and in terms of the molar extinction coefficient $\epsilon$ it is $\sigma(cm^2) = 3.82 \times 10^{-21}\epsilon$. For a vibrational relaxation time of $10^{-12}$ second of the pumped state the light intensity must be kept less than $2.6 \times 10^{32}/\epsilon$ photon $sec^{-1}$ $cm^{-2}$. This inequality is met for cw excitation sources except in the extreme conditions of exceptionally high power (say 10 W) focused to a spot having a few microns diameter into a material having a molar extinction coefficient of ca. $10^5$. It is evident that with a conventional 10-ns pulse having an energy of 10 mJ focused to a diameter of 0.2 mm $I(= 2.7 \times 10^{27}$ photon $sec^{-1}$ $cm^{-2})$ is perilously close to $1/\tau_v\sigma$ for moderately large extinction coefficients, and for picosecond laser pulses equations (22) and (23) may be inappropriate unless the pulse energies are kept in the microjoule regime. (However, see Section 5, which deals with coherent effects.) The solution to these equations for a system having $N$ molecules per cm$^3$ is

$$n_2(t) = \sigma \int_{-\infty}^{t} d\gamma' \, I(\gamma')(N - n_2) \exp\left[-(t - \tau')/\tau\right] \qquad (24)$$

Generally, this can be solved iteratively to obtain successive approximations for $n_2$. For the case where $I(\tau)$ is a square pulse of duration $\tau_p$ and intensity $I_0$ the exact solution for $n_2(t)$ is

$$n_2(t) = \left\{\frac{N}{1 + 1/I_0\sigma\tau_p}\right\} \left[\exp\left(I_0\sigma + 1/\tau\right)\tau_p - 1\right] \exp\left[-(I_0\sigma + 1/\tau)\right]t \qquad (25)$$

A common situation with short laser pulses is $\tau_p < \tau$, so that the fraction of molecules that become excited after an irradiation time of $\tau_p$ is given by

$$n_2/N = \left[1 - \exp\left(-I_0\sigma\tau_p\right)\right]/(1 + 1/I_0\sigma\tau_p) \qquad (26)$$

For high light intensity (or $I_0\sigma\tau_p \gg 1$) the ratio becomes unity, representing the situation known as complete *bleaching* of the sample. When a sample is significantly bleached by the exciting light, the fluorescent signal is no longer proportional to the light intensity of the pump: At the extreme limit of complete bleaching, increases of the intensity have no further effect on the fluorescence signal. If the lifetimes of the initially pumped levels are *not* sufficiently small that their populations may be neglected, then, from (20) and (21), excitation by a steady intensity for time $\tau_p > \tau$ results in a fraction of excited molecules given by

$$n_2/N = \left[1 - \exp\left(-2I_0\sigma\tau_p\right)\right]/(2 + 1/I_0\sigma\tau_p) \qquad (27)$$

which tends to $1/2$ as $I_0\sigma\tau_p$ becomes greater than unity. This equalization of populations of ground and excited states is termed *saturation*. This effect is common in gases and low-temperature solids where, in a good approximation, the populations of only two molecular levels are influenced by the laser field. However, for conventional condensed phase systems the two levels are often coupled to many other molecular states by various rapid relaxation pathways so that bleaching – and not saturation – occurs in strong light fields.

Complex molecules of the aromatic and heteroaromatic type have the characteristic that the energy separations between neighboring electronic excited states are significantly less than the separation of the excited states from the ground state. Thus the conventional electronic spectra of such molecules do not usually begin until the near ultraviolet (ca. 4 eV), whereas the excited-state spectra always occur in the near infrared or visible region of the spectrum. On the other hand, the first ionization potentials of these molecules are in the range 7–9 eV, so that many conventional excited-state spectra must involve transitions to states above the ionization potential. Furthermore, bond dissociation energies of ca. 5 eV are readily achieved by the successive absorption of two photons from near UV or visible lasers. These remarks underline the possible importance of nonlinear processes in laser-excited experiments, since most molecules have states at twice and three times the incident laser energy.

In a situation such as that described in Figure 1 the efficiency of some chemical process having a rate constant $k_{3c}$ and requiring the absorption of two photons may become significant. The critical factor is whether or not the exciting light pulse is much shorter than the lifetime of the state 2, pumped in the one-photon absorption step. Certainly this is a common situation when picosecond pulses are used in experiments, and under these circumstances the populations created initially by the laser pulse can be seen to be in the ratio $n_1 : n_2 : n_3 = 1 : (N_p/A)\sigma_1 : (N_p/A)^2 \sigma_1 \sigma_2$, where $(N_p/A)$ is the number of photons per unit area of the pulse, and $\sigma_1$ and $\sigma_2$ are the absorption coefficients for the $1 \rightarrow 2$ and $2 \rightarrow 3$ transitions. Care must be taken during experiments to keep the dimensionless quantity $(N_p/A)\sigma$ less than unity, otherwise the population is unavoidably concentrated in level 3 and multiphoton processes may occur with high efficiency. Even with moderate pulse energy and average laser spot sizes, $(N_p/A)\sigma$ is greater than unity. For example, with $10^{14}$ photons focused to a 0.25 mm spot, and $\sigma_1 = \sigma_2 = 4 \times 10^{-17}$ cm$^2$, we find $(N_p\sigma/A)^2 = 3$, so that 53% of the total population in the three levels immediately after pumping resides in level 3! Most picosecond experiments utilize pulses carrying more photons than used in this example. Thus photochemical processes, transient species, and fluorescence decays must be dominated initially by properties of states other than those excited by one photon absorption. The situation is totally altered when systems are pumped by laser pulses that are longer than the relaxation times. When the relaxation time of state 2 is $\tau$ and the width of the

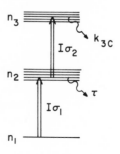

FIGURE 1. Schematic diagram of a molecular system having two excited states (2 and 3) both of which undergo relaxation. $\sigma_1$ and $\sigma_2$ are the absorption coefficients for transitions $1 \rightarrow 2$ and $2 \rightarrow 3$, respectively; $I$ is the light intensity (photon s$^{-1}$ cm$^{-2}$), so that $I\sigma$ is the rate constant for transitions between levels.

pulse $\tau_p$ is greater than $\tau$, then the ratio of the populations in levels 3 and 2 immediately after excitation is no longer $(N_p/A)\sigma$ but is reduced to $(N_p/A)\sigma(\tau/\tau_p)$.

# 5. Coherent Interactions of Molecules and Light

The equations given in the previous section are not useful for describing coherent properties of the system. By a coherent effect we mean one that requires that a macroscopic polarization wave be present in the sample. When an electromagnetic field $E(t) = E_0 \cos \omega t$ is incident on a sample, it induces a dipole moment, or coherent polarization, $P(t)$. The generation of electromagnetic fields by the sample (for example, the emergence of light at the other end of the cell) is understood to arise from the existence of $P(t)$. In effect $P(t)$ is a polarization wave with components at frequency $\omega$ and is therefore a true source of electromagnetic fields. Obviously the coherent polarization of the medium is built up from the dipoles induced in separate molecules oscillating in a reinforcing manner. Thus $P(t)$ may be diminished if molecules are caused to undergo interactions which are different in their strength or timing for the different molecules in the ensemble. An example of such an interaction is the random buffetting of molecules by surrounding solvent. So-called random stochastic interactions of this type give rise to an irreversible diminution of the coherent polarization and hence convert the energy of the electromagnetic field into a population of excited molecular states. *It follows that the rate of absorption of light by a sample to produce a population is determined by a time characteristic of coherence-loss or dephasing processes.* The characteristic coherence-loss time in a homogeneous system is usually given the symbol $T_2$. The $T_2$ is a function of both the population decay time (or fluorescence lifetime) *and* a pure dephasing time, since the coherent polarization is clearly diminished by both of these processes. When $T_2$ is very small compared with the laser pulse width, it is not necessary to consider coherent processes, since all the coherence is lost on a time scale that is short compared with the quantities of experimental interest. This will normally be the situation in solutions at ambient temperatures that are optically pumped with laser pulses longer than ca. 1 ps. In such circumstances the $\sigma$ in equations (22) and (23) automatically incorporates the value of $T_2$. Since the linewidth of the optical absorption is $1/T_2$, the absorption coefficient at the peak is simply proportional to $T_2$. Since $T_2$ changes with temperature, so also does $\sigma$ and the pumping dynamics.

In situations where $\tau_p < T_2$ defines the interesting experimental region, the equations developed in Section 4.1 are not even approximately valid. The inequality $\tau_p < T_2$ can be achieved in liquid solutions by employing subpicosecond pulses, and in low-temperature systems where $T_2$ often becomes longer than conventional fluorescence lifetimes. In this case the level populations must be calculated from coupled differential equations involving both the coherent amplitude and the populations, rather than just the populations as in the rate-equation approach of equations (20) to (23). These coupled equations including the coherences are called the optical Bloch equations, and they are quite analogous to the Bloch equations used to describe coherence dynamics in NMR. In the regime where coherences are important, the incident light pulse may cause the system to oscillate between its ground and excited state so that the fluorescence from the system also undergoes an oscillation. There are many optical techniques based on coherent effects that will surely become of increasing importance in biology, so a few review articles on these and related topics are included in the bibliography.

## 5.1. The Distinctions between Fluorescence and Resonance Raman Effects

The study of fluorescence emission and lifetimes becomes more complex when laser sources are used in experiments. How does one know that the observed radiation is due to fluorescence and not to the Raman effect? The answer to this question is concerned with the fundamental understanding of luminescence phenomena. The Raman process is a direct transition between the initial state, usually the ground state, and the final state, usually a vibrational state, accompanied by the omnidirectional emission of photons having energy equal to that of the incident field less the molecular transition energy. In such scattering it does not matter what is the value of the laser frequency in relation to the molecular resonance frequency. These excited states near the laser frequency play no role in the Raman process other than to influence the scattered light intensity associated with particular modes. The Raman scattering is most efficient when the laser is closest to resonance with electronic transitions. In fact, with a small frequency mismatch $\Delta$, the scattered light intensity is proportional to

$$|\mu_{00}'\mu_{0'v}|^2/(\Delta^2 + \Gamma^2) \tag{28}$$

where $\Gamma = 1/T_2$ is the halfwidth of the optical transition. The transition-moment factor is precisely the same as that involved in fluorescence spectra, so obviously there are no differences to be expected in the spectroscopic selection rules for the two processes. In relation (28) the quantity $2\Gamma$ is the full width at half maximum of the optical-absorption spectrum. The Raman lines themselves have a width determined not by $\Gamma$ but by the relaxations in the vibrational states of the molecule. Thus they have widths similar to those found in the infrared spectrum. The fluorescence lines, however, have a width that includes $\Gamma$. Thus the actual lineshape of emission from a sample is a composite of a narrow (Raman) and a broad (fluorescence) part, each centered at the same frequency for the case of excitation at exact resonance ($\Delta = 0$). A reliable pictorial model is presented in Figure 2 in which the excitation process is shown as occurring in two steps. The first step produces a polarization $P(t)$ in the medium. The polarized system can spontaneously radiate, and this is termed Raman scattering. It can be dephased to yield a conventional population $n_2(t)$ of excited states. The emission of this population is termed fluorescence. As before, dephasing is meant to imply the going out of phase of the oscillating-induced dipoles on different molecules in the ensemble. The polarization is made up from those induced dipoles that maintain an in-phase relationship to the driving field. The relative proportions of dephased to in-phase oscillating dipoles is determined in the steady state by the rate of the random stochastic process acting on the molecular-transition frequency. Such $T_2$ effects are responsible for the larger part of optical spectral linewidths $\Gamma$ at normal temperatures, as we have seen.

The dephasing occurs simply because a particular molecule, which at one time has a shift $\Delta$ of its peak with respect to the light frequency, undergoes an interaction with the medium that shifts its transition frequency to another value of $\Delta$. After some time the transition frequency will shift again, and so on. At

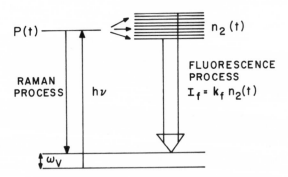

FIGURE 2. Schematic of the transition from the *polarized state* (in which all molecules have induced dipoles oscillating in-phase with the driving electromagnetic field), and the *population* $n_2(t)$ of phase-uncorrelated states. The Raman process originates from the ensemble of in-phase oscillating induced dipoles signified by $P(t)$. The energy of this polarization state is strictly the incident light frequency times $h$. The Raman line occurs at $h\nu - \omega_v$. The interactions between molecules and the medium generates the population $n_2(t)$ that spans a range of energies determined by this coupling. It is $n_2(t)$ that gives rise to the fluorescence. Both of these processes always occur as discussed in the text.

any time this molecule could emit light at some frequency near to the transition, but it no longer contributes to the Raman scattering (recall that Raman lines have a specific frequency shift with respect to the laser). The interactions of molecules with the medium are random both in time and in the strength of the coupling. Downward shifts are nearly as likely as upward shifts, so the net result of these fluctuations is to produce a broad emission band centered at the transition frequency. The polarization can also be dephased by relaxation to lower energy states, say for a molecule in solution, so the $\Gamma$ includes effects due to $\tau_v$.

The fraction of the total emission occurring as Raman is the ratio of the rate constants for spontaneous emission $(1/\tau_r)$ and dephasing $(1/T_2)$, namely $T_2/\tau_r$. Obviously this ratio is very small indeed, except at low temperatures and for excitation conditions where vibrational relaxation is slowed down. We now recapitulate the various types of emission that can occur in a conventional system. The emission originating from molecules contributing to the polarization is Raman scattering; emission from the same electronic state of the molecule that is interacting with the photon but after being influenced by fluctuations in the medium is (resonance type) fluorescence. Emission from an ensemble of molecular states reached from the driven state after some net exchange of energy with the medium is termed *relaxed fluorescence*. Such emission arises from all levels involved in the evolution of the system toward thermal equilibrium. Thus the relaxed emission may be *"hot luminescence"* — luminescence from the equilibrium distribution of excited states. Typical values of $\Gamma = 1/2\pi c T_2$ at normal temperatures are probably $\sim 15\ \mathrm{cm}^{-1}$. This corresponds to a dephasing time of $(1/2\pi c\Gamma) = 0.32\ \mathrm{ps}$. A typical radiative lifetime is $10^{-8}\ \mathrm{s}$, so that the fraction of emission occurring as Raman is $3.2 \times 10^{-5}$. If the absorption cross section at the laser frequency is $\sigma$, then the resonance Raman cross section is just $\sigma_R = 1.6 \times 10^{-5}\,\sigma$. This determines the total amount of Raman scattering into all frequencies and all angles; that is, if light intensity of $I_0$ is incident on an absorbing sample of length $L$, the total resonant Raman intensity $I_R$ is given by $I_R = I_0(1 - \exp(-\sigma_R NL)) \simeq I_0 \sigma_R NL$. The resonance Raman efficiency can obviously be greatly increased by going to lower temperatures, since at absolute zero where there are no thermally induced fluctuations we should find no resonance fluorescence. The value of $\Gamma$ is now determined by vibrational and electronic relaxation processes. A typical time scale for these radiationless transitions is ca. $10\ \mathrm{ps}$, so that the Raman cross section is increased by two orders of magnitude on the basis of the earlier example. The ratio of the Raman to relaxed fluorescence remains very small in this example, and of course there is no alteration in the amount of relaxed fluorescence with temperature — only the ratio of Raman to resonance emission changes.

In general a careful study of luminescence indicates the presence of emission occurring from vibrationally excited molecules in addition to expectations based on a thermal equilibrium. This hot luminescence can be useful since it provides information about vibrational-relaxation pathways and relaxation times. The nonequilibrium luminescence becomes more important as shorter and shorter time resolution is used in luminescence experiments. In principle the radiative lifetime can be used as an internal clock to gauge the time that optically excited systems take to relax to equilibrium. The basic point is that, to a good approximation, the radiative lifetime is independent of the nature of the vibrational excitation. In such a circumstance the variations of intensity of different hot luminescence bands arises from their radiationless decay. The rule works best for large molecules where optical excitations involve relatively low average vibrational energies compared with the dissociative energies. The other important factor is that the displacements of the nuclei resulting from electronic excitation remain relatively small so that the electronic transition moment is nearly the same at all points of the potential surface that are explored. Under these circumstances each excited level has the same rate constant for the decay by radiative emission into all possible lower-energy states.

The hot luminescence is extremely weak when systems are excited by conventional light sources. This is presumably because of the very large number of levels that are excited by broad-band sources: The fractional population of any particular level is then very small. Furthermore the vast majority of excited-state levels radiate at frequencies that nearly coincide with the equilibrium emission. In complex molecules this is a result of there being normally a small number of Franck–Condon active modes in comparison with those that are inactive. To a first approximation the inactive ones each have emission spectra that are independent of the specific type of mode excited.

The situation is different when lasers are used for excitation. Narrow-band excitation may result in the significant population of relatively few modes so that the emission spectrum is now expected to be different from the equilibrium fluorescence. These effects are particularly evident in low-temperature systems where the discrete spectra allow the laser to be tuned into specific levels. For example, with azulene in a naphthalene crystal quite different spectra were observed for each excitation wavelength using a cw laser. The results were utilized to estimate the vibrational relaxation of a number of levels of the excited state. In order that hot-luminescence studies can be used to determine relaxation times, it is necessary to obtain the fluorescence quantum yields for each level. In many respects, with current short-pulse laser technology it is just as straightforward to measure the lifetimes of the unrelaxed levels directly. In the next section the topic of picosecond fluorescence decays is discussed.

It is frequently asserted that Raman processes only occur while the driving light field is present on the molecular system. One can see that this distinction between Raman and fluorescence is not necessarily useful. The model of

Figure 2 shows clearly that if the dephasing interactions are *not* present on the time scale of interest, then the population $n_2(t)$ is never formed! This situation can only prevail in the limit that the stochastic interactions between the molecule and its surroundings are sufficiently slow that the decay of the polarization (i.e., the light absorption) only occurs by spontaneous molecular processes at a rate governed by $\tau$. Actually the emitted light *intensity* measures the number of molecules excited, and this decays at twice the rate that the polarization amplitude decays, so that the limit of interest is $T_2 = 2\tau$. In this limit, the so-called coherent limit, there is only Raman scattering in the sense that the term is used here. Experimentally the coherent limit can be approached in molecular beams where there are no collisions, or in condensed phases at sufficiently low temperatures where there are no excitations. Obviously, as one approaches the coherent limit, it becomes easier to observe the Raman scattering when the driving light pulse has passed by. For example, suppose $\tau$ is $10^{-8}$ s and $T_2$ is $2\tau$. Excitation of a sample by a picosecond pulse generates a polarization which remains intact long after the pulse has traversed the sample cell. In fact the polarization in this case is exponentially diminished with a time constant $2 \times 10^{-8}$ s. The cause of the diminution of the polarization is omnidirectional light emission — spontaneous emission with a time constant of $10^{-8}$ s. Thus, the spontaneous emission originating from the coherently driven molecules may arise during or after the application of the light field depending on the parameters of the system. In solutions at ambient temperature, $T_2 \ll \tau$ for most examples, so that Raman scattering can only occur during the light pulse if $\tau_p > T_2$. If $T_2 \sim 0.3$ ps or less, it is apparent that conventional excitation sources satisfy these inequalities. However, the situation can change dramatically as samples are cooled or when a feintosecond laser pulse is used.

## 6. Ultrafast Fluorescence Decay

The quantum yield of fluorescence is intended as a measure of the fraction of excited molecules that emit a photon in the spectrum of interest. It is a time-averaged quantity. Thus if in equation (24) we choose the linear regime where $N \approx n_1$, we can obtain a first approximation to $n_2(t)$ by putting $n_2 = 0$ on the right-hand side, leading to

$$n_2^{(1)}(t) = N\sigma \int_{-\infty}^{t} d\tau' I(\tau') \exp\left[-(t - \tau')/\tau\right] \qquad (29)$$

This simply says that the number of excited states produced by time $t$ is the sum (integral) of increments of population created by light absorption at each time until $t$, modified by the exponential decay of whatever is formed. If an ultra-short pulse $I(\tau')$ is used (a so-called $\delta$-function pulse with a time width approaching zero on the experimental scale, and much shorter than all processes of interest), we may replace the integral by its value at one time $\tau' = \tau_0$. Generally $\tau_0$ is chosen to be zero, so that the value of $n_2(t)$ becomes

$$n_2(t) = N\sigma I_0 \exp(-t/\tau) \tag{30}$$

Obviously $N\sigma I_0$ is just the number of excited molecules generated by the pulse at $t = 0$ assuming linear absorption. The quantum yield of fluorescence $\phi_F$ is then given by

$$\phi_F = \int_0^{\infty} n_2(t)\,dt/N\sigma I\tau_r = \tau/\tau_r \tag{31}$$

This is a well-known result. The development is used to emphasize that the fluorescence yield is a time-integrated quantity whereas the decay of the fluorescence is a time-gated experiment. The rate of emission of photons at time $t$ $(dn_2/dt)$ is simply $n_2(t)/\tau_r$, which is then directly proportional to the fluorescent signal observed at that time. This means that at $t = 0$ the rate of emission of photons does not depend on the actual lifetime $\tau$ of the state but is given by $N\sigma I_0/\tau_r$. Thus the rate of emission of photons at zero time (emitted light intensity at zero time) depends only on the fraction of the incident light intensity absorbed per unit pathlength (i.e., $N\sigma$) and on the *radiative* lifetime.

The considerations of the previous paragraph are indeed important in the planning of picosecond pulse experiments. In a cw experiment the crucial factor that decides feasibility is the quantum yield. If the emission is to be time gated, the key factor is the radiative lifetime. For example, consider a molecular transition having a radiative lifetime of 100 ns and a quantum yield of $10^{-4}$. Clearly the cw fluorescence experiment will present a relatively weak signal. On the other hand, the peak signal in a picosecond time-gated fluorescence experiment in this case is not any different than if the quantum yield were unity! Indeed the fluorescence light signal obtained just after excitation with a 1-ps pulse is not more than 1% different for $\phi_F = 10^{-4}$ than for $\phi_F = 1$. Obviously this implies that time-gated fluorescence methods such as streak cameras offer many new opportunities for fluorescence research.

## 7. The Effects of Inhomogeneous Distributions

The spectrally narrow-band laser allows the study of different microscopic portions of systems that are inhomogeneous. In many complex molecular systems, especially at lowered temperatures, all molecules in a sample may not experience the same environment, so that there are variations expected in the spectra of different molecules. These spectral distinctions between molecules may persist for periods that are long compared with the observation time.

Conventional broad-band optical excitation of such systems might simultaneously excite all the various types of molecules, but a narrow-band laser may only excite a special subset whose energies are shifted into resonance with the laser by their interactions with the medium. Thus the fluorescence spectrum may change as the excitation frequency is shifted, and it may have considerably narrower bands than those obtained by conventional techniques. This site-selection technique can be extremely useful in simplifying the spectra of molecules in complex environments. Of course it is only expected to work effectively in systems where the sites remain energetically distinguishable throughout the period of the experiment. This situation can certainly be achieved in most systems by lowering the temperature sufficiently. However, it may also be realized in systems which, due to their size and complexity, undergo structural changes on a time scale that is slow compared with the lifetime of the fluorescence of the chromophores associated with the structure. The fluorescence lines in these experiments can be even narrower than conventional Raman or infrared transitions of the same system. If the laser is sufficiently narrow, the inhomogeneous broadening of the vibrational transitions is also overcome.

Obviously luminescence line-narrowing can prove extremely useful since sharper spectra are obtained and more detailed structural information is made available. However, not all laser-excited fluorescence in inhomogeneous systems will be line-narrowed. In this case the difference between the laser and conventional excitation is not so much the bandwidth — since conventional sources can also be narrowed — but the tunability as well. Thus, the laser has a large number of photons per second within variable narrow-frequency ranges. This property is a key one for line-narrowing, because of the composition of broad-band optical spectra. At the low-energy side of an optical spectrum, near the 0–0 transition, only a few molecular states are involved in the transitions. For example, at $77°K$ only the zero-point level of the ground state and few vibrational levels of the excited state are important. Of course each of these few molecular transitions is associated with excitation of the medium-molecule states (so-called lattice states). The lattice states are excited whenever the interaction of a molecule and its surroundings is modified by electronic excitation of the molecule. In higher-energy regions of the optical spectrum any particular wavelength of light would excite many different molecular transitions because of the congestion of molecular states each with its associated lattice states. Thus different subsets of molecules can only be expected to be distinguished from one another in the low-energy region of the spectrum. The Franck–Condon progression of lattice modes usually peaks $50–100 \, cm^{-1}$ from the lattice-free transition ("zero-phonon transition"), so that even in the 0–0 region one may only achieve significant fluorescence line-narrowing for excitation at the longest wavelength edge of the absorption spectrum.

Besides the analysis of line-narrowed or site-selective fluorescence, the occurrence of inhomogeneous distributions also gives rise to the phenomenon of optical hole burning. The saturation or bleaching of only a narrow energy distribution of sites leaves the absorption spectrum with "holes" (i.e., narrow-energy regions where the transmission is increased) that can be utilized to investigate the spectra of complex systems in more detail. The time available to the investigator to study these holes depends on the occurrence and lifetime of bottleneck states in the molecular relaxation. Hole-burning spectroscopy has found recent application to biological chromophores. More permanent holes can be generated by both photochemical and nonphotochemical means in many organic systems. In photochemical hole burning the narrow-band laser converts a small subset of molecules into products, thereby leaving a hole in the absorption profile. Nonphotochemical hole-burning mechanisms are not as clear, but frequently it is found that continued irradiation of an inhomogeneous system results in the formation of holes in the absorption profile even though no photodamage occurred. Apparently the absorption frequencies of molecules in the subset that absorb the light are shifted.

ACKNOWLEDGMENT

This research was supported by the National Science Foundation and the Institute of General Medical Sciences of the National Institutes of Health.

## Bibliography

Owing to the generality of the present article, the bibliography was chosen to cover both background and more advanced material pertaining to the various subjects discussed. The citations are mainly to review articles and books where these are available. The structure of the bibliography follows that of the article, each section being identified with the same arabic numeral as used in the text. I am indebted to my graduate students F. Doany, E. Heilweil, P. McCarthy, and J. Trout for compiling the bibliography.

### 1. General References on Radiationless Processes

Atkinson, G. H., Parmenter, C., and Schuyler, M. W., Single vibronic level fluorescence, in *Creation and Detection of the Excited State*, Vol. 3 (W. R. Ware, ed.), Marcel Dekker, New York (1974).
Avouris, P., Gelbart, W., and El-Sayed, M. A., Nonradiative electronic relaxation under collision-free conditions, *Chem. Rev.* 77 (1977).
Bondybey, V. E., Experimental studies of nonradiative processes in low-temperature matrices, *Adv. Chem. Phys.* 47, 521 (1981).
Brus, L. E., and Bondybey, V. E., Spectroscopic and time-resolved studies of small molecule relaxation in the condensed phase, in *Radiationless Transitions* (S. H. Lin, ed.), Academic Press, New York (1980).

El-Sayed, M. A., Double resonance and the properties of the lowest excited triplet state of organic molecules, *Annu. Rev. Phys. Chem.* **26**, 235 (1975).

Fong, Francis, *Theory of Molecular Relaxation: Applications in Chemistry and Biology,* Wiley, New York (1975).

Freed, K. F., The theory of radiationless transitions in polyatomic molecules, *Top. Curr. Chem.* **31**, 105 (1972).

Freed, K. F., Radiationless transitions in molecules, *Acc. Chem. Res.* **11**, 74 (1978).

Freed, K. F., Energy dependence of electronic relaxation processes in polyatomic molecules, *Top. App. Phys.* **15**, 1 (1976).

Henry, B. R., and Kasha, M., Radiationless molecular electronic transitions, *Ann. Rev. Phys. Chem.* **19**, 161 (1968).

Henry, B. R., and Siebrand, W., Radiationless transitions, in *Organic Molecular Photophysics,* Vol. 1 (J. B. Birks, ed.), Wiley, New York (1973), pp. 153–237.

Hochstrasser, R. M., and Weisman, R. B., Relaxation of electronically excited molecular states in condensed media, in *Radiationless Transitions* (S. H. Lin, ed.), Academic Press, New York (1980), p. 317.

Hochstrasser, R. M., Analytical and structural aspects of vibronic interactions in the ultraviolet spectra of organic molecules, *Acc. Chem. Res.* **1**, 266 (1968).

Jortner, J., and Mukamel, S., Molecular radiationless processes in *Molecular Energy Transfer* (R. Levine and J. Jortner, eds.), Wiley, New York (1976).

Jortner, J., Rice, S. A., and Hochstrasser, R. M., Radiationless transitions in photochemistry, *Adv. Photochem.* **7**, 149 (1969).

Jortner, J., and Mukamel, S., Preparation and decay of excited molecular states, in *The World of Quantum Chemistry* (R. Daudel and B. Pullman, eds.), Reidel, Boston (1974).

Leach, S., Electronic spectroscopy and relaxation processes in small molecules, in *Molecular Energy Transfer* (R. Levine and J. Jortner, eds.), Wiley, New York (1976).

Lee, E. K. C., and Loper, G. L., Experimental measurement of electronic relaxation of isolated small polyatomic molecules from selected states, in *Radiationless Transitions* (S. H. Lin, ed.), Academic Press, New York (1980).

Lin, S. H., Some considerations of theory and experiment in ultrafast processes, in *Radiationless Transitions* Academic Press, New York (1980).

Rhodes, W., Dynamic aspects of molecular excitation by light, in *Radiationless Transitions* (S. R. Lin, ed.), Academic Press, New York (1980).

Rice, S. A., Some comments on the dynamics of primary photochemical processes, in *Excited States* Vol. 2 (E. C. Lim, ed.), Academic Press, New York (1975).

Robinson, G. W., Molecular electronic radiationless transitions from excited states, in *Excited States* Vol. 1, (E. C. Lim, ed.), Academic Press, New York (1974).

Schlag, E. W., Schneider, S., and Fischer, S. F., Lifetimes in excited states, *Annu. Rev. Phys. Chem.* **22**, 465 (1971).

Yardley, J. T., Dynamic properties of electronically excited molecules, in *Chemical and Biochemical Applications of Lasers* Vol. 1 (C. B. Moore, ed.), Academic Press, New York (1974).

## 2. General Aspects of Spontaneous Emission, and Optical-Parameter Interrelationships

Becker, R. S., *Fluorescence*, Wiley, New York (1969).

Condon, E. U., and Shortley, G. H., *The Theory of Atomic Spectra,* Cambridge University Press (1963), pp. 79–11.

Corney, Alan, *Atomic and Laser Spectroscopy*, Clarendon Press, Oxford (1977), pp. 93–117.

Fowler, W. B., and Dexter, D. L., Relation between absorption and emission probabilities in luminescent centers in ionic solids, *Phys. Rev.* **128**, 2154–2166 (1962).

Hochstrasser, R. M., and Prasad, P. N., Optical spectra and relaxation in molecular solids, Vol. 1 (E. C. Lim, ed.), Academic Press, New York (1974) pp. 79–128.

Hochstrasser, R. M., Some principles governing the luminescence of organic molecules, in *Excited States of Proteins and Nucleic Acids* (R. F. Steiner and I. Weinryb, eds.), Plenum Press, New York (1960), pp. 1–29.

Kasha, M., Theory of molecular luminescence, in *Proceedings of the International Conference on Luminescence*, Budapest (1966), pp. 166–182.

Loudon, R., *The Quantum Theory of Light*, Clarendon Press, Oxford (1973).

Mccomber, J. D., *The Dynamics of Spectroscopic Transitions*, Wiley, New York (1966).

Mitchell, A. C. G., and Zemansky, M. W., *Resonance Radiation and Excited Atoms*, Cambridge University Press, London (1961).

Stenholm, S., Quantum theory of electromagnetic fields interacting with atoms and molecules, *Phys. Rep. C* **6**, 1 (1973).

Strickler, S. J., and Berg, R. A., Relationship between absorption intensity and fluorescence lifetime of molecules, *J. Chem. Phys.* **37**, 814–822 (1962).

## 3. Transition Moments, Their Orbital Sources, and Polarization Properties

Birks, J. B., ed., *Organic Molecular Photophysics*, Vols. 1 and 2, Wiley, New York (1973–1975).

Kasha, M., Paths of Molecular Excitation, *Radiat. Res.* **2**, 243 (1960).

Kasha, M., in *Light and Life* (W. M. McElroy and B. Glass, eds.), The John Hopkins Press, Baltimore (1961), pp. 31–64.

Kasha, M., Characterization of electronic transitions in complex molecules, *Discuss. Farad. Soc.* **9**, 14–19 (1950).

Lim, E. C., Vibronic interactions and luminescence in aromatic molecules with nonbonding electrons, *Excited States*, Vol. 3 (E. C. Lim, ed.), Academic Press, New York (1977), p. 305.

Longuet-Higgins, H. C., Theory of molecular energy levels, *Adv. Spectrosc.* **2**, 429–472 (1961).

McGlynn, S. P., Azumi, T., and Kinoshita, M., *Molecular Spectroscopy of the Triplet State* Prentice-Hall, Englewood Cliffs, New Jersey (1969).

Simpson, W. T., *Theories of Electrons in Molecules*, Prentice-Hall, Englewood Cliffs, New Jersey (1961), pp. 171–178.

Turro, N. J., *Modern Molecular Photochemistry*, Benjamin/Cummings, Menlo Park, California (1978).

## 4. Luminescence Using Lasers: Properties of Lasers; Nonlinear Processes and Bleaching Kinetics

Allen, L., and Eberly, J. H., *Optical Resonance and Two Level Atoms*, Wiley-Interscience, New York (1975).

Bernheim, R. A., *Optical Pumping*, W. A. Benjamin, New York (1965).

Corney, A., *Atomic and Laser Spectroscopy*, Clarendon Press, Oxford (1977).

Diels, J. C., Feasibility of measuring phase relaxation time with subpicosecond pulses, *IEEE J. Quantum Electron.* QE-16, 1020 (1980).

Happer, W., Optical pumping, *Rev. Med. Phys.* 44, 169 (1972).

Mourou, G., Drovin, B., Gergeron, M., and Denariez-Roberga, M. M., Kinetics at bleaching in polymethine cyanine dyes, *IEEE J. Quantum Electron.* QE-9, 787 (1973).

Sargent, M., III, Scully, M. O., and Lamb, W. E., Jr., *Laser Physics,* Addison-Wesley, London (1974).

Spaeth, M. L., and Sooy, W. R., Fluorescence and bleaching of organic dyes for a passive Q-switched laser, *J. Chem. Phys.* 48, 2315 (1968).

Steinfeld, J. I., *Molecules and Radiation: An Introduction to Modern Molecular Spectroscopy,* M.I.T. Press, Cambridge (1978).

Walther, H., Atomic and molecular spectroscopy with lasers, in *Laser Spectroscopy of Atoms and Molecules* (H. Walther, ed.), Springer, Berlin (1976).

Yariv, A., *Quantum Electronics,* 2nd edn., Wiley, New York (1975).

## 5. Coherent Interactions. Dephasing and the Distinction between Fluorescence and Resonant Raman Scattering

Behringer, J., Experimental resonance Raman spectroscopy, *Mol. Spectrosc.* 3, 163 (1975).

Behringer, J., The relation of resonance Raman scattering to resonance fluorescence, *J. Raman Spectrosc.* 2, (3) 275 (1974).

Berne, B. J., and Pecora, R., *Dynamic Light Scattering with Applications To Chemistry, Biology, and Physics,* Wiley, New York (1976).

Bloembergen, N., *Nonlinear Optics,* W. A. Benjamin, New York (1977).

Burns, M. J., Lin, W. K., and Zewail, A. H., Nonlinear laser spectroscopy and dephasing of molecules: An experimental and theoretical overview, in *Spectroscopy and Excitation Dynamics of the Molecular Condensed Phase,* a series in *Modern Problems in Solid State Physics* (V. M. Agranovich and R. M. Hochstrasser, eds.), North-Holland, Amsterdam (1982).

Harris, C. B., and Breiland, W. G., Coherent spectroscopy in electronic excited states, *Laser and Coherent Spectroscopy* (J. I. Steinfeld, ed.), Plenum Press, New York (1978), p. 373.

Hochstrasser, Robin M., and Nyi, Corinne A., Resonance fluorescence, Raman and relaxation of single vibronic levels in the condensed phase: Azulene in naphthalene, *J. Chem. Phys.* 70, 1112 (1979).

Hochstrasser, R. M., Novak, F., and Nyi, C. A., Statistical aspects of resonance light scattering in the condensed phase, *Israel J. Chem.* 16, 250–257 (1977).

Laubereau, A., and Kaiser, W., Vibrational dynamics of liquids and solids investigated by picosecond light pulses, *Rev. Mod. Phys.* 50, 607 (1978).

Macomber, J. D., A bibliography of transient effects in the resonant elastic response of matter to an intense light pulse, *IEEE J. Quantum Electron.* QE-4, 1 (1968).

McGurk, J. C., Schmaltz, T. O., and Flygare, W. H., A density matrix, Bloch equation description of infrared and microwave Transient Phenomena, *Adv. Chem. Phys.* 21, 1 (1974).

Novak, F. A., Friedman, J. I., and Hochstrasser, R. M., Resonant scattering of light by molecules: Time-dependent and coherent effects, in *Laser and Coherent Spectroscopy* (J. I. Steinfeld, ed.), Plenum, New York (1978), p. 451.

Shorygin, P. P., New possibilities and chemical applications of Raman spectroscopy, *Russ. Chem. Rev.* 47, 1697 (1978).

Steinfeld, J. I., Optical analogs of magnetic resonance spectroscopy, *Chemical and Biochemical Applications of Lasers* (C. B. Moore, ed.), Academic Press, New York (1974).

Zewail, A. H., Optical Dephrasing: Principles of and probings by coherent laser spectroscopy, *Acc. Chem. Res.* **13**, 369 (1980).

## 6. Picosecond Fluorescence Concepts

Eisenthal, K. B., Picosecond spectroscopy, *Annu. Rev. Phys. Chem.* **28**, 207 (1977).

Eisenthal, K. B., Studies of chemical and physical processes with picosecond lasers, *Acct. Chem. Res.* **8**, 118 (1975).

Hochstrasser, R. M., and Weisman, R. B., Relaxation of electronically excited molecular states in condensed media, in *Radiationless Transitions* (S. H. Lin, ed.), Academic Press, New York (1980) p. 317.

Kaufman, K. J., and Rentzepis, P. M., Picosecond spectroscopy in chemistry and biology, *Acct. Chem. Res.* **8**, 407 (1975).

Laubereau, A., and Kaiser, N., Picosecond spectroscopy of molecular dynamics in liquids, *Annu. Rev. Phys. Chem.* **26**, 83 (1975).

Pellegrino, F., Kolpaxis, A., and Alfano, R. R., Time-resolved picosecond laser spectroscopy, S.P.I.E. Volume 148, *Computers in Optical Systems* (1978).

## 7. Effects of Inhomogeneous Distribution on Fluorescence

deVries, H., and Wiersma, D. A., Photophysical and photochemical molecular hole burning, *J. Chem. Phys.* **72**, 1851 (1980).

Friedrich, J., Scheer, H., Zichendraht-Wandelstadt, B., and Haarer, D., High-resolution optical studies on C-phycocyanin via photochemical hole burning, *J. Am. Chem. Soc.* **103**, 1030 (1981).

Kohler, B. E., Site selection spectroscopy, in *Chemical and Biochemical Applications of Lasers* (C. B. Moore, ed.), Academic Press, New York (1979).

Letokhov, V. S., and Chelotayev, V. P., *Principles of Nonlinear Laser Spectroscopy,* Izd. Nauka, Moscow (1975).

Mouron, G., and Denairez-Roberga, M. M., Polarization of fluorescence and bleaching of dyes in a high-viscosity solvent, *IEEE J. Quantum Electron.* QE-9, 787 (1973).

Personov, R. I., Site selection spectroscopy of complex in solutions and its applications, in *Spectroscopy and Dynamics of Molecules in the Condensed Phase,* (V. Agranowich and R. M. Hochstrasser, eds.), North-Holland, Amsterdam (1982).

Small, G. J., Persistent Nonphotochemical Hole Burning in Organic Glasses, in *Spectroscopy and Dynamics of Molecules in the Condensed Phase* (V. Agranowich and R. M. Hochstrasser, eds.), North-Holland, Amsterdam (1982).

Stoneham, A. M., Shapes of inhomogeneously broadened lines in solids, *Rev. Mod. Phys.* **41**, 82 (1969).

Szabo, A., Observation of hole burning and cross relaxation effects in ruby, *Phys. Rev. B.* **11**, 4512 (1975).

# 2

# Covalent Fluorescent Probes

## RICHARD P. HAUGLAND

## *1. Introduction*

In the application of fluorescence techniques to biomolecules it is frequently necessary to introduce fluorophores with experimentally advantageous fluorescence properties. These extrinsic fluorescent probes can be of two kinds. Non-covalent probes form a reversible association with the biomolecule by a combination of hydrophobic, dipole–dipole, and ionic interactions. The covalent fluorescent probes described in this chapter form much stronger chemical bonds with specific atoms on the biomolecule and usually are essentially irreversibly bound. The two types of probes are complementary and each has advantages, limitations, and specific applications to fluorescence studies.

## *2. Primary Considerations in Fluorescent Labeling of Biomolecules*

In the covalent fluorescent modification of a biomolecule (protein, cell membrane, nucleic acid, polysaccharide, or related material) three chemical moieties must be considered: first, potential reactive groups on the biomolecule; second, a chemically reactive "handle" which is usually part of the fluorescent probe; third, a fluorescent chromophore with spectral characteristics appropriate to the problem being investigated and the fluorescence technique employed. Extensive research with non-fluorescent probes has fortunately led to classification of many potentially reactive residues on biomolecules and identified

RICHARD P. HAUGLAND • Molecular Probes, Inc., Junction City, Oregon 97448.

several chemical reactions for their selective chemical modification. Research primarily in the last decade has developed a broad arsenal of fluorescent probes that take advantage of these selective coupling reactions for biomolecule modification. Rather than being an exhaustive review of all of the fluorescent probes available and of all of the applications of covalent probes, this chapter will concentrate on probes with unique advantages, on improved methods of conjugation, on obtaining selective chemical modification, and on matching the probe to its fluorescence application.

## 2.1. Reactive Sites in Biomolecules

Covalent fluorescent probes have been used most extensively to modify proteins, although methods to label nucleic acids, polysaccharides, lipids, and other biomolecules have also been developed. The predominance of probe–protein studies is directly due to the ease of modification of the frequently large number of reactive peptide side chains. Indeed, it is frequently difficult to specifically modify *nonprotein* components of complex systems when proteins are present.

In the language of organic chemistry the reactive residues on proteins are predominantly nucleophiles, and the reactions with probes are mainly examples of nucleophilic substitution or addition reactions. Nucleophiles are atoms or groups of atoms with excess electron density in their bonding or unshared electron pairs. When several nucleophilic residues are present in one biomolecule or complex and a limited amount of probe is used, the sites of modification are determined by the competing reaction rates for each site. Several important factors influence the reaction rates. Primary among these is the "nucleophilicity" of the atoms at the site. High nucleophilicity is associated with base strength, and consequently aliphatic amines such as the *epsilon*-amino group of lysine in peptides ($pK_a$ about 10), which are the strongest bases in most biomolecules, are rapidly modified by several reagents. Other factors effect nucleophilicity, and the weaker base mercaptide (thiolate) of cysteine ($pK_a$ about 8) usually undergoes nucleophilic reactions faster than lysine. A second major consideration in determining relative reactivity is the concentration of the reactants. With most of the commonly modified residues in proteins, the concentration of reactive sites is determined not just by the amino acid composition but also by the acidity. This is especially important in reaction of the *epsilon*-amino group of lysine. at pH 7, less than 1% of the lysines are in the form of unprotonated nucleophiles ($R–NH_2$). Since the protonated lysines ($R–NH_3^+$) are *totally* unreactive, the effective concentration of lysines is considerably lowered and the overall rate of modification is very slow at low pH. Since the *alpha*-amino group of peptides is a much weaker base ($pK_a$ about 7), a significant fraction remains unprotonated and the amino terminus can be selectively modified at physiological pH.

Due to its having a $pK_a$ lower than lysine and the high intrinsic nucleophilicity of the mercaptide anion (RS$^-$) cysteine is rapidly modified by many reagents, fluorescent and otherwise. It remains quite reactive in even the acidic (R–SH) form. Reactions in which phenols such as tyrosine are the nucleophiles are also highly pH sensitive and are fastest above the phenol ionization constant. The heterocyclic ring of histidine undergoes some reactions usually as a relatively poor nucleophile near or below its $pK_a$ of about 7 where other potential nucleophiles are less reactive. Carboxyl residues in proteins and other biomolecules are much poorer nucleophiles yet, especially when exposed in aqueous solution. Modification of carboxyl groups usually requires reactions in which it functions as other than a nucleophile. One common amino acid whose rate of side-chain modification is not highly pH dependent is methionine. Recent examples of methionine alkylation in noncysteine-containing proteins have used a relatively low pH to suppress reaction with other nucleophiles to give specific, albeit sometimes slow, modification of methionine.

Since most biochemical modification reactions are carried out in aqueous solution, specific reaction of the weakly nucleophilic hydroxyl groups such as in serine, threonine, polysaccharides, and nucleic acids is usually impractical due to competitive reaction with water. The guanidinium group of arginine remains protonated and generally unreactive up to pH 12 to 13; however, some special reagents effect its virtually specific modification at a lower pH. While no fluorescent reagents currently exist, the indole side chain of tryptophan remains a potential site of specific modification, most likely by an electrophilic reagent.

While occurring only rarely in proteins, phosphate esters are common in other biomolecules. Being poor nucleophiles, they are unlikely to be a good candidate for fluorescent modification, except possibly enzymatically. The other common functional groups in biomolecules such as amides, esters, ethers and hydrocarbons and the common nucleotide bases are generally too unreactive for one to give serious consideration to their modification by ordinary organic reactions in aqueous solutions. Two common and reactive organic functional groups, aldehydes and ketones, are rare in large biomolecules; however, their formation by specific oxidation of glycols in polysaccharides and polyribonucleotides has provided a new entry into these otherwise difficult to modify biomolecules.

Due to their relative polarity or charges, most potential nucleophiles are exposed to an aqueous environment and, unless protected by other factors such as a probe-impermeant membrane, have no difficulty in coupling with an appropriately reactive fluorophore. Macromolecular folding causes certain normally nucleophilic residues to be restrained from achieving complete aqueous solvation. Nucleophilic organic reactions commonly proceed fastest in a polar yet aprotic environment, and the lower solvation and possible catalysis by spatially adjacent residues may account for the unusually reactive nucleophilic reactions in the active-site serine proteases and in several proteins with "fast thiols". These regions of enhanced reactivity have frequently been found in relatively large clefts associated with enzyme active sites. Steric hindrance caused by macromolecular folding can also slow or eliminate modification by the frequently bulky fluorescent probes. Although many probes are uncharged and readily pass through cell membranes or even label proteins from within the membrane, some probes are essentially impermeant and restricted to external reactions.

## 2.2. Reactive Handles on Fluorescent Probes

In most cases, chemically reactive functional groups have been attached to the fluorophore. These reactive handles are predominantly electron-deficient moieties that interact with the electron surplus nucleophiles on the biomolecule. The activation energy to form a covalent bond must be relatively low since most biomolecule modifications must occur at an appreciable rate below $40°C$ to be useful. This is usually achieved by using a relatively unstable (high-energy) reactive group on the probe. While reagent instability accelerates the modification, it also introduces limitations such as highly reactive functional groups like isocyanates and acid chlorides being destroyed by water before they have a chance to react with other nucleophiles. Intrinsic reagent stability also affects storage, reproducibility, and selectivity of labeling. With most useful probes, a compromise between high reactivity and stability in storage must be reached, but all covalent probes should be treated as unstable, particularly when in solution in nucleophilic solvents such as water or alcohols.

For a given type of nucleophilic reaction with a single type of reactive group, such as the reaction of protein thiols with a maleimide, the residue in a set that reacts fastest by having the lowest activation energy is the site modified selectively when a limited amount of probe is used. In an irreversible competitive reaction between two similar sites for a limited amount of probe, a 1 kcal/mole difference in activation energy results in about five times as much modification of the more reactive site. A difference of 2 kcal/mole results in a 28-fold selectivity. The activation energy can be decreased by either catalysis, which *lowers* the peak energy required for the reaction, or unequal solvation at the two different sites, which *raises* the energy of the reactants and especially that of the nucleophile.

The factors of relative nucleophilicity, effective nucleophile concentration, accessibility, and solvation all apply to the residue on the biomolecule to be modified. With the exception of variations of pH and temperature, there is usually little one can do to alter these intrinsic reactivity factors for the sites on a given biomolecule. The factors that will determine which classes of sites are modified (for instance, thiols versus amines) are related to the reactive handles on the probes. Fluorescent probes and the reactions they undergo with biomolecules can be classified as one of the following types: (1) alkylating, (2) acylating, (3) aldehyde and ketone reactive, (4) photoactivatable, or (5) miscellaneous.

The alkylating reagents include the functional groups haloacetyl (such as iodoacetamides), maleimides, aziridines, epoxides, and alkyl and aryl halides. All preferentially react with cysteine thiols in proteins if they are present but may also react with histidine, tyrosine, lysine, methionine, and possibly carboxylic acids.

Acylating probes are reactive derivatives of carboxylic, sulfonic, phosphoric, or boric acids and include anhydrides, acid halides, "active" esters, imidoesters, and isothiocyanates. Their reaction products with cysteine, histidine, or carboxylic acids in proteins are frequently hydrolytically or thermodynamically unstable, and they do not react with methionine. Stable products are formed with lysine or tyrosine at elevated pH, while several are selective or specific for serine proteases.

Aldehyde and ketone reagents are almost all derivatives of hydrazine or, potentially, hydroxylamine. Photoactivatable fluorescent probes are chemically nonreactive until photolyzed to a reactive intermediate that can covalently insert into convenient nearby bonds. Miscellaneous reagents include disulfides and mercurials and those used in nonnucleophilic or enzyme-assisted modification of biomolecules.

## 2.3. Fluorophores

With the possible exception of some polyenes such as parinaric acid and retinol and some rare earth chelates, all fluorophores of practical use for labeling biomolecules are derivatives of aromatic compounds. These usually have from one to five conjugated rings. The usual range of useful spectral absorption and emission for extrinsic fluorescent probes is from 300 to about 600 nm although more sensitive phototubes are extending the long wavelength into the near infra-red. Limitations of greater light scattering at shorter wavelengths or intrinsic biomolecule absorption and fluorescence may favor use of the long-wavelength probes. The use of long-wavelength lasers for excitation may limit the choice of probes.

For many applications of fluorescence the most important consideration is the ultimate intensity detected by the recorder. Two noninstrumental factors effect this intensity. These are the quantum yield and the molar absorptivity (extinction coefficient) at the excitation wavelength. Both properties are intrinsic to the conjugated fluorophore at a given light intensity, and it is essentially the product of these two factors that determines the ultimate sensitivity obtainable. The vast majority of organic compounds has a quantum yield less than 0.01 and are useless as extrinsic fluorescent probes. Essentially all

## Table 1. Covalent Fluorescent Probes

| Fluorophore | Absorption maximum | A | Emission maximum | Q | τ | Probe number | Covalent derivatives | References |
|---|---|---|---|---|---|---|---|---|
| Salicylate | 313 | L | 410 | H | S | 1 | 4-Maleimide | 1 |
| | 343 | L | 435 | H | S | 2 | 5-Maleimide | 2 |
| | 320$^a$ | L | 410$^a$ | M | S | 3 | 4-Iodoacetamide | 3, 4 |
| | 323 | L | 405 | 0.44 | 2.0 | 4 | 5-Iodoacetamide | 5, 6 |
| | 320$^a$ | L | 400$^a$ | H | S | 5 | Hydrazide | 7, 8 |
| Anthranilate | 315–330 | L | 400–420 | H | S | 6 | Isatoic anhydride | 2 |
| | 342 | 4200 | 422 | 0.53 | 8 | 7 | p-Nitrophenyl | 9–11 |
| | 330$^a$ | L | 420$^a$ | H | S | 8 | Anthraniloyl aziridine | 2 |
| | 330$^a$ | L | 420$^a$ | H | S | 9 | Hydrazide | 2 |
| | 350 | L | 430 | M | S | 10 | 5-Azidoisophthalic acid | 12 |
| | 330$^a$ | L | 420$^a$ | H | S | 11 | Succinimidyl ester of N-methyl | 13 |
| | 334 | 3600 | 438 | 0.5 | S | 12 | Carboxymethylisatoic anhydride | 14 |
| 1-Alkylthio isoindoles | 334 | 5700 | 450 | 0.33–0.47 | 18 | 13 | o-Phthalaldehyde/ mercaptoethanol | 15, 16 |
| Pyrrolinones | 390 | 5300 | 475 | 0.25 | 1.7– 11.7 | 14 | Fluorescamine® | 15–19 |
| | 390 | L | 475 | L | S | 15 | Methoxydiphenylfuranone | 20, 21 |
| Bimanes | 385 | 4200 | 484 | 0.2 | ? | 16, 17 | Monobromo- and dibromobimanes | 22 |
| Benzoxazole | 308 | 32000 | 370 | 0.17 | 0.88 | 18 | Benzoxazolephenylmaleimide | 23 |
| Benzimidazole | 313 | 28000 | 361 | 0.5 | S | 19 | Benzimidazolephenylmaleimide | 24–26 |
| Benzofurazan (NBD) | 480 | 26000 | 510–545 | L–M | S | 20 | NBD chloride/amines | 27–29 |
| | 425 | 13000 | 540 | L | S | 20 | NBD chloride/thiols | 30–32 |
| | 382 | 11400 | Weak | | | 20 | NBD chloride/phenols | 33, 34 |
| | 475 | 25200 | 520–540 | L | S | 21, 22 | NBD disulfides | 30, 35, 36 |
| | 480 | 25000 | 530 | L | S | 23 | Iodoacetate | 37, 38 |
| | 480$^a$ | 25000* | 530$^a$ | L | S | 24 | Aziridine | 2, 39 |
| | 480$^a$ | 25000 | 530$^a$ | L | S | 25 | NBD methylaminoacetaldehyde | 2 |
| | 470$^a$ | 25000* | 520$^a$ | L–M | S | 26 | NBD azide | 2 |
| Naphthalenes | 340 | 4500 | 500–550 | 0.05–0.5 | 5–15 | 27 | 1,5-Dansyl chloride | 40–42 |
| | 378 | 3000 | 475 | 0.76 | 25–29 | 28 | 2,5-Dansyl chloride | 43 |
| | 359 | 5700 | 435 | 0.71 | 13.2 | 29 | 2,6-Dansyl chloride | 43 |
| | 321 | 23000 | 415–460 | Variable | S–M | 30 | Mansyl chloride | 44, 45 |
| | 315, 360 | 20000 | 430 | Variable | 17 | 31 | TNS Chloride | 46 |
| | 340 | 6100 | 495 | 0.68 | 19.4 | 32 | 1,5-IAEDANS | 5, 6, 47, 48, 49, 50 |
| | 349 | 6500 | 480 | 0.52 | 12.6 | 33 | 1,8-IAEDANS | 47 |
| | 357 | L | 412 | H | 12.0 | 34 | 2,6-IAEDANS | 51 |
| | 330 | 20000$^a$ | 450 | Variable | S–M | 35 | 2,6-ANS iodoacetamide | 37, 52–54 |
| | 330 | 20000$^a$ | 380–440 | Variable | S–M | 36 | 2,6-ANS maleimide | 52, 55 |
| | 355 | 13,180 | 448 | Variable | Variable | 37 | 1-Anilinonaphthyl-4-maleimide | 26, 56 |
| | 360 | L | 410–440 | 0.1–0.4 | S–M | 38–40 | Dimethylaminonaphthyl-maleimides | 57 |
| | 345 | 3980 | 480–535 | 0.1–0.5 | S–M | 41 | Dansyl aziridine | 58–56 |
| | 328–350 | 7950 | 520 | 0.61 | 11.5 | 42 | Dansyl cystine | 5, 6, 62 |
| | 333 | 4000 | 535 | M | M | 43 | Dansyl cystine mercurial | 63–65 |
| | 340 | 4500$^a$ | 500–550 | M | M | 44, 45 | Dansyl imido esters | 66 |
| | 340 | 4480 | 530 | M | M | 46 | Succinimidyl dansyl glycinate | 13, 67 |
| | 325 | L | 510 | M | M | 47 | Dansyl hydrazine | 68–71 |
| | 430 | ? | 540 | 0.21 | M | 48 | Lucifer Yellow carbohydrazide | 72–74 |
| | 340 | 4500 | 510 | 0.85 | 18.5 | 49 | Monodansyl cadaverine | 75–78 |
| | 335 | 4000 | 500–560 | M | M | 50, 51 | Dansyl chloromethyl ketones | 79 |
| | 340 | L | 450 | M | M | 52 | Dimethylaminonaphthylisothio-cyanate | 80 |
| | 350 | L | 520 | M | M | 53 | Dansyl fluoride | 81, 82 |
| | 340 | L | 550 | M | M | 54 | Dansylamidophenylboronic aci | 83, 84 |
| | 340 | 4500$^a$ | 500–550 | M | M | 55 | Dansyl azide | 2 |
| | 322 | 1700 | 467 | 0.29 | 20.2 | 56 | 3-Azidonaphthalene-2,7-disulfonate | 85, 86 |
| | | | | | | 57 | Azidonaphthalenesulfonyl chloride | 87 |
| Coumarins | 379 | 26000 | 457 | 0.67 | S | 58 | Maleimide DACM | 88–90 |
| | 387 | 30200 | 465 | H | S | 59 | Coumarin phenyl maleimide CPM | 91 |
| | 390 | 30000$^a$ | 460 | H | S | 60 | Coumarin phenyl isothiocyanate | 2 |
| | 315 | 12500 | 400 | M | 1.8 | 61 | 7-Methoxy-4-bromomethyl-coumarin | 92–94 |
| | 325 for OH | 14400 | 390 for OH | ? | S | 62 | 7-Hydroxycoumarin-acethydrazide | 95 |
| | 371 for O⁻ | 19100 | 454 for O⁻ | H | S | | | |
| Stilbenes | 345–360 | 30000$^a$ | 430 | M–H | 1.2 | 63 | Acetamide isothiocyanate SITS | 96, 97 |
| | 334 | 51000 | 435 | M | 0.81 | 64 | Benzamide isothiocyanate BIDS | 98 |
| | 340 | 34000 | 430 | M–H | S | 65 | Diisothiocyanate DIDS | 99–101 |
| | 335 | 30000$^a$ | 430 | M–H | S | 66 | Dimaleimide | 2 |

*Table 1. (Continued)*

| Fluorophore | Absorption maximum | A | Emission maximum | Q | τ | Probe number | Covalent derivatives | References |
|---|---|---|---|---|---|---|---|---|
| | 335 | 30000$^a$ | 430 | M–H | S | 67 | Dimaleimide disulfonate | 2 |
| | 364 | 20000 | 460 | 0.4 | 5 | 68 | Dimethoxystilbenemaleimide | 2, 102 |
| | 384 | 25000$^a$ | 460 | H | S | 69 | Dimethoxystilbeneiodoacetamide | 2 |
| | 350 | 221800 | 480 | 0.32 | 1.5 | 70 | Dimethylaminostilbenemaleimide | 103 |
| Carbazoles | 370 | M | 420 | M | 22.7 | 71 | 3-Azido-N-ethylcarbazole | 2 |
| Phenanthridine | 504 | 4000 | 600 | Variable | 6–30 | 72, 73 | Ethidium mono- and diazides | 104–106 |
| Anthracenes | 390 | L | 450 | H | M | 74 | Anthracene-9-maleimide | 107 |
| | 387 | L | 450 | 0.6 | 29 | 75 | Anthracene-2-isocyanate | 208, 109 |
| | 390 | L | 420–450 | 0.1 | 20 | 76 | Succinimidyl 9-anthroate | 110 |
| | 393 | 8000 | 465 | 0.017 | M | 77 | Carboxaldehyde carbohydrazone | 111, 112 |
| Acridines | 360 | 12000 | 440 | 0.17 | S | 78 | 9-Maleimide | 113 |
| | 440 | M | 470 | 0.04 | 3.5 | 79 | 9-Hydrazine | 5 |
| | 445 | 4600 | 516 | 0.021 | S | 80 | Proflavinmonosemicarbazide | 111 |
| Fluorescein | 490–495 | 75000 | 520 | 0.5 | 4.2 | 81, 82 | 5- and 6-isothiocyanates | 109, 114, 115 |
| | 495 | 77000 | 521 | H | 4$^a$ | 83, 84 | 5- and 6-iodoacetamides | 2, 48, 49, 116, 117 |
| | 488 | 55000 | 520 | H | 4$^a$ | 85 | 5-Maleimide | 2, 23, 118 |
| | 492 | 82000 | 513 | H | 4$^a$ | 86, 87 | 5- and 6-dichlorotriazines | 2, 119, 120 |
| | 492 | 69000 | 518 | 0.55 | 2.9 | 88 | 5-Thiosemicarbazide | 2, 121–123 |
| | 490$^a$ | 70000$^a$ | 520$^a$ | 0.5$^a$ | 3$^a$ | 89 | Fluorescein alanine hydrazide | 124 |
| | 495 | H | 518 | 0.12 | 2.5 | 90 | Difluorescein cystine | 5, 6, 125, 126 |
| Fluorescein | 499 | 78000 | 525 | ? | S | 91 | Fluorescein mercuric acetate | 127, 128 |
| Eosins | 525 | 83000 | 560 | 0.2 | 0.9 | 92, 93 | 5- and 6-isothiocyanates | 2, 129–131 |
| | 525 | 80000$^a$ | 560$^a$ | L–M | S | 94 | 5-Iodoacetamide | 130, 132 |
| | 525 | 80000$^a$ | 560$^a$ | L–M | S | 95 | 5-Maleimide | 130, 133, 134 |
| | 525 | 80000$^a$ | 560$^a$ | L–M | S | 96 | 5-Dichlorotriazine | 2 |
| | 525 | 91000 | 542 | 0.07 | S | 97 | 5-Thiosemicarbazide | 122, 135 |
| Rhodamines | 550 | H | 585 | 0.7(?) | 3 | 98 | Rhodamine B isothiocyanate | 109 |
| | 515, 550 | H | 580 | 0.28 | S | 99 | Tetramethyl isothiocyanate TRITC | 136–138 |
| | 585 | H | 610 | M–H | S | 100 | XRITC | 139 |
| | 550 | H | 580 | M–H | S | 101 | Tetramethyl iodoacetamide | 140 |
| | 535 | H | 580 | L | 1 | 102 | Lissamine Rh sulfonyl chloride | 109, 141, 142 |
| | 596 | 80000$^a$ | 615 | H | S | 103 | Texas Red sulfonyl chloride | 2 |
| | 568 | H | 600$^a$ | ? | S | 104 | Rosamine isothiocyanate | 143 |
| | 550$^a$ | H | 585$^a$ | M$^a$ | S | 105, 106 | Rhodamine hydrazides | 124 |
| Pyrenes | 343 | 38000 | 380 | 0.18 | 40, 120 | 107 | 1-Pyrenemaleimide | 23, 144–146 |
| | 357 | 16900? | ..$^a$ | M–H | 30+ | 108 | 1-Pyrenemethyl iodoacetate | 147 |
| | 342, 365 | 22400 | 386, 407 | 0.08–0.41 | 4–10 | 109 | 1-Pyreneiodoacetamide | 2, 208 |
| | 386 | M | ? | M | M | 110 | 1-Pyrene isothiocyanate | 2 |
| | 346 | 40500 | 380, 400 | H | 90+ | 111 | Pyrene Butyryl sulfate | 148 |
| | 345 | 40500 | 378, 398 | 0.6 | 75–127 | 112 | Pyrene butyric acetic anhydride | 149 |
| | 345 | 40500 | 380, 400 | 0.6$^a$ | 100$^a$ | 113 | p-Nitrophenyl pyrene butyrate | 2 |
| | 345 | 40500 | 380, 400 | 0.6$^a$ | 100$^a$ | 114 | Succinimidyl pyrene butyrate | 2 |
| | 373 | H | 463 | ? | 3–20 | 115 | Succinimidyl pyrene acrylate | 2 |
| | 348 | 36500 | 397 | 0.07–0.6 | M–Lg | 116 | 1-Pyrenebutyryl hydrazide | 111, 150 |
| | 350 | 30000$^a$ | 380, 398 | M | M–Lg | 117 | 1-Pyrenesulfonyl hydrazine | 2 |
| | 350 | 30000$^a$ | 380, 400 | H | 30$^a$ | 118 | 1-Pyrenesulfonyl chloride | 2, 151 |
| | 349 | 39000 | 380 | 0.3 | 80, 160 | 119 | Pyrenebutylphosphono fluoridate | 152, 153 |
| | ? | ? | ? | ? | ? | 120 | Isothiocyanopyrene trisulfonate | 154 |
| | 340 | H | 440 | H | M–Lg | 121 | 1-Azidopyrene | 155, 156 |
| | 346 | H | 380 | H | M–Lg | 122 | 1-Pyrene sulfonyl azide | 157, 158 |
| Chrysenes | 360 | H | 400 | H | M–Lg | 123 | 6-Chrysene maleimide | 2 |

$^a$ This author's estimates.

of the commercial dyes such as azo and quinone dyes are weakly fluorescent or nonfluorescent due to facile nonradiative pathways for depopulation of the excited states. Those chromophores with detectable fluorescence may vary considerably in molar absorptivity, which is a factor just as important as quantum yield in determining the ultimate intensity. In general, chromophores with extended conjugation and resonance have more allowed transitions and higher extinction coefficients for their longest-wavelength absorptions. Compounds with one or two aromatic rings have low absorptivity (under 10,000

$cm^2/mmol$, while fluorophores with several rings and especially those with two equivalent resonance forms such as rhodamines and symmetric carbocyanines can have extinctions up to $200,000\ cm^2/mmol$, although values of 10,000 to $50,000\ cm^2/mmol$ are more common.

Relatively few fluorophores have been coupled to reactive handles to make them useful as covalent fluorescent probes. The approximate spectral properties of a large number of covalent conjugates are arranged in Table 1 in order of increasing molecular size of the basic fluorophore. Estimates of the spectral properties are for the protein conjugates rather than for the probes themselves since in several cases (for example, most "fluorescent" maleimides and dansyl chloride) the probes are not fluorescent until after conjugation, and absorption of the adduct is considerably different from that of the probe. The spectral data is derived from many sources and different instruments. The emission wavelengths may or may not be corrected for variations in the sensitivity of the photomultiplier with wavelength. In many cases, the wavelength and intensity of emission and, to a lesser extent, of absorption vary with the solvent or specific environment and/or pH. This is sometimes indicated in Table 1. In some cases where specific data are incomplete for molar absorptivity and quantum yield, this author's estimates are given. The approximate ranges for absorptivity are: "low (L)", less than $10,000\ cm^2/mmol$; "medium (M)", $10,000-25,000\ cm^2/mmol$; and "high (H)", above $25,000\ cm^2/mmol$. For quantum yields the approximate ranges are: "low", 0–0.1; "medium", 0.1–0.3; and "high", 0.3–1.0. Again the latter may vary considerably and are frequently difficult to measure in biomolecule conjugates. Fluorescence lifetimes ($\tau$) have been determined on only a limited number of probes and are indicated in Table 1 where published. Estimates are given in other cases with the following probable ranges: "short (S)", 9–10 ns; "medium (M)", 10–20 ns; and "long (Lg)", greater than 20 ns. Like the quantum yields, the values for fluorescence lifetime show wide variations with solvent or environment in some cases.


## 3. Covalent Labeling of Biomolecules

An experimental protocol for covalent labeling must be developed for each probe with special regard to intrinsic factors of stability, solubility, and degree or location of modification desired. Optimum methods have been developed for only a few probes in limited situations, so that initial use of probes becomes somewhat a matter of guesswork. There are, however, some guidelines and generalizations that can be made to aid the initial use of a new probe or an old probe in a new situation. Further refinements in such matters as pH, temperature, and incubation time can then be made to optimize labeling.

## 3.1. Solubilization

Before chemical reactions can occur, it is almost invariably necessary to dissolve the probe in a solvent or buffer. Although biomolecules can be labeled in biphasic mixtures such as crystals, amorphous solids, suspensions, or even *in vivo*, better reproducibility in labeling usually occurs when they are dissolved in true solutions where standard kinetics controls the rate of modification. Unfortunately, many spectrally useful probes are almost totally water insoluble. The usual solution is to add the probe from a concentrated stock solution in an organic solvent to the aqueous biomolecule. Probably the best single organic solvent to use is *N,N*-dimethylformamide (DMF). This highly polar aprotic solvent dissolves almost all fluorescent probes yet is not nucleophilic and is compatible with many biomolecules. Dimethyl sulfoxide (DMSO) has been used similarly, but some probes such as dansyl chloride react exothermically with DMSO to give decomposition products.[159] Traces of dimethyl sulfide in DMSO may also react with iodoacetamides. Other solvents used include the lower alcohols, acetone, tetrahydrofuran (THF), dioxane, and acetonitrile, all of which are completely water miscible. Being weak nucleophiles, alcohols should not be used with most acylating reagents. Acetone can potentially form Schiff bases, especially at higher pH, and is, of course, not compatible with aldehyde/ketone reagents. THF and dioxane frequently contain high levels of organic peroxides that can oxidize thiols and other groups. Acetonitrile is excellent as long as it dissolves the probe. Due to the highly reactive nature of most probe molecules, fresh stock solutions should be prepared frequently for maximum reproducibility of labeling.

The natural "soapiness" of many biomolecules in solution is a great asset in preparing and maintaining concentrations of otherwise water-insoluble probes orders of magnitude higher than in pure water, and it is usually advantageous to dilute the probe into the concentrated biomolecule solution rather than into pure buffer whenever high local concentrations of organic solvents can be avoided by rapid mixing.

It is sometimes necessary to avoid use of organic solvents. A number of acidic probes such as fluorescein and salicylic acid derivatives and IAEDANS (32)* are not appreciably water soluble unless the pH is adjusted above 7. The crystals of several probes are very slow to dissolve. Dissolution is much faster if the probe is dissolved in a volatile solvent such as ether and mixed with the buffer, followed by partial evaporation, or the volatile solvent is evaporated on the surface of the container used for labeling. Sonication is another method of dispersing a water-insoluble probe and increasing the surface area of the probe in contact with the aqueous biomolecule.

---

* Numbers refer to compounds in Table 1.

An even higher surface area for labeling with water-insoluble probes can be achieved by adsorbing the probe on Celite (Johns–Manville brand of diatomaceous earth).[160] This is best done by evaporating the probe from an organic solvent onto Celite that has been previously dried at 200°C so that the probe-to-Celite ratio is about 1 to 10. While this sometimes gives rapid labeling with water-insoluble probes in aqueous suspension, the location of the labeling may be different from that done in homogeneous solution. For instance, α-chymotrypsin labeled with dansyl chloride in solution rapidly inactivates the enzyme by reaction with the buried active site serine, while dansyl chloride on Celite predominantly modifies surface lysines. Conversely, Celite-bound fluorescent probes may be useful in preserving biological activity or in achieving mainly surface labeling of cells.

Another alternative for increasing aqueous probe solubility is entrapment of the probe in β-cyclodextrin (cycloheptaamylose).[161,162] This partially water soluble clathrate forms a moderately polar cavity around the probe. This has been used to solubilize dansyl chloride and fluorescamine and should be suitable for most two-ring fluorophores.

In some cases aqueous probe concentration can be increased without biomolecule denaturation by use of low concentrations of detergents such as sodium dodecyl sulfate (SDS) or the easily removable nonionic surfactants such as the Triton, Brij, or alkyl sugars. Similarly, dansyl chloride has been ultrasonically dispersed in lecithin–cholesterol micelles and these used to label erythrocyte membranes without damage.[163]

## 3.2. Removal of Excess Probe

In most cases, removal of the nonconjugated probe is necessary after labeling. This can range from very easy to very difficult. With water-soluble biomolecules the most rapid method is chromatography on a molecular-weight exclusion column such as Sephadex G-25, with the low-molecular-weight probe being tightly bound and the biomolecule passing through with the void volume. Dialysis or membrane filtration can also be used with water-soluble probes. Water-insoluble biomolecules can be directly centrifuged and washed; however, this is frequently not sufficient to remove water-insoluble probes that are concentrated in hydrophobic regions of the biomolecules. A good method for assisting removal of water-insoluble probes is to convert the unreacted probe to a more water-soluble derivative. This also quenches the reaction and can assist in kinetic studies of the rate of modification. For alkylating fluorophores ("sulfhydryl reagents") the following are suggested: thioglycolic acid, thiomalic acid, dithiothreitol (DTT), mercaptoethanol, cysteine, or glutathione. For acylating reagents, buffered solutions of glycine, lysine, glutamic acid, taurine, or glucosamine are useful. Pyruvic acid, acetone dicarboxylic acid, or potentially glucose can assist removal of excess aldehyde/ketone reagents. These quenching reagents are usually used at 1–10 mM at the highest practical pH. Another general method useful for water-soluble biomolecules is to adsorb the non-conjugated probe on activated charcoal followed by centrifugation or filtration. This is frequently necessary to remove the last traces of fluorescein from FITC conjugates and to remove rhodamines from their conjugates since they are typically strongly adsorbed noncovalently.

## 3.3. Other Considerations in Labeling

The rate of labeling in homogeneous solution is usually either second order (i.e., rate $= k$ [biomolecule] [probe]) at low probe concentration or pseudo-first order (rate $= k'$ [biomolecule]) when the ratio of probe to biomolecule is high. In either case the biomolecule and probe concentrations should be kept as high as possible for the fastest labeling. The high biomolecule concentration also helps to solubilize the probe. At low biomolecule concentrations, rates of decomposition or reaction of the probe with water may exceed the modification rate, leading to a low degree of labeling, particularly when the probe concentration is also low.

Probably a useful beginning point for modification of a biomolecule with a new probe is to use the highest practical concentration of biomolecule with about 0.01–0.1 mM probe at the optimum pH for the type of reaction (discussed below) for either 2 h at room temperature or overnight on ice. Alternatively, the kinetics of modification can be followed by removing aliquots from the reaction, freeing the aliquots from excess probe, diluting, and measuring the incorporated fluorescence. Direct and continuous determination of modification rates by following changes in fluorescence with time in the reaction mixture is frequently not possible due to the photosensitivity of many probes and the high absorbance and scattering of most reaction mixtures. The labeling reagents may also be intrinsically fluorescent.

One of the most difficult problems in labeling with fluorescent probes is in precise determination of the degree of conjugation. Probably the most precise method is with radioactive fluorescent probes, but currently only three (dansyl chloride, NBD chloride, and 1,5-IAEDANS) are commercially available. The molar absorptivity (extinction coefficient) can often be inferred from model compounds and measured for biomolecule conjugates after correction for light scattering. For a pure water-soluble biomolecule of known extinction, this can give the degree of modification. The precision of using the extinction coefficient of model compounds is quite variable. The extinction coefficient of FITC-labeled proteins is usually about 10% less than model compounds, but the value for NBD chloride derivatives of amines varies from 18,000 to 30,000, depending on solvent. Even more variable as a quantitative measure of conjugation is the fluorescence intensity of the probe. For many probes the quantum yield varies an order of magnitude or greater depending on the environment and, indeed only by knowing the degree of modification can one accurately estimate the average quantum yield.

When labeling with any reactive probe it is advisable to remove oxidation-protecting reagents such as mercaptoethanol or DTT unless these have been shown to be nonreactive with the probe. These can be readded to quench the reaction. An inert gas atmosphere may be useful as an alternative if necessary. Also to be avoided in some cases is tris-hydroxymethylaminomethane (TrIS) and related buffers which introduce a high concentration of potentially reactive nucleophiles. Phosphate, borate, and carbonate buffers are generally nonreactive. The presence of chelators in the buffer may actually enhance modification by removing heavy metal contaminants from thiols.

Many reactive handles on the covalent fluorescent probes and especially iodoacetamides and azides are very light sensitive. Some fluorophores are also light sensitive. Labeling in the dark is therefore always recommended. Most common fluorophores and reactive groups are quite oxygen stable.

# 4. Selective Modification Reactions

This section will consider techniques and probes that have been developed with some emphasis on achieving selective modification in complex systems. While there is some structural and reactivity overlap among the major classes of biomolecules (proteins, nucleic acids, polysaccharides, and lipids), most reactions have been studied on the isolated materials.

## 4.1. Thiol Modification

Free thiols (sulfhydryl groups, mercaptans) occur primarily as cysteine in proteins. On rare occasions they occur as thio derivatives of nucleic acids. Due to high nucleophilicity they are readily modified by all of the alkylating reagents. They are also specifically modified by organic mercurials and exchange reactions by fluorescent disulfides. When present, aqueous exposed sulfhydryl groups can be modified with a high degree of selectivity with respect to other protein nucleophiles. In some cases environment-enhanced reactivity permits selective labeling of a single thiol in proteins containing several cysteines. A good example is myosin where the two homologous thiols (one on each head) are at least 90% modified by IAEDANS (32) before any of the other 42 thiols are significantly reacted.[164] To selectively label cysteine, a pH of 7.0–8.0 is usually optimum. Above this pH the reaction rate with tyrosine and lysine becomes more significant. Below this pH the thiol is primarily in the protonated form and the reaction with histidine may be of comparable rate.

Three major classes of reactive alkylating groups have been used for selective thiol modification: haloacetyl, maleimides, and aziridines. Haloacetyl probes, of which the fluorescent iodoacetamides are most common, have been widely used. The major probes to be used have been 1,5-IAEDANS at shorter wavelengths and 5-iodoacetamidofluorescein (83) at longer wavelengths. Other fluorescent iodoacetyl derivatives from Table 1 are 3, 4, 23, 33, 34, 35, 69, 84, 94, 101, 108, and 109. Both IAEDANS and iodoacetamidofluorescein are appreciably water soluble above pH 7 but, like all iodoacetamides, are rapidly photolyzed by ultraviolet light. The covalent conjugates are of much greater photolytic stability. The fluorescein derivative absorbs more strongly (by a factor of at least 10) and has emission of high quantum yield that is not particularly environment (but somewhat pH) sensitive. The 1,5-IAEDANS lifetime is relatively long (up to 20 ns), making it useful for rotational correlation and segmental flexibility studies of proteins of molecular weight up to $5 \times 10^5$ daltons.[165,166] Pyrene methyl iodoacetate (108) has an even longer lifetime (30 to 60 ns). The spectral overlap of the fluorescence of 1,5-IAEDANS and absorbance of 5-iodo-acetamidofluorescein permits efficient fluorescence-energy transfer between sites labeled with this pair with an approximate distance of 50% quenching ($R_0$) of 5 nm.[48,49,50,167] Actin labeled with 5-iodoacetamidofluorescein retains full biological activity when injected into living cells and becomes regionally concentrated during different stages of cell division.[117,168,169] Laser photobleaching of 5-IAF-labeled myosin fragments has been used to estimate translational diffusion coefficients and to determine the apparent association constant for myosin binding to actin.[170]

Several probes, including two iodoacetyl derivatives, have the unique feature of giving appreciably fluorescent conjugates only with proteins having a suitably situated nucleophilic residue located in a cavity in which aqueous solvation is partially excluded. This double requirement of both a nucleophile and a hydrophobic cavity is satisfied by only a limited number of proteins. Surface-labeled sites exposed to water have a low quantum yield and contribute little to the overall fluorescence intensity. These covalent hydrophobic probes are predominantly derivatives of anilinonaphthalene sulfonate (ANS) or nitrobenzoxadiazole amines (NBD) the fluorescence intensity of which is strongly quenched in aqueous solution. The iodoacetyl derivatives of ANS (IAANS, 35) and of NBD (IANBD, 23) and the maleimide derivatives of ANS (Mal-ANS, 36) and anilino-naphthalene (ANM, 37) are predominantly sulfhydryl reactive while toluidinyl-naphthalene sulfonyl chloride (31) and mansyl chloride (30) are amine and phenol selective. NBD chloride (20) and all other NBD derivatives share this same high degree of environmentally sensitive fluorescence. Since their fluorescence intensity is very environment sensitive, these probes also tend to respond to conformational changes induced by substrate, modifier, or inorganic ion binding. Creatine kinase is labeled specifically by IAANS at a thiol adjacent to the active site. The fluorescence intensity of the conjugate is strongly enhanced by purine nucleotides but partially quenched by pyrimidine nucleotides or anions.[52] Mal-ANS showed little response to these same modifiers, apparently because its fluorescence was already maximally enhanced by the polarity of the environment sensed. Troponin C labeled with IAANS shows strong fluorescence enhancement on $CA^{2+}$ binding,[54] while divalent cations influenced the rate and extent of modification with Mal-ANS.[55] IANBD shows appreciable fluorescence on only the two fast reacting thiols of myosin[37] but is also useful as a long-wavelength fluorescence energy acceptor.[171] The fluorescence intensity of proteins labeled with covalent hydrophobic probes is sensitive to the folding of the protein and is eventually reduced on exposure to water during denaturation.

Most of the maleimide derivatives of fluorophores have the desirable property of being nonfluorescent until after conjugation of a reactive nucleophile across the maleimide double bond. Since fluorescence of the unreacted probe usually contributes insignificantly to the total fluorescence, the rate of modification can readily be followed. The fluorescence properties of the conjugate are determined by the chromophore attached to the maleimide. In addition to the maleimide covalent hydrophobic probes Mal-ANS and ANM mentioned above, some of the more useful "fluorescent" maleimides are: derivatives of salicylic acid (1 and 2) that are highly water soluble and reported to be membrane impermeant,[1] maleimides of stilbene and coumarin (58, 59, 68, 70) with both high absorbance and quantum yield, pyrene maleimide (107) with a fluorescence lifetime up to 100 ns, and fluorescein and eosin maleimides (85 and 95) with the longest-wavelength fluorescence and high intensity. Since the fluorescence intensity is proportional to the degree of conjugation for a nonenvironment-sensitive probe, the intensity relative to some calibration standard can be used to both locate and quantitate sulfhydryl groups. Perhaps the best probe for this purpose is 5-maleimidylsalicylic acid (2) which is very water soluble and not particularly environment sensitive. This reagent, and potentially several other fluorescent maleimides, can also be used to locate thiol-containing peptides separated by electrophoresis or as a sensitive sulfhydryl spray reagent on chromatograms. The coumarin maleimides DACM (58) and CPM (59) (the latter recently developed by T. O. Sippel of the University of Michigan) appear to be the reagents of choice for histochemical demonstration of thiols. CPM has the advantage over DACM of not requiring long reaction times for maximal fluorescence.

The chemical instability of the maleimide ring attaching group introduces some limitations in its use. Ring-opening hydrolysis of the anhydride-like maleimide ring occurs at an appreciable rate (half opened in minutes) above pH 8.[172] Since selective modification of thiols is usually conducted below this pH and the maleimide reagent is usually used in excess, this is often not critical. A more significant problem exists when the protein thiol adduct of the maleimide ring opens after conjugation to the biomolecule, since the fluorescence properties of the closed-ring thiol adduct and the open-ring adduct (an amide of thiomalic acid) are sometimes significantly different. An interesting application of this ring instability is the potential cross-linking of the maleimide adduct with a spatially adjacent lysine residue to form a diamide derivative. This has been reported to occur for the pyrene maleimide derivative of serum albumin.[145] Pyrene maleimide also shows the interesting feature of excimer fluorescence when two probes are on spatially adjacent thiols, as occurs in tropomyosin.[173]

A potential use of the new stilbene dimaleimides **66** and **67** is for cross-linking spatially adjacent thiols. It would be expected that only the adducts in which *both* maleimides are reacted will be fluorescent so that only intra- or interprotein cross-linked molecules will be observable. Presumably other dimaleimides with different distances between reactive groups can be developed to span longer or shorter lengths.

Aziridines can be divided chemically into "activated" and "unactivated" classes. The activated aziridines are amides of carboxylic acids such as anthranilic acid (**8**) or of sulfonic acids such as dansyl (**41**). The unactivated aziridines are *N*-alkyl- or *N*-arylethylenimine derivatives such as NBD-aziridine (**24**). Activated aziridines undergo ring-opening nucleophilic reactions with thiols or amines in basic solution, while the unactivated aziridines are very stable in base and undergo ring-opening reactions only with acid catalysis. The latter may be most useful for carboxylic acid modification. By far the most widely used fluorescent aziridine has been dansyl aziridine, particularly with the muscle proteins actin, myosin, and troponin C. On actin, one cysteine is modified with high selectivity,[174] while on the cysteine-free troponin C the site of reaction is a methionine residue.[59,60,175]

Two reactions of thiols are specific for thiols in the presence of all other nucleophiles found in biomolecules. These are the disulfide interchange reaction and reaction with organic mercurials. Both reactions are reversible, and the label can be displaced by the common thiols such as 2-mercaptoethanol, DTT, cysteine, or reduced glutathione. They can also be transferred from one cysteine to another within or between proteins, which can make their localization difficult. Didansyl cystine (**42**), difluorescein cystine (**90**), and some NBD disulfides (**21** and **22**) have all been used as disulfide probes. Organomercurials in which the mercury is directly attached to the aromatic ring usually show heavy-atom fluorescence quenching, but a mercurial derived from dansyl cysteine (**43**) shows strong fluorescence enhancement after formation of a protein-thiol adduct; fluorescein mercurial has also been used as a sulfhydryl reagent and, being a dimercurial, may be a cross-linking reagent.[127,128]

Two types of cysteine reactive handles that are best termed "arylating reagents" rather than "alkylating reagents" are NBD chloride (**20**) and several dichlorotriazine probes (**86, 87, 96**). These rapidly form moderately stable adducts with thiols. NBD chloride is similar in reactivity to fluorodinitrobenzene (FDNB, Sanger's reagent) in reacting with either thiols or amines. Absorption of the NBD thiol adduct is at about 425 nm while the amine adduct is near 475 nm, so that the type of residue modified can readily be determined. The fluorescence of the NBD thiol adduct is very weak, especially in an aqueous environment. With 2-mercaptoethylamine (cysteamine) and NBD chloride, the initial site of reaction is the thiol. This is followed by an intramolecular transfer of the benzoxadiazole to the amine. An identical S to N transfer appears to occur in actin where initial reaction of a thiol is transferred to a spatially adjacent lysine. The NBD chromophore can also be removed from some protein thiol (but not protein amine) adducts by treatment with extrinsic thiols.

Dichlorotriazines are the reactive acid chlorides of cyanuric acid. These probes are usually prepared by replacement of one chlorine of cyanuric chloride by a fluorescent amine derivative. The two remaining chlorides can be successively replaced by nucleophiles, including thiols and amines. In the case of a protein, reaction of the third chlorine would give a cross-linked product, although this reaction is usually slow. The dichlorotriazinylamino derivatives of fluorescein (**86** and **87**) have been proposed as alternatives to fluorescein isothiocyanate for immunofluorescence tracing since they are equally fluorescent and the conjugation proceeds with greater ease and reproducibility at a lower pH.[119,120] Several reactive triazines, including bifunctional dichlorotriazines of diaminostilbenedisulfonic acid, have been described in the patent literature as cloth-reactive fluorescent brightening agents and may be useful as fluorescent cross-linking reagents or covalent probes.

Thiol residues do react with common acylating reagents; however, few stable conjugates are formed. The reaction with thiols can, however, consume reagent and sometimes modify the protein without formation of a fluorescent conjugate. An example is the reaction of dansyl chloride with the active-site adjacent thiol of creatine kinase, which results in net oxidation of the thiol (probably to a sulfenic acid) without incorporation of the probe. Reaction of dansyl chloride with cysteamine requires two equivalents of dansyl chloride, and the product is a disulfide rather than the original thiol. Similar reactions probably occur frequently in thiol-containing proteins modified with dansyl chloride. Isothiocyanates, including the widely used fluorescein isothiocyanate, also react with thiols to give unstable dithiourethanes.[176] These can either revert back to the isothiocyanate, transfer the acyl group to an amine, or undergo hydrolysis with loss of reagent. Thio esters can also be formed from reactive derivatives of carboxylic acids. Again these tend to transfer the acyl group to other nucleophiles, particularly amines.

In summary, modification of thiols in proteins is usually the most facile reaction to achieve with a variety of fluorescent alkylating reagents, with the most stable adducts formed from iodoacetamides and aziridines. Maleimide derivatives have spectral advantages for following the modification but are susceptible to unpredictable ring-opening reactions. Modification by several other classes of compounds is partially reversible.

## 4.2. Methionine Modification

Dialkyl sulfides such as methionine are generally nonreactive; however, particularly when protected from aqueous solvation in a protein, they are susceptible to alkylation to a sulfonium salt. Although no instances of reaction of fluorescent maleimides with methionine have been reported, aziridines and iodoacetamides undergo the reaction. The ternary salts formed, however, may not be completely stable since trialkyl sulfonium salts are themselves alkylating reagents for other nucleophiles.[177] Furthermore, iodoacetamide has been suggested as a reagent for the specific cleavage of peptides at methionine residues,[178] and this may be a useful technique for locating the probe in the protein. Methionine modification in the presence of cysteine is unlikely to be successful unless the cysteine residues are protected as disulfides such as by prior reaction with dithiobis-2-nitrobenzoic acid (DTNB, Elman's reagent). In the presence of other potential nucleophiles such as histidine, tyrosine, and lysine, advantage can be taken of the pH insensitivity of methionine modification to suppress the reaction of other nucleophiles by working at the lowest practical pH.

## 4.3. Histidine Modification

Little systematic work has been done on the selective modification of histidine by fluorescent reagents although it has occasionally been found that the imidazole ring of histidine has been alkylated by iodoacetamides. In several enzymes, the presence of a catalytically active histidine near the active site has been indicated. In chymotrypsin this proximity and catalytic preference for aromatic substrates has been used to effect the specific alkylation of histidine with a series of chloromethyl ketones.[79] Similar use of reactive substrate analogs may enable labeling of other catalytically important histidine residues in proteins that do not contain cysteines. The histidine of lysozyme has been selectively labeled at pH 6.2 by a dichlorotriazinyl spin label which indicates that this reactive handle on the fluorescent probes **86, 87,** and **96** may also be useful for histidine modification.[179] N-acyl imidazoles are generally very unstable hydrolytically and can themselves be used as modification reagents. Dansyl imidazole has been prepared and found to be similarly reactive.[180]

## 4.4. Amine Modification

Selective modification of the epsilon amino groups of lysine in proteins is usually easy provided the pH can be raised above 9 where the concentration of unprotonated amine starts to become significant. The protonated aliphatic amine is devoid of reactivity, and modification by any reagent will be very slow below this pH. In contrast, the $pK_a$ of the alpha amino group of peptides is near 7, so that selective modification of the N-terminus can usually be accomplished if it is free. Amino acyl derivatives of transfer ribonucleic acids can also be selectively modified at this pH. Phosphatidylethanolamine is also a relatively weak base and is frequently modified where it is a component of cell membranes. Probably the four most widely used covalent fluorescent probes are dansyl chloride, fluorescein isothiocyanate, o-phthalaldehyde, and Fluorescamine (trademark of Roche Diagnostics), all of which are predominantly amine reactive and all of which can be used under approximately the same conditions. Dansyl chloride (27) forms highly stable sulfonamide derivatives with amines. The spectral properties of the conjugates have both good and bad features. The absorptivity is very low (typically 4500 cm²/mmol) compared with fluorescein (about 75,000 cm²/mmol). The quantum yield is also quite environment sensitive and can be low (less than 0.1) for sites exposed to water. The Stokes shift of emission is extremely high with emission up to 200 nm red shifted from absorption. The fluorescence lifetime, while only moderate, is longer than most covalent probes. The reactivity of the reagent is quite high, which means conjugation is facile, but the reagent also hydrolyzes at an appreciable rate. Storage can also be a problem. Reaction with tyrosine to form O-dansyl esters is sometimes predominant, and these esters are also stable. As mentioned above, dansyl chloride has a tendency to react oxidatively with thiols. Under certain conditions it also reacts with amino acids to give aldehydes.[181] Despite these shortcomings, dansyl chloride will likely continue to be widely used, particularly for detection purposes.

Fluorescein isothiocyanate is almost an ideal covalent fluorescent probe. Its only spectral drawbacks are a small Stokes shift of about 25 nm, which necessitates sharp cutoff filters or monochromators, and a tendency toward photobleaching. Its advantages are high extinction coefficient, high quantum yield that is almost invariant with environment, and water solubility for FITC. The spectra are far removed from the usual naturally fluorescent components of biomolecules although serum with high bilirubin interferes. Frequently the limiting factor in detecting fluorescein fluorescence is the Raman scattering peak buried under the emission. In the synthesis of FITC, two isomers are formed, first called isomer I and isomer II, which correspond to fluorescein-5-isothiocyanate and fluorescein-6-isothiocyanate, respectively. These differ in the point of attachment of the isothiocyanate to the single benzene ring. The 5-isomer (isomer I) of FITC has been most widely used although the spectral properties of the two isomers are quite similar and some commercial preparations are a mixture of the two isomers. In contrast to the original isocyanates,[182] the isothiocyanates undergo solvolysis only slowly in aqueous or alcoholic solutions. The thiourea adduct formed by reaction of an isothiocyanate with an amine is somewhat unstable to hydrolysis to a urea and hydrogen sulfide, particularly at extremes of pH. FITC conjugates do not survive the conditions of protein hydrolysis for amino acid analysis. Fluorescein conjugates are susceptible to concentration quenching whenever two fluorescein chromophores are spatially adjacent. This can even occur at longer distances by singlet–singlet energy transfer. Fluorescein fluorescence is also strongly quenched by binding of its antibody, which can be useful for determining whether the label is on the outside surface of a cell.[183] As an isothiocyanate, FITC and certain other fluorescent isothiocyanates have been suggested but not widely used as alternatives to phenyl isothiocyanate in the Edman degradation of peptides for sequence analysis.[114,154]

Rhodamine isothiocyanates have the advantage over FITC of greater photolytic stability; however, they have a lower quantum yield, are usually a mixture of the 5- and 6-isomers, and are difficult to dissolve in the conjugation medium; it is also difficult to remove excess isothiocyanate after conjugation. The new XRITC (100) absorbs and emits at longer wavelengths than either rhodamine B (tetraethyl) or tetramethylrhodamine isothiocyanates.

o-Phthalaldehyde and Fluorescamine are both reagents that have predominantly been used to quantitate primary aliphatic amines. o-Phthalaldehyde (13) is reacted with amines in the presence of a mercaptan (usually 2-mercaptoethanol) to give a fluorescent isoindole derivative.[15,16] Modifications of the method permit detection of the secondary amino acid, proline. The isoindole conjugates are not completely stable and the fluorophore is destroyed with time so that, as a probe, phthalaldehyde may not be useful. Fluorescamine is an unusual probe in that neither it nor its hydrolysis product is fluorescent but its amine conjugates are. Conjugation is usually by vortexing a solution of the biomolecule with Fluorescamine dissolved in acetone. The hydrolysis rate of Fluorescamine in water is so fast that conjugation must be complete in minutes. From the published spectral data, it would be expected that the sensitivity of each reagent for amines would be about equal, but at least one paper[15] claims o-phthalaldehyde detection of amines to be 5–10 times as sensitive as Fluorescamine.

Several other lysine-reactive reagents have been developed for specific applications and as alternatives to those mentioned above. The 2,5- and 2,6-isomers of dansyl chloride (28 and 29) show much greater Stokes shift than the "normal" 1,5-isomer, with the 2,5-isomer showing an exceptionally long fluorescence lifetime. Pyrene sulfonyl chloride (118) has a quantum yield similar to dansyl chloride but significantly stronger absorbance, making its detectability greater. It also has a longer fluorescence lifetime.

The two very water-soluble isothiocyanate derivatives of stilbene, SITS and DIDS (63 and 65), have been used most frequently for their specific anion-transport inhibition[97] although they are also of potential use as the short-wavelength "third probe" for multiply labeled fluorescence-activated cell sorting, due to its fluorescence being spectrally well separated from other probes used. Coumarin phenyl isothiocyanate (60) is a new and both intensely absorbing and strong emitting isothiocyanate.

Several methods exist for formation of simple amide derivatives between fluorescent carboxylic acids and biomolecule amines. Since carboxylic acids and amines form salts rather than covalent bonds, all of these methods depend on some method of "activation" of the carboxy probe to provide the necessary reaction mechanism. These probes differ in the nature of the "leaving group" and include most of the methods used by synthetic-peptide chemists. Carboxylic acid chlorides are usually not used due to their high reactivity, while simple esters usually are not sufficiently reactive. Exceptions are esters of p-nitrophenol and N-hydroxysuccinimide. Nitrophenyl esters of aliphatic acids such as pyrene butyric acid (113) readily acylate aliphatic (but not aromatic) amines. They are also good substrates for some enzymatic esterases such as chymotrypsin. The reactivity of p-nitrophenyl esters of aromatic acids depends in part on other substituents on the aromatic ring. Electron-withdrawing substituents increase reactivity while donating groups decrease reactivity. The effect is so strong that p-nitrophenyl anthranilate (7) irreversibly acylates the active-site serine of chymotrypsin but shows no tendency to react with the epsilon amino group of lysines. Succinimidyl esters of carboxylic acids (11, 46, 76, 114, 115) show high specificity for amine modification, while the much more reactive acyl imidazoles also acylate tyrosine and hydroxyl groups. Competitive hydrolysis by water can be a problem with acyl imidazoles.

Fluorescent carboxylic acids can also be made protein reactive by conversion to anhydrides. Usually the symmetric anhydride is not employed since only half of an anhydride is incorporated when it reacts with nucleophiles. Instead "mixed anhydrides" are usually employed. Examples are mixed anhydrides with ethyl chloroformate, sulfur trioxide (111), or trifluoroacetic anhydride. Carbodiimides and particularly the water-soluble carbodiimides have also been used to activate carboxylic acids through formation of a mixed anhydride with the carbodiimide being converted to a urea.[184]

In the modification of amines with all of the reagents so far described, the positive charge associated with the epsilon amine of lysine is lost and in some cases converted to a negative charge. This change can sometimes have a negative effect on the stability, solubility, and/or activity of a biomolecule. Two types of reactions can be used to avoid loss of the positive charge. The first utilizes reaction of the lysine with an imido ester to give a positively charged amidine derivative. Unfortunately most imido esters are somewhat unstable in storage, including the only fluorescent derivatives so far described.[66] The reversible reaction of aldehydes and ketones with amines to form Schiff bases can be used for the highly specific labeling of amines if the Schiff base is reduced with sodium borohydride or (preferably) sodium cyanoborohydride. This reductive alkylation reaction retains the basic character of the amine and produces a highly stable derivative that survives amino acid analysis conditions. The first fluorescent reagent to take advantage of this reaction for covalent modification of amines is an NBD acetaldehyde derivative (25).

## 4.5. Tyrosine Modification

Tyrosine in proteins can be modified by many fluorescent reagents but usually with low selectivity in the presence of other common amino acids such as cysteine and lysine. Dansyl chloride gives stable sulfonate derivatives, FITC reacts to some degree, and iodoacetamides react readily at elevated pH. A spin-labeled imidazole[185] gave tyrosine-specific modification in nucleosomes and particles, suggesting that fluorescent imidazoles may do likewise. An early paper of interest[186] on specific tyrosine modification used the facile nitration of tyrosine with tetranitromethane to form nitrotyrosine followed by dithionite reduction to aminotyrosine. Being weakly basic, the aromatic amine could be modified at a low pH with dansyl chloride. Diazonium salts readily couple with tyrosine at neutral pH to form azo dyes. While azo dyes are not usually fluorescent, the azo dye derived by coupling diazotized o-aminophenol to tyrosine forms a strongly fluorescent chelate with such cations as aluminum or magnesium.[187]

## 4.6. Carboxylic Acid Modification

Due to low nucleophilicity of the carboxylate anion, direct modification of carboxylic acids in biomolecules is usually difficult. Some solutions are possible, but few have been explored using fluorescent probes. The first is reversing the carbodiimide-mediated coupling of an amine with a carboxylic acid to form an amide by using a fluorescent amine derivative to attack a carbodiimide-activated protein carboxyl group. Since a large number of carboxyl groups are usually present, selectivity of modification depends on the ability of the carbodiimide to reach the carboxyl group to form the mixed anhydride derivative. For surface carboxyl residues a polar water-soluble carbodiimide is probably most suitable, but for the rarer buried carboxyl groups a nonpolar carbodiimide such as dicyclohexyl carbodiimide (DCC) or the oxidation–reduction coupling reagent EEDQ may be better. Choice of the amine is also very important. To avoid protein cross-linking, a fluorescent amine with a low $pK_a$ (weak base) is necessary since a significant concentration of free base can be maintained at a pH of 5 to 7 where essentially all of the lysines are in the unreactive charged form. Glycine amides and esters, hydrazides, and hydroxylamines all meet these requirements by having a $pK_a$ near 7.[188]

Fluorescent reagents for direct modification of carboxyl groups have not been demonstrated; however, some are suggested through analogy to carboxylic acid derivatization reagents. A new reagent is 9-diazomethylanthracene.[189] Like diazomethane, this reagent esterifies carboxylic acids at room temperature. It is sufficiently stable to be stored for a period in the freezer, but its use with proteins has not yet been reported. Halomethylcarbonyl compounds (for example, phenacyl bromides) are common carboxylic acid derivatization reagents, and the dansyl chloromethyl ketones (50, 51) have this structure. Aziridines such as NBD aziridine (24) react with carboxylic acids at a low pH; however, this reaction has not yet been reported for proteins.

## 4.7. Modification of Other Residues in Proteins

Although guanidines such as arginine form fluorescent adducts with ninhydrin and o-phenanthrenequinone,[190] the extremely basic conditions required preclude their use at physiological pH. It has been reported, however, that aromatic glyoxal derivatives react reversibly with arginine in the stoichiometry of two glyoxals per arginine.[191,192] While this has been done with chromophoric glyoxal derivatives, it has not yet been reported for fluorescent glyoxal derivatives.

One of the most promising methods for specific modification of biomolecules is enzyme-mediated fluorescent labeling. The most widely used system so far has been guinea pig liver transglutaminase-catalyzed incorporation of dansyl cadaverine (49) into proteins.[76,77,193] Protein glutamine residues are specifically converted to fluorescent glutamic acid amides. This has been reported for actin, nitrate reductase, rhodopsin, spectrin in erythrocyte ghosts, casein, and sarcoplasmic reticulum membrane proteins. Other possibilities for enzyme-mediated fluorescence labeling include acyl transferases utilizing fluorescent Co-A derivatives and enzymes that can transfer fluorescent analogs of nucleotides, sugars, or amino acids.

The nonreactivity of serine and threonine side-chain hydroxyl groups to either alkylating or acylating reagents is not unexpected due to their low nucleophilicity. The major exceptions are the serine residues at the active sites of chymotrypsin, trypsin, thombin, and the cholinesterases. Several examples of very specific fluorescent labeling of these enzymes have been recorded. Mentioned above was the specific modification of chymotrypsin with p-nitrophenyl anthranilate to give a stable acyl enzyme.[9] Dansyl fluoride (53) is the fluoride analog of dansyl chloride, but, unlike that reagent, dansyl fluoride does not react with amines and is very stable to hydrolysis. It reacts with chymotrypsin at the active-site serine[82] and similarly with thrombin[10] and cholinesterase[81] but not with the serine-blocked enzymes. The fluorescent pyrene phosphonofluoridate (119) also specifically modifies the serine of acetylcholinesterase.[152,153]

## 4.8. Modification of Nucleic Acids, Polysaccharides, and Glycoproteins

While many fluorescent reagents exist for modification of proteins, most other biomolecules have an almost total absence of reactive nucleophiles with which to react with the probe. Some specialized methods and reagents have been developed for their selective modification. Enzyme-mediated fluorescence labeling may also become an important method of labeling these biomolecules.

A common feature in polysaccharides, glycoproteins, and RNA (but not DNA) is the vicinal glycol (two hydroxyl groups on adjacent carbons). These are oxidized by periodate at neutral pH to aldehydes, ketones, or mixtures thereof. Cysteine and methionine are also oxidized by periodate but usually at a slower rate than glycols. In RNA, only the terminal free ribonucleotide can be oxidized. After removal of the excess periodate salt, the aldehydes (or rarely ketones) can then be condensed with a fluorescent amine, hydrazine, or hydroxylamine derivative.[194] In the case of reaction with an amine, the resulting Schiff base is usually reduced with sodium borohydride or preferably sodium cyanoborohydride to increase the stability of the linkage. Hydrazone and thiosemicarbazone derivatives are usually sufficiently stable without reduction. One advantage of the reductive method is that it can be done with radioactive borohydride or cyanoborohydride for tracing the reacted sites. Dansyl hydrazine (47) has been the most widely used probe for labeling aldehydes, but several other hydrazine derivatives of pyrene (116, 117), fluorescein (88, 89), eosin (97), acridine (79, 80), anthracene (77), coumarin (62), and tetramethylrhodamine[195] have been developed for aldehyde labeling with spectral characteristics determined by the fluorophores. The same reagents can be used to label the natural aldehydes that occur to some extent in sugars and steroidal ketones.

A new set of fluorescent boronic acid derivatives (Fluoraboras®) shows great promise for labeling glycols and related derivatives without periodate oxidation.[83,84] Boronic acids form derivatives of varying stability with 1,2- or 1,3-glycols and 1,2- or 1,3-amino alcohols, including catecholamines and collagen peptides containing hydroxylysine. The two reactive alcohols or amino alcohol need not come from the same residue of the biomolecule but from any two spatially adjacent residues within a bond's length of the boron. The dansyl derivative (54) has been reported to agglutinate red blood cells and to be of use as a vital stain.[84]

A novel method of labeling the oxidized 3′-terminus of RNA was introduced by Reines and Cantor.[111] This consisted of periodate oxidation of the glycol to the dialdehyde, reaction with carbohydrazide (other dihydrazides such as succinic dihydrazide should also work), then reaction with a fluorescent aldehyde. Similar approaches may be used to introduce sulfhydryl residues into cell-surface carbohydrate residues using thiol hydrazide derivatives such as 2-

acetamido-4-mercaptobutyric acid hydrazide[196] or 2-mercaptoacethydrazide. After introduction of the thiol, one of the many thiol-selective reagents can be used to achieve specific carbohydrate modification.

Except for fluorescent lectins, direct modification of carbohydrate hydroxyl groups by fluorescent probes without oxidation does not appear to have been reported. A spin-labeled acyl imidazole derivative has been reported to specifically label the 2′ hydroxyl of nucleotide homopolymers in aqueous solution.[197] With protein-containing systems, however, the same spin-labeled imidazole gave specific tyrosine modification.[198]

Direct modification of the bases of polynucleotides has not been very successful due to lack of nucleophilic groups. Two exceptions, however, exist. The first is modification of the cytidine base by transamination reactions either catalyzed by bisulfite[199] or using hydrazide derivatives at pH 4.[200] This method may also be suitable for introduction of a reactive thiol at cytidine residues by use of 2-aminoethanethiol (cystamine) followed by reaction with a fluorescent thiol reagent. A second unexploited method is the use of glyoxal derivatives to specifically modify guanosine residues at neutral pH.[201,202] While chromophoric glyoxal derivatives have been reported,[192] fluorescent glyoxals have not.

## 4.9. Modification by Photoactivated Fluorescent Probes

A unique method of achieving fluorescent labeling of biomolecules has recently been developed. The probes are chemically stable in the absence of light but are activated by absorbed light to a highly reactive and short-lived species that can react, usually nonspecifically, with sites that can be reached by diffusion before the reactive intermediate is consumed by reaction with water or by other mechanisms. Most molecules employed have been organic derivatives of azides ($R-N_3$) such as the fluorescent azides 10, 26, 55, 56, 57, 71, 72, 73, 121, and 122. On photolysis these yield reactive nitrenes. These electron-deficient atoms can undergo several reactions, including insertion into carbon–hydrogen bonds or carbon–carbon double bonds, to give covalent adducts.[203] In most cases the azides are nonfluorescent until conjugated to the biomolecule. These probes provide the only method for the potential labeling of lipids using some of the lipid-soluble probes. While several of the azide derivatives may be expected to partition into the lipid component of membranes, diffusion of the nitrene during its lifetime may result in eventual modification of the protein component in contact with the lipid, as apparently occurs with 1-pyrenesulfonyl azide (122).[157,158]

The azide 3-azidonaphthalene-2,7-disulfonic acid (ANDS, 56) is an excellent topographical probe. Unlike most of the other fluorescent photoaffinity reagents, ANDS is essentially membrane impermeant due to two sulfonic acid groups. Photolysis in the presence of intact human erythrocytes gives fluorescent labeling of only those proteins on the external surface. These can be readily identified by gel electrophoresis.[85,86]

While photoaffinity fluorescent labeling is frequently very nonspecific, in certain cases a high degree of specificity can be conferred by making the azide part of a substrate or modifier. Such is the case with 8-azido-1-$N^6$-ethenoadenosine-3',5'-cyclic monophosphate[204] and the similar azidoetheno-ATP analog.[205,206] Each of these nucleotide analogs can be photolyzed from the nonfluorescent azides to give highly specific fluorescent labeling of the nucleotide binding sites. In a similar manner, the mono and diazide derivatives of ethidium bromide (72 and 73) can be photolyzed to give fluorescent nucleic acid derivatives,[104-106] and an azidodecamethonium analog gives a fluorescent-labeled acetylcholine receptor.[207]

## References

1. E. Mercado, G. Carvajal, A. Reyes, and A. Rosado, *Biol. Reprod.* **14**, 632 (1976).
2. R. P. Haugland, *Handbook of Fluorescent Compounds,* Molecular Probes, Inc., Junction City, Oregon (1981).
3. J. J. Holbrook, P. A. Roberts, and R. B. Wallis, *Biochem. J.* **133**, 165 (1973).
4. R. B. Wallis and J. J. Holbrook, *Biochem. J.* **133**, 173 (1973).
5. C.-W. Wu and L. Stryer, *Proc. Natl. Acad. Sci. USA* **69**, 1104 (1972).
6. C.-W. Wu and L. Stryer, *J. Supramol. Struct.* **1**, 348 (1973).
7. J. Burns and P. B. Neame, *Blood* **28**, 674–682 (1966).
8. H. Hahn der Dorsche, *Acta Histochem. Suppl.* **19**, 49–59 (1977).
9. R. P. Haugland and L. Stryer, in *Conformation of Biopolymers* (N. Ramachandran, ed.), Academic Press, New York (1967), pp. 321–335.
10. L. J. Berliner and Y. L. Shen, *Thrombosis Res.* **12**, 15 (1977).
11. R. Walenga, Y. J. Vanderhock, and M. B. Feinstein, *J. Biol. Chem.* **255**, 6024–6027 (1980).
12. W. E. White, Jr., and K. L. Yielding, *Biochem. Biophys. Res. Commun.* **52**, 1129–1133 (1973).
13. P. W. Schiller and A. N. Schechter, *Nucleic Acids Res.* **4**, 2161–2167 (1977).
14. T. M. Jovin, P. T. Englund, and A. Kornberg, *J. Biol. Chem.* **248**, 3173 (1973).
15. J. R. Benson and P. E. Hare, *Proc. Natl. Acad. Sci. USA* **72**, 619–622 (1975).
16. R. F. Chen, C. Scott, and E. Trepman, *Biochim. Biophys. Acta* **576**, 440–455 (1979).
17. M. Weigele, S. DeBernardo, and W. Leimgruber, *Biochem. Biophys. Res. Commun.* **50**, 352–356 (1973).
18. R. E. Stephens, *Anal. Biochem.* **84**, 116–126 (1978).
19. S. De Bernardo, M. Weigele, V. Toome, K. Manhart, W. Leimgruber, F. Böhlen, S. Stein, and S. Udenfriend, *Arch. Biochem. Biophys.* **163**, 390–399 (1974).
20. J. Wideman, L. Brink, and S. Stein, *Anal. Biochem.* **86**, 670–678 (1978).
21. S. Chen-Kiang, S. Stein, and S. Udenfriend, *Anal. Biochem.* **95**, 122–126 (1979).
22. N. S. Kosower, E. M. Kosowerm, G. L. Newton, and H. M. Ranney, *Proc. Natl. Acad. Sci. USA* **76**, 3382–3386 (1979).
23. A. Rao, P. Martin, R. A. F. Teithmeier, and L. C. Cantley, *Biochemistry* **18**, 4505–4516 (1979).
24. Y. Kanaoka, M. Machida, K. Ando, and T. Sekine, *Biochim. Biophys. Acta* **207**, 269–277 (1970).

25. T. Sekine, T. Ohyashiki, M. Machida, and Y. Kanaoka, *Biochim. Biophys. Acta* **351**, 205–213 (1974).
26. T. Iio and H. Kondo, *J. Biochem.* **86**, 1883–1886 (1979).
27. P. B. Ghosh and M. W. Whitehouse, *Biochem. J.* **108**, 155–156 (1968).
28. R. S. Fager, C. B. Kutina, and E. Abrahamson, *Anal. Biochem.* **53**, 290–294 (1973).
29. K. Nitta, S. C. Bratcher, and M. J. Kronman, *Biochem. J.* **177**, 385–392 (1979).
30. G. Allen and G. Lowe, *Biochem. J.* **133**, 679–686 (1973).
31. D. J. Birkett, N. C. Price, G. K. Radda, and A. G. Salmon, *FEBS Lett.* **6**, 346–348 (1970).
32. K. Nitta, S. C. Bratcher, and M. J. Kronman, *Biochem. J.* **177**, 385–392 (1979).
33. A. A. Aboderin, E. Boedefeld, and P. L. Luisi, *Biochim. Biophys. Acta* **328**, 20–30 (1973).
34. R. H. Sigg, P. L. Luisi, and A. A. Aboderin, *J. Biol. Chem.* **252**, 2507–2514 (1977).
35. K. Brocklehurst, J. P. G. Malthouse, and M. Shipton, *Biochem. J.* **183**, 223–231 (1979).
36. K. Brocklehurst, J. A. L. Herbert, R. Norris, and. H. Suschitzky, *Biochem. J.* **183**, 369–373 (1979).
37. R. P. Haugland, *J. Supramol. Struct.* **3**, 338–347 (1975).
38. J. S. Franzen, P. S. Marchetti, and D. S. Feingold, *Biochemistry* **19**, 6080–6089 (1980).
39. J. David Johnson, personal communication.
40. W. R. Gray, *Meth. Enzymol.* **11**, 139–151 (1967).
41. N. Seiler, *Methods Biochem. Anal.* **18**, 259–337 (1970).
42. B. A. Davis, *J. Chromatogr.* **151**, 252–255 (1978).
43. B. K. Fung and L. Stryer, *Biochemistry* **17**, 5241–5248 (1978).
44. R. P. Cory, R. R. Becker, R. Rosenbluth, and I. Isenberg, *J. Am. Chem. Soc.* **90**, 1643–1647 (1968).
45. M. Onodera, H. Shiokawa, and T. Takagi, *J. Biochem.* **79**, 195–202 (1976).
46. F. C. Greene, *Biochemistry* **14**, 747–753 (1975).
47. E. N. Hudson and G. Weber, *Biochemistry* **12**, 2250–2255 (1973).
48. R. Takashi, *Biochemistry* **18**, 5164–5169 (1979).
49. H. Eshaghpour, A. E. Dietrich, C. R. Cantor, and D. M. Crothers, *Biochemistry* **19**, 1797–1805 (1980).
50. D. J. Marsh and S. J. Lowey, *Biochemistry* **19**, 774–784 (1980).
51. A. Steinemann, J. Bietenhader, and M. Dockter, *Anal. Biochem.* **86**, 303–309 (1978).
52. R. P. Haugland, *J. Supramol. Struct.* **3**, 192–199 (1975).
53. N. C. Price, *Biochem. J.* **177**, 603–612 (1979).
54. J. D. Johnson, J. H. Collins, S. P. Robertson, and J. D. Potter, *J. Biol. Chem.* **255**, 9635–9640 (1980).
55. S. S. Gupte and L. K. Lane, *J. Biol. Chem.* **254**, 10362–10369 (1979).
56. Y. Kanaoka, M. Machida, M. Machida, and T. Sekine, *Biochim. Biophys. Acta* **317**, 563–568 (1973).
57. M. Machida, M. Bando, Y. Migita, M. I. Machida, and Y. Kanaoka, *Chem. Pharm. Bull.* **24**, 3045–3057 (1976).
58. W. H. Scouten, R. Lubcher, and W. Baugman, *Biochim. Biophys. Acta* **336**, 421–426 (1974).
59. J. D. Johnson and A. Schwartz, *J. Biol. Chem.* **253**, 6451–6458 (1978).
60. J. D. Johnson, S. C. Charlton, and J. D. Potter, *J. Biol. Chem.* **254**, 3497–3502 (1979).
61. E. P. Lankmayr, K. W. Budna, K. Mueller, and F. Nachtman, *Fresenius Z. Anal. Chem.* **295**, 371–374 (1979).

62. H. C. Cheung, R. Cooke, and L. Smith, *Arch. Biochem. Biophys.* **142**, 333–339 (1971).
63. P. C. Leavis and S. S. Lehrer, *Biochemistry* **13**, 3042–3048 (1974).
64. W. E. Harris and W. L. Stahl, *Biochim. Biophys. Acta* **426**, 325–334 (1976).
65. W. E. Harris and W. L. Stahl, *Biochim. Biophys. Acta* **485**, 203–214 (1977).
66. H. J. Schramm, *Hoppe-Seyler's Z. Physiol. Chem.* **356**, 1375–1379 (1975).
67. L. A. Aleksandrova, *Bull. Acad. Sci. USSR, Div. Chem. Sci.* **26**, 583–589 (1977).
68. R. Chayen, R. Dvir, S. Gould, and A. Habell, *Anal. Biochem.* **42**, 283–286 (1971).
69. P. Weber, F. W. Harrison, and L. Hof, *Histochemistry* **45**, 271–277 (1975).
70. P. Weber and L. Hof, *Biochem. Biophys. Res. Commun.* **65**, 1298–1302 (1975).
71. G. Abraham and P. S. Low, *Biochim. Biophys. Acta* **597**, 285–291 (1980).
72. W. W. Stewart, *Cell* **14**, 741–759 (1978).
73. T. A. Reaves, Jr., and J. N. Hayward, *Proc. Natl. Acad. Sci. USA* **76**, 6009–6011 (1979).
74. C. Bowman and H. Tedeschi, *Science* **209**, 1251–1252 (1980).
75. P. Stenberg, J. L. G. Nilsson, O. Erikkson, and R. Lunden. *Acta Pharm. Suec.* **8**, 415–422 (1971).
76. A. Dutton and S. J. Singer, *Proc. Natl. Acad. Sci. USA* **72**, 2568–2571 (1975).
77. J. S. Pober, V. Iwanij, E. Reich, and L. Stryer, *Biochemistry* **17**, 2163–2169 (1978).
78. L. Lorand, G. E. Siefring, Y. S. Tong, J. Bruner-Lorand, and A. J. Gray, *Anal. Biochem.* **93**, 453–458 (1979).
79. G. S. Penny and D. F. Dyckes, *Biochemistry* **19**, 2888–2894 (1980).
80. H. Inchikawa, T. Tanimura, T. Nakajima, and Z. Tamura, *Chem. Pharm. Bull.* **18**, 1493–1495 (1970).
81. C. M. Himel, W. G. Aboud-Saad, and S. Uk, *J. Agric. Food Chem.* **19**, 1178–1185 (1971).
82. W. L. C. Vaz and G. Schoellmann, *Biochim. Biophys. Acta* **439**, 194–205 (1976).
83. T. J. Burnett, H. C. Peebles, and J. H. Hageman, *Biochem. Biophys. Res. Commun.* **96**, 157–162 (1980).
84. P. M. Gallop and M. A. Paz, *Fed. Proc.* **39**, 1603 (1980).
85. M. E. Dockter, *J. Biol. Chem.* **254**, 2161–2164 (1979).
86. R. B. Moreland and M. E. Dockter, *Anal. Biochem.* **103**, 26–32 (1980).
87. T. T. Ngo and C. F. Yam, *Gen. Pharmacol.* **11**, 193–196 (1980).
88. M. Machida, N. Ushijima, T. Takahashi, and Y. Kanaoka, *Chem. Pharm. Bull.* **25**, 1289–1294 (1977).
89. M. Machida, M. I. Machida, T. Sekine, and Y. Kanaoka, *Chem. Pharm. Bull.* **25**, 1678–1684 (1977).
90. K. Yamamoto, Y. Okamoto, and T. Sekine, *Anal. Biochem.* **84**, 313–318 (1978).
91. T. O. Sippel, *J. Histochem. Cytochem.* **29**, 314–316 (1981).
92. C.-H. Yang and D. Soll, *Proc. Natl. Acad. Sci. USA* **71**, 2838–2842 (1974).
93. C.-H. Yang and D. Soll, *Biochemistry* **13**, 3615–3621 (1974).
94. W. Dunges, *Anal. Chem.* **49**, 442–445 (1977).
95. J. Krejcoves, J. Drobnik, J. Jok, and J. Kalal, *Collect. Czech. Chem. Commun.* **44**, 2211–2220 (1979).
96. H. Maddy, *Biochim. Biophys. Acta* **88**, 390–399 (1964).
97. P. A. Knauf and A. Rothstein, *J. Gen. Physiol.* **58**, 190–210 (1971).
98. A. Rao, P. Martin, R. A. F. Reithmeier, and L. C. Cantley, *Biochemistry* **18**, 4505–4516 (1979).
99. Z. I. Cabantchik and A. Rothstein, *J. Membr. Biol.* **15**, 207–226 (1974).
100. Z. I. Cabantchik, P. A. Knauf, and A. Rothstein, *Biochim. Biophys. Acta* **515**, 239–302 (1978).

101. S. Dissing, A. J. Jesaitis, and P. A. G. Fortes, *Biochim. Biophys. Acta* **553**, 66–83 (1979).
102. K. A. Angelides, private communication.
103. D. A. Howlaka and G. G. Hammes, *Biochemistry* **16**, 5538–5545 (1977).
104. P. H. Bolton and D. R. Kearns, *Nucleic Acids Res.* **5**, 4891–4903 (1978).
105. D. E. Graves, L. W. Yielding, C. L. Watkins, and K. L. Yielding, *Biochim. Biophys. Acta* **479**, 93–104 (1977).
106. C. E. Cantrell, K. L. Yielding, and K. M. Pruitt, *Mol. Pharmacol.* **15**, 322–330 (1979).
107. C. Graue and M. Klingenberg, *Biochim. Biophys. Acta* **546**, 539–550 (1979).
108. L. F. Fieser and H. J. Creech, *J. Am. Chem. Soc.* **61**, 3502–3506 (1939).
109. R. F. Chen, *Arch. Biochem. Biophys.* **133**, 263–276 (1969).
110. A. Carmel, M. Zur, A. Yaron, and E. Katchalski, *FEBS Lett.* **30**, 11–14 (1973).
111. S. A. Reines and C. R. Cantor, *Nucleic Acids Res.* **1**, 767–786 (1974).
112. D. C. Cottel and D. C. Livingstone, *J. Histochem. Cytochem.* **24**, 956–958 (1976).
113. M. Machida, T. Takahashi, K. Itoh, T. Sekine, and Y. Kanaoka, *Chem. Pharm. Bull.* **26**, 596–604 (1978).
114. H. Maeda, N. Ishida, H. Kawauchii, and K. Tuzimura, *J. Biochem.* **65**, 777–783 (1969).
115. G. Steinbach, *Acta Histochem.* **49**, 19–34 (1974).
116. D. L. Taylor and Y.-L. Wang, *Proc. Natl. Acad. Sci. USA* **75**, 857–861 (1978).
117. D. L. Taylor and Y.-L. Wang, *Nature (London)* **284**, 405–410 (1980).
118. S. K. Curtis and R. R. Cowden, *Histochemistry* **68**, 23–28 (1980).
119. D. Blakeslee and M. G. Baines, *J. Immunol. Methods* **13**, 305–320 (1976).
120. D. Axelrod, *Proc. Natl. Acad. Sci. USA* **77**, 4823–4827 (1980).
121. O. W. Odom, Jr., D. J. Robbins, J. Lynch, D. Dottavio-Martin, G. Kramer, and B. Hardesty, *Biochemistry* **19**, 5947–5954 (1980).
122. B. D. Wells and C. R. Vantor, *Nucleic Acids. Res.* **8**, 3229–3246 (1980).
123. C. C. Lee, B. D. Wells, R. H. Fairclough, and C. R. Cantor, *J. Biol. Chem.* **256**, 49–53 (1981).
124. M. Wilchek, S. Spiegel, and Y. Spiegel, *Biochem. Biophys. Res. Commun.* **92**, 1215–1222 (1980).
125. J. R. Bunting and R. E. Cathou, *J. Mol. Biol.* **77**, 223–235 (1973).
126. J. R. Bunting and R. E. Cathou, *J. Mol. Biol.* **87**, 329–338 (1974).
127. F. Karush, N. R. Klinman, and R. Marks, *Anal. Biochem.* **9**, 100–114 (1964).
128. K. Mihashi, M. Nakabayashi, H. Yoshimura, and H. Ohnuma, *J. Biochem. (Tokyo)* **85**, 359–366 (1979).
129. R. J. Cherry, A. Cogoli, M. Opplinger, G. Schneider, and G. Semenza, *Biochemistry* **15**, 3653–3656 (1976).
130. R. J. Cherry, *Biochim. Biophys. Acta* **559**, 289–327 (1979).
131. R. H. Austin, S. S. Chan, and T. M. Jovin, *Proc. Natl. Acad. Sci. USA* **76**, 5650–5654 (1979).
132. E. Nigg, M. Kessler, and R. M. Cherry, *Biochim. Biophys. Acta* **550**, 328–340 (1979).
133. E. A. Nigg and R. J. Cherry, *Proc. Natl. Acad. Sci. USA* **77**, 4702–4706 (1980).
134. J. C. Skou and M. Esmann, *Biochim. Biophys. Acta* **601**, 386–402 (1980).
135. R. J. Cherry, E. A. Nigg, and G. S. Beddard, *Proc. Natl. Acad. Sci. USA* **77**, 5899–5903 (1980).
136. L. C. Felton and C. R. McMillion, *Anal. Biochem.* **2**, 178–187 (1961).
137. D. Kramer, H. Klapper, and F. Miller, *Spectrosc. Lett.* **1**, 23–26 (1968).
138. J. W. Goding, *J. Immunol. Methods* **13**, 215–226 (1976).
139. Thomas M. Chused, unpublished.

140. A. Levi, Y. Shechter, E. J. Neufeld, and J. Schlessinger, *Proc. Natl. Acad. Sci.* **77**, 3469–3473 (1980).
141. D. L. Taylor, Y.-L. Wang, and J. M. Heiple, *J. Cell. Biol.* **86**, 590–598 (1980).
142. Y.-L. Wang and D. L. Taylor, *J. Histochem. Cytochem.* **28**, 1198–1206 (1980).
143. F. Borek and A. M. Silverstein, *Arch. Biochem. Biophys.* **87**, 293–297 (1960).
144. J. K. Weltman, R. P. Szaro, A. R. Frackelton, R. M. Dowben, J. R. Bunting, and R. E. Cathou, *J. Biol. Chem.* **248**, 3173–3177 (1973).
145. C.-W. Wu, L. R. Yarbrough, and F. Y.-H. Wu, *Biochemistry* **15**, 2863–2868 (1976).
146. T. Kouyama and K. Mihashi, *Eur. J. Biochem.* **105**, 279–287 (1980).
147. T. Kouyama and K. Mihashi, *Eur. J. Biochem.* **114**, 33–38 (1981).
148. A. B. Rawitch, E. Hudson, and G. Weber, *J. Biol. Chem.* **244**, 6543–6547 (1969).
149. J. A. Knopp and G. Weber, *J. Biol. Chem.* **244**, 6309–6315 (1969).
150. P. Koenig, S. A. Reines, and C. R. Cantor, *Biopolymers* **16**, 2231–2242 (1977).
151. C. S. Owen, *J. Membr. Biol.* **54**, 13–20 (1980).
152. H. A. Berman and P. Taylor, *Biochemistry* **17**, 1704–1713 (1978).
153. H. A. Berman, J. Yguerabide, and P. Taylor, *Biochemistry* **19**, 2226–2235 (1980).
154. G. Braunitzer, B. Schrank, S. Petersen, and U. Petersen, *Hoppe-Seyler's Z. Physiol. Chem.* **354**, 1563–1566 (1973).
155. T. Yamaoka, H. Kashiwagi, and S. Nagakura, *Bull. Chem. Soc. Jpn.* **45**, 361–365 (1972).
156. D. Neiva-Gomez and R. B. Gennis, *Proc. Natl. Acad. Sci. USA* **74**, 1811–1815 (1977).
157. J. M. Gonzalez-Ros, V. Sator, P. Calvo-Fernandez, and M. Martinez-Carrion, *Biochem. Biophys, Res. Commun.* **87**, 214–220 (1979).
158. J. J. Zakowski and R. R. Wagner, *J. Virol.* **36**, 93–102 (1980).
159. R. E. Boyle, *J. Org. Chem.* **31**, 3880–3882 (1966).
160. H. Rinderknect, *Experientia* **16**, 430–431 (1960).
161. T. Kinoshita, F. Iinuma, and A. Tsuji, *Anal. Biochem.* **61**, 632–637 (1974).
162. K. Nakaya, M. Yabuta, F. Iinuma, T. Kinoshita, and Y. Nakamura, *Biochem. Biophys Res. Commun.* **67**, 760–766 (1975).
163. R. Schmidt-Ullrich, H. Knufermann, and D. F. H. Wallach, *Biochim. Biophys. Acta* **307**, 353–365 (1973).
164. R. Takashi, J. Duke, K. Ue, and M. Morales, *Arch. Biochem. Biophys.* **175**, 279–283 (1976).
165. R. A. Mendelson, M. F. Morales, and J. Botts, *Biochemistry* **12**, 2250–2255 (1973).
166. T. Ikkai, P. Wahl, and J.-C. Auchet, *Eur. J. Biochem.* **93**, 397–408 (1979).
167. J. M. Vanderkooi, A. Ierokomas, H. Nakamura, and A. Martonosi, *Biochemistry* **16**, 1262–1267 (1977).
168. J. Wehland and K. Weber, *Exp. Cell. Res.* **127**, 397–408 (1980).
169. Y.-L. Wang and D. L. Taylor, *J. Cell. Biol.* **82**, 672–679 (1979).
170. J. Borejdo, *Biopolymers* **18**, 2807–2820 (1979).
171. J. S. Franzen, P. S. Marchetti, and D. S. Feingold, *Biochemistry* **19**, 6080–6089 (1980).
172. M. Machida, M. I. Machida, and Y. Kanaoka, *Chem. Pharm. Bull.* **25**, 2739–2743 (1977).
173. P. Graceffa and S. S. Lehrer, *J. Biol. Chem.* **255**, 11296–11300 (1980).
174. T.-I. Lin, *Arch. Biochem. Biophys.* **185**, 285–299 (1978).
175. J. D. Johnson, J. H. Collins, and J. Potter, *J. Biol. Chem.* **253**, 6451–6458 (1978).
176. D. Podhradsky, L. Drobnica, and P. Kristian, *Experientia* **35**, 154–155 (1979).
177. H. R. Horton and W. P. Tucker, *J. Biol. Chem.* **245**, 3397–3401 (1970).

178. W. B. Lawson, E. Gross, C. M. Foltz, and B. Whitkop, *J. Am. Chem. Soc.* **84**, 1715–1718 (1962).

179. G. I. Likhtenshtein and Yu. D. Akhmedov, *Mol. Biol.* **4**, 551–559 (1970).

180. L. A. Aleksandrova, A. A. Kraevskii, A. A. Arutyunyan, V. A. Spivak, and B. P. Gottikh, *Izv. Akad Nauk SSSR, Ser. Khim.*, 1321–1324 (1973).

181. D. J. Neadle and R. J. Pollitt, *Biochem. J.* **97**, 607–608 (1965).

182. A. H. Coons and M. H. Kaplan, *J. Exp. Med.* **91**, 1–13 (1950).

183. R. M. Watt and E. W. Voss, Jr., *Immunochemistry* **14**, 533–541 (1977).

184. S. Bauminger and M. Wilchek, *Meth. Enzymol.* **70**, 151–159 (1980).

185. A. I. Petrov and B. I. Sukhorukov, *Nucleic Acids Res* **8**, 4221–4234 (1980).

186. R. A. Kenner and H. Neurath, *Biochemistry* **10**, 551–557 (1971).

187. G. G. Guilbault, *Practical Fluorescence. Theory, Methods, and Techniques*, Marcel Dekker, New York (1973), pp. 230–235.

188. K. L. Carraway and D. E. Koshland, Jr., *Methods Enzymol.* **25B**, 616–623 (1972).

189. S. A. Barker, J. A. Monti, S. T. Christian, F. Benington, and R. D. Morin, *Anal. Biochem.* **107**, 116–123 (1980).

190. K. Yasuda and W. S. Chilton, *Anal. Biochem.* **74**, 609–614 (1976).

191. K. Takahashi, *J. Biol. Chem.* **243**, 6171–6179 (1968).

192. C. L. Borders, Jr., L. J. Pearson, A. E. McLaughlin, M. E. Gustafson, J. Vasiloff, F. Y. An, and D. J. Morgan, *Biochim. Biophys. Acta* **568**, 491–495 (1979).

193. D. L. Gard and E. Lazarides, *J. Cell. Biol.* **81**, 336–347 (1979).

194. P. C. Zamecnik, M. L. Stephenson, and J. F. Scott, *Proc. Natl. Acad. Sci. USA* **46**, 811–822 (1960).

195. J. G. L. Bauman, J. Wiegant, P. Borst, and P. van Duijn, *Exp. Cell. Res.* **128**, 485–489 (1980).

196. K. E. Taylor and Y. C. Wu, *Biochem. Int.* **1**, 353–358 (1980).
     *Biochem. Int.* **1**, 353–358 (1980).

197. A. I. Petrov and B. I. Sukhorukov, *Nucleic Acids Res.* **8**, 4221–4234 (1980).

198. D. C. F. Chan, *Biophys. J.* **32**, 454 (1980).

199. D. E. Draper and L. Gold, *Biochemistry* **19**, 1774–1781 (1980).

200. J. Summerton and P. A. Bartlett, *J. Mol. Biol.* **122**, 145–162 (1978).

201. R. Wagner and H. G. Gassen, *Biochem. Biophys. Res. Commun.* **65**, 519–529 (1975).

202. R. Wagner and R. A. Garrett, *Nucleic Acids Res.* **5**, 4065–4075 (1978).

203. J. V. Staros, *Trends Biochem. Sci.*, 320–322 (1980).

204. G. Dreyfuss, K. Schwartz, E. R. Blout, J. R. Barrio, F.-T. Liu, and N. J. Leonard, *Proc. Natl. Acad. Sci. USA* **75**, 1199–1203 (1978).

205. H.-J. Schäfer, P. Schuerich, G. Rathgeber, and K. Dose, *Nucleic Acids Res.* **5**, 1345–1351 (1978).

206. H.-J. Schafer, P. Scheurich, G. Rathgeber, and K. Dose, *Anal. Biochem.* **104**, 106–111 (1980).

207. V. Witzemann and M. A. Raftery, *Biochemistry* **16**, 5862–5868 (1977).

208. T. Kouyama and K. Mihashi, *Eur. J. Biochem.* **114**, 33–38 (1981).

# 3

# Nanosecond Pulse Fluorimetry of Proteins

TSUNG-I LIN AND ROBERT M. DOWBEN

## 1. Introduction

During the past decade, studies of proteins using nanosecond pulse fluorescence spectroscopy have proliferated. In large part, this is the result of advances in instrumentation and the availability of commercial nanosecond spectrofluorimeters as well as the development of many new fluorescent probes. Methodologically, these studies fall into two main categories: (1) lifetime measurements, to be addressed in this chapter, and (2) time-resolved fluorescence anisotropy, the topic of the next chapter. The application of lifetime-resolved emission spectroscopy is still in its infancy, and only very few studies can be found in the literature. From the instrumentation point of view, anisotropy measurements are similar to lifetime measurements; in the former, decay of polarization is measured, while in the latter, decay of total fluorescence is measured. The information obtained from these two types of measurements is vastly different. Lifetime experiments tell us about the dynamic properties of the fluorescent moiety itself, while anisotropy measurements provide information regarding the shape and hydrodynamic properties of the proteins to which the fluorescent moiety is attached. An exhaustive review of the literature has not been attempted; rather representative results are described in this chapter.

In general, the applications of nanosecond lifetime measurements in protein studies fall into one or a combination of the following categories: (1) lifetime

TSUNG-I LIN AND ROBERT M. DOWBEN ● Department of Pathology, Baylor University Medical Center, Dallas, Texas 75246.

changes as indicative of conformational changes in proteins, (2) lifetime changes as a means of evaluating protein–ligand and protein–protein binding interactions, (3) studies of the heterogeneity or multiplicity of probe binding sites, (4) quenching experiments, and (5) energy-transfer experiments for inter- and intramolecular distance measurements.

Studies which fall into the first and second categories above usually involve only a single decay lifetime measurement. The experimental design is quite straightforward and the results yield information basically similar to those that can also be obtained from quantum yield studies. The measurement of multiple lifetimes is more complex and requires quite sophisticated mathematical analysis. Lifetime-quenching experiments are a valuable tool for probing accessibility of various protein functional sites and dynamic fluctuation of protein structures in solution. The energy transfer type experiments remain at present the only practical tool for measuring molecular distances between two specific functional sites in single or complex protein systems in solution. In this review, only studies involving proteins will be discussed; studies on membranes and membrane proteins (which are also quite numerous) will be omitted.

## 2. Instrumentation and Data Analyses

Before discussing fluorescent studies in protein systems, let us briefly consider the instrumentation and data analysis required for these experiments. Nanosecond pulse fluorimetry preferably utilizes the single-photon counting technique for measuring the fluorescence decay of excited samples, which is based on statistical sampling. The single-photon counting method offers several advantages over the phase-shift method. The former method provides all three of the following modes of measurements: lifetime, time-resolved anisotropy decay (see the next chapter), and lifetime-resolved emission spectroscopy. On the other hand, the phase-shift method only determines lifetime. In the single-photon counting method, the decay profile of fluorescence emission is measured directly, whether it be monoexponential or multiexponential. In contrast, in the phase-shift method no decay profile is measured and only one lifetime can be determined readily.

### 2. 1. Single-Photon-Counting Nanosecond-Pulse Fluorimeter

### 2.1.1. Principle of Operation

Fluorescence decay of the excited fluorophore molecules follows single or multiple exponential functions. The statistical-sampling technique requires that only one single photon be detected per excitation pulse. The probability of detecting a single photon is proportional to the number of excited states in the sample. The number of excited states as a function of time depends on the decay time of the sample. Thus, the distribution of the detected events as a function of time will be a direct measure of the decay-time profile of the sample. In its simplest monoexponential function, the decay curve is described by the equation

$$N(t) = N_0 \exp(-t/\tau) \tag{1}$$

where $\tau$ is defined as the lifetime of the fluorescence decay, and $N_0$ and $N(t)$ are, respectively, the number of photon events detected at times zero and $t$.

A large statistical sampling of the time distribution of detected photon events is required in order to reconstruct a true and accurate decay-time spectrum. Since only one event is detected per pulse, it is necessary that the sample be excited and single-photon events be detected many thousands of times during an experiment at a repetition rate as fast as the experiment permits.

The basic setup for the nanosecond pulse fluorescence decay measurement is described here, using commercially available components. A block diagram of the system is depicted in Figure 1. It should be noted that with a few additional components, the system in Figure 1 can be converted to measure time-resolved anisotropy decay (see the next chapter) or lifetime-resolved spectra.

## 2.1.2. Pulse Light Sources

The light pulser in Figure 1 is a high-voltage discharge gas flash lamp, containing two tungsten electrodes to form a spark gap, between which the gas is charged with a high voltage by a circuit operating typically at 9–10 kV. A gas selected to give the desired light spectrum is flushed through the gap via two inlet and outlet tubes (about 6 mm in diameter); the gas can be air, nitrogen, hydrogen, deuterium, krypton, etc. at pressures up to 3 atm. Air and nitrogen give good spectral output in the near visible to blue region, with distinct sharp peaks and a maximum at 337 nm (Table 1). Hydrogen and deuterium give continuous spectral output with strong intensity in the UV region. Depending on the gas used, the gap distance, and the operating voltage, the pulse duration is typically 2–6 ns FWHM (full width at half maximum intensity). The repetition rate is adjustable by varying the gap distance (0.2–2 mm) between the electrodes and changing the operating voltage. Typically, a repetition rate of 15–40 KHz can be obtained when the electrodes are smooth and polished. Changes in operating voltage and gap distance affect not only repetition rate but also pulse duration width and light output. We obtain good results using unregulated air (without forced circulation) pulsing at 16–22 kHz. The pulse width is typically 0.8–1.5 ns FWHM. Higher operating voltage and narrower gap distance usually gives higher repetition rates (up to about 46 kHz) but lower light output (about $5 \times 10^8$ photons/flash). However, since in the single-photon counting method only one photon is detected per flash, the low photon output hardly affects the

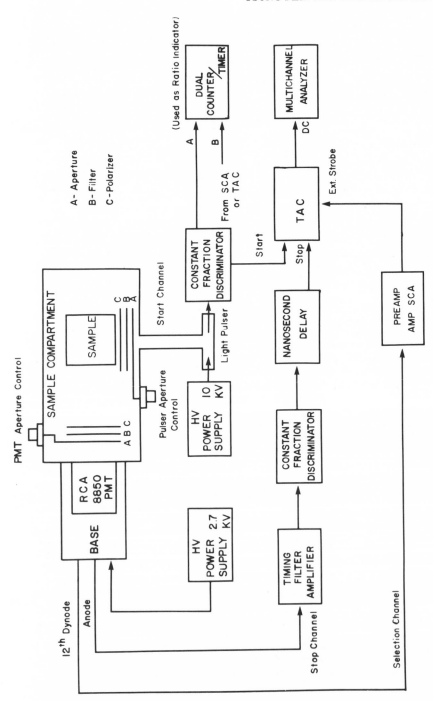

FIGURE 1. A block diagram for a single-photon-counting nanosecond-pulse fluorimeter.

experiment unless the sample has extremely low quantum efficiency. Because in experimental strategy it is often desirable to run at the highest repetition rate possible, it is sometimes advantageous to use a different type of flash lamp fired by a gated thyratron,[1] which is available commercially.

The pulsed light enters the sample compartment via a small quartz window (about 6 mm in diameter) and illuminates the sample cuvette seated in the thermostat-controlled sample block. Situated between the pulser and the sample block are a photoaperture which controls the amount of light entering the sample compartment and a filter which selects the excitation wavelength for the fluorescent sample. Similarly, a photoaperature control and a filter are used in the emission path. The detector, a 2-in., end-on RCA 8850 photomultipler (PMT) is oriented at 90° to the exciting light path. Although use of a monochromator for selecting excitation or emission wavelength offers the best possible spectral resolution, quite often a pair of good band-pass filters with narrow bandwidth (typically FWHM 10 nm) will do a satisfactory job with less loss of light. This is so because a majority of fluorophores have a well-defined emission band and air flash lamp has sharp spectral peaks. However, it is critical to minimize the reflected light from the filters (the filter should be of a low-fluorescent grade). This can be accomplished by enclosing the filter in a black filter holder box with 5-mm-diameter holes in the front and back plates. Scattered light is also minimized by blackening the sample holder and inside of the compartment. A nonfluorescent (photographic quality) black paint must be used.

*Table 1. Comparison of Characteristics of Air-Gap Flash Lamp and Pulsed Laser Light*

| Characteristics | Laser | Air flash lamp |
|---|---|---|
| Excitation wavelengths (nm) | 515–560, 293–305, 260–275, 570–650 (tunable wavelengths); 514, 488, 458 (discrete wavelengths) | 294–407, broad band with discrete lines at 316, 337 (max.), and 358 |
| Pulse width (ns) | < 0.035 (tunable wavelength); 0.25 (discrete wavelength) | 0.8–2.2 |
| Pulse power (W) | > 150 (tunable wavelength); 0.05–0.40 (discrete wavelengths) | $5 \times 10^{-6}$; broad band |
| Pulse shape and stability | Uniform and stable | Nonuniform and somewhat erratic |
| Repetition frequency | 10 kHz to 10 MHz; switch selectable | 5–40 kHz; free running |
| Optical property | Monochromatic and polarized | Finite spectral lines and unpolarized |

## 2.1.3. Pulsed Laser Light Source

For the past few years, the availability of lasers has provided alternative light sources with significant advantages, including monochromicity, high intensity, high repetition rates, short pulse duration times (in subpicosecond range), and stable pulse profiles. A synchronously pumped tunable dye laser which gives subnanosecond pulse width has been constructed in our laboratory.[2] A block diagram of this system is depicted in Figure 2. By gain modulation, the system uses the mode-locked 514.5 nm line of a 5-W argon-ion laser to pump a rhodamine 6G tunable dye laser. Other laser lines and laser dyes can be used to cover other spectral regions. The pumping laser cavity was extended to 2.025 m, and an acousto-optic modulator was inserted to obtain mode locking, driven at a frequency of 38.401 MHz, resulting in a 76.802-MHz pulse train output with a 13.020-ns spacing between mode-locked pulses. The cavity of the dye laser was extended so that it was equal to and resonant with that of the pumping laser, maximizing the coupling efficiency. A Brag cell was mounted in the dye-laser cavity and was used to dump the mode-locked pulses by intercavity diffraction at a gated 470-MHz acoustic frequency. The associated electronics accurately synchronize the acoustic coupling pulse with the arrival of a mode-locked dye pulse. The tuning wedge inside the cavity permits adjustment of the mode-locked rhodamine 6G laser over the range 545–655 nm. The characteristics of the air-gap discharge lamp and pulsed laser light sources are compared in Table 1.

## 2.1.4. Associated Electronics

The heart of a nanosecond pulse fluorimeter is the high-speed electronics which provides the following basic functions:

(1) Start channel — timing of the start of nanosecond light pulse;
(2) Stop channel — timing of the detection of a fluorescence emission photon;
(3) Selection channel — selection of true single-photon events;
(4) Data acquisition — storage of the accumulated photon events in their appropriate time channels in a multichannel analyzer (MCA).

With reference to Figure 1, assume that a light flash has occurred; the electronics in the start channel uses the current pulse directly from the pulser to trigger a fast constant-fraction discriminator whose output signal is fed into the start channel of a time-to-pulse-height converter (TAC). In the meantime the flash light excites the sample which emits many fluorescence photons. The photon striking the photomultiplier cathode leads to an electron current pulse on the anode. The pulse from the anode of the PMT is fed into a fast constant-fraction discriminator whose output is connected to stop the TAC. Constant-fraction discriminators are used because they give the best timing precision. A

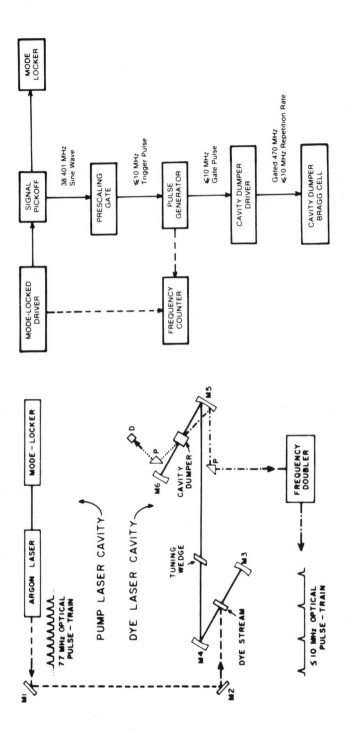

FIGURE 2. Left: optical configuration of a synchronously pumped tunable dye laser system. Right: associated electronics for the mode-locked dye laser. From Koester and Dowben.[a]

delay unit is connected between the discriminator and TAC for calibration of the timebase. In some commercial systems, a timing filter amplifier NIM module is also connected between the PMT anode and the discriminator to amplify anode signal. We found this amplification stage undesirable because the relatively long (4.5 ns) risetime of the NIM module output compared to the less than 2-ns risetime of the PMT anode pulse led to greater timing uncertainty.[2] Instead of using an NIM amplifier, we operate the PMT at 2.75 kV so that the resulting anode pulse is sufficiently large to trigger the discriminator directly.

Since in the single-photon counting method, only single-photon events are selected for constructing a decay histogram, multiphoton events must be rejected. Statistically, multiphoton events have a higher incident rate when data are collected at a higher rate. In order to obtain high collecting rates while collecting only single-photon events, an "energy window" technique[3] is employed to eliminate the multiphoton events. Multiphoton events can be identified in a selected PMT tube which has a narrow pulse height distribution for single photons such as an RCA 8850 tube. A single-channel analyzer (SCA) is connected to the 12th dynode of the PMT. This linear channel has a sufficiently long time constant, about 600 ns, so that the amplifier output amplitude is much larger for multiphoton events than for single photons. The lower- and upper-level voltage discriminators in the SCA are adjusted so that only the single photon impinging on the PMT can produce an output pulse to the TAC.

When the TAC successively registers a start and a stop pulse, it produces an output voltage whose magnitude is proportional to the time lapse between the start and stop signals. The TAC output is strobed out to the amplitude-to-digital converter (ADC) of the multichannel analyzer (MCA) triggered by the SCA pulse which signals the detection of only the single-photon events. The ADC in the MCA then determines the proper address in proportion to the amplitude of the TAC to which that count is to be registered. The address number in the MCA is therefore in direct proportion to the time lapse of a photon event. After the output in the TAC is strobed to the MCA, or if no signal from the SCA is received within a preset time, the TAC resets itself and waits for the next sequence of events. The above sequence is repeated many thousand times. When sufficient counts have been accumulated in the MCA, a histogram of counts (ordinate) versus time (abscissa) is produced.

## 2.1.5. Calibration of Timebase in the MCA

The measured time span of fluorescence decay is adjusted by the setting on the TAC. Therefore, it is necessary to determine the timebase between the channels in the MCA. This can be done by the following procedure. An excitation pulse profile is collected on the MCA using a light-scattering solution of Ludox-30 (DuPont Chemicals, Inc.). The peak channel which registered the maximal counts (PCN) is identified. A precalibrated time delay $t_d$ is introduced to the stop of the TAC which delays the PCN by a number of channels in proportion to the delay time. The timebase per channel $t_b$ is thus equal to $t_d/(PCN_2 - PCN_1)$. It is assumed that both TAC and MCA have linear responses. Sometimes this assumption cannot be met when full 512 or 1024 addresses of channels are used. Several calibrated delay times are used. A plot of $t_d$ vs. PCN is made from which a straight line can be obtained by least-squares fitting. The slope of this line equals $t_b$. Routinely we use an NIM which provides a fixed delay time in increments of 1 ns in the range 1–31 ns for short timebase calibration and a combination of a 31-ns-delay NIM with two 60-ns calibrated delay cables in series for long timebase determinations.

## 2.1.6. Sample Preparation and Experimental Setup

In general the preparation of a sample is similar to that used in other types of fluorescence measurements. Since lifetime is sensitive to oxygen quenching, samples usually are bubbled with nitrogen, covered with a stopper, and sealed with parafilm. For samples that are photolabile, it is essential to minimize light exposure during preparation. Ideally, thermostated water is circulated in the sample compartment. However, in a well air-conditioned room, experiments sometimes are done at ambient temperature. Ordinarily, the sample is measured first in order to set an appropriate time range for the TAC. The time range usually covers about five times the longest lifetime. When a sample exhibits two or more components whose lifetimes differ by more than an order of magnitude, it is recommended that two separate measurements be made, one for the short-lifetime component and the other for the long-lifetime component. The time span on the MCA is adjusted by varying the gain setting on the TAC over the range 50 ns to 80 μs. After the TAC is set properly, a fluorescence decay profile of the sample is collected on the MCA. The MCA has a full address of 1024 channels which can be selected for halves or quarters. We usually use 512 channels for each decay profile. The zeroth channel is reserved for recording the collecting time. The first 30–40 channels are for collecting background counts. Counting continues until a preset time or a preset count of peak channels is reached on the MCA. We typically collect $10^6$–$10^7$ total counts with $10^4$–$10^5$ counts in the peak channels. However, data containing as few as 2000 counts in the peak channel can sometimes be analyzed. The excitation profile can be collected using scattered light either at the excitation wavelength or at the emission wavelength. The excitation or emission filter is removed from its path, and the photoapertures are readjusted to give approximately the same photon counting rate as in the collection of sample emission decay. Collecting excitation profile at the excitation wavelength takes much less time than that taken at the emission wavelength. However, since the PMT response is wavelength dependent, if the excitation and emission wavelength used in the measurement are separated by more than 100 nm, using an excitation profile collected at the excitation

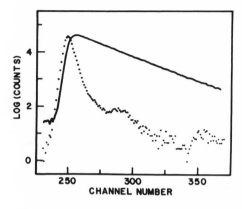

FIGURE 3. A typical excitation pulse profile of the laser system (dots) and the fluorescence emission decay curve of rhodamine B (solid curve). Timebase is 64.1 ps/channel. From Koester and Dowben.[2]

wavelength may cause "zero time shift" errors (see Section 2.2.1.4). On the other hand, it may not be possible to collect an excitation profile at the emission wavelength in the visible region since most flash lamps have low output in the visible region. Furthermore, the lamp may have a wavelength-dependent output.

After two profiles are collected, the MCA is full and its data must be transferred either directly to a computer or stored on tape. Depending on the light source, the quantum efficiency of the sample, and the collecting efficiency of the instrument, obtaining an emission decay profile can take 5 min to several hours. Collection of an excitation profile usually takes a few minutes. Typical excitation and emission decay profiles are displayed on Figure 3.

Three instrument parameters are continuously monitored during the experiment. They are (1) the pulser rate $C_1$, (2) the PMT anode photon counting rate $C_2$, and (3) the single-photon detection rate $C_3$. A stable pulser rate is important particularly for long experiments. An unstable pulser rate may produce an unstable excitation profile owing to lamp drift and zero-time jittering. The ratio $C_2/C_1$ is roughly the photon counting efficiency. We usually set this ratio to between 5 to 25%. When a higher collecting efficiency is desired, $C_2/C_1$ is increased to about 40% and $C_3/C_2$ is maintained at about 40%.

## 2.2. Fluorescent Decay Data Analysis

In a nanosecond pulse fluorimetry experiment, a fluorescent sample is excited with a very brief pulse of light. Many of the excited molecules emit photons; with proper adjustments, only single photons are selected for each light pulse by the instrument. The time between the start of an excitation pulse and the detection of a single-photon emission is measured, and the event is recorded. This process is repeated thousands of times until a satisfactory histogram is obtained representing a fluorescence decay profile $F(t)$ of the sample. In general, fluorescence decay follows an exponential time function. When the sample contains more than one component that does not interact, the impulse response function $I(t)$ can be assumed to be

$$I(t) = \sum_i^N \alpha_i \exp(-t/\tau_i) \tag{2}$$

where $N$ is the number of decay components, $\alpha_i$ and $\tau_i$ are, respectively, the amplitude and lifetime of the $i$th component, and $N$, $\alpha$, and $\tau$ are positive numbers and are measureable physical parameters. The impulse response function $I(t)$ thus gives an explicit description of the fluorescent decay behavior of the sample. Experimentally, the excitation pulse $E(t)$ always has a finite pulse width, and the observed fluorescence decay $F(t)$ is a convolution integral of $E(t)$ and $I(t)$ defined by

$$F(t) = \int_0^t E(t-s)I(s)\,ds \tag{3}$$

From equations (2) and (3) one needs to extract $N, \alpha_i$, and $\tau_i$. This can be done by one of several numerical analytical methods[4] or by curve fitting techniques.[5]

### 2.2.1. Methods of Moments

In our laboratory, the method of moments (MOM) originally developed by Isenberg and his colleagues[6, 7] is used with some modifications. Improvements have also been made in error analysis.

Using the moments of $F(t)$ and $E(t)$ defined respectively as

$$\mu_k = \int_0^\infty t^k F(t)\,dt \qquad k = 1, 2, \ldots \tag{4a}$$

$$m_k = \int_0^\infty t^k E(t)\,dt \qquad k = 1, 2, \ldots \tag{4b}$$

we can determine the moment $G_k$ of $I(t)$ defined as

$$G_k = \int_0^\infty t^k I(t)\,dt \qquad k = 1, 2, \ldots \tag{5}$$

It can be shown that these moments are related by equation (6)

$$\frac{\mu_k}{k!} = \sum_{s=1}^{k+1} \frac{G_s m_{k+1-s}}{(k+1-s)!} \qquad k = 1, 2, \ldots \tag{6}$$

Thus, $G_s$ moments higher than $s$ can be calculated successively from $\mu_k$ and $m_k$.

From equation (6), for successive values of $k$, we obtain a set of equations:

$$\mu_0 = G_1 m_0 \tag{7a}$$

$$\mu_1 = G_1 m_1 + G_2 m_0 \tag{7b}$$

$$\mu_2/2! = G_1 m_2/2! + G_2 m_1 + G_3 m_0 \tag{7c}$$

$$\cdot$$
$$\cdot$$
$$\cdot$$

If $N$ components exist, we have $2N$ unknown parameters $\alpha_i$ and $\tau_i$, and therefore we need to calculate $2N$ of $G_s$. It can be shown that moments $G_k$ are related to the set $\alpha_i$, $\tau_i$ by the equation

$$G_k = \sum_{i=1}^{N} \alpha_i \tau_i^k \tag{8}$$

Consider the following polynominal:

$$P_N(x) = \begin{bmatrix} 1 & x & x^2 \dots x^N \\ G_1 & G_2 & G_3 \dots G_{N+1} \\ G_2 & G_3 & G_4 \dots G_{N+2} \\ \cdot & \cdot & \cdot \dots \cdot \\ \cdot & \cdot & \cdot \dots \cdot \\ \cdot & \cdot & \cdot \dots \cdot \\ G_N & G_{N+1} & G_{N+2} \quad G_{2N} \end{bmatrix}$$

$$= \begin{bmatrix} 1 & 0 & 0 & \dots & 0 \\ 0 & \alpha_1\tau_1 & \alpha_2\tau_2 & \dots & \alpha_N\tau_N \\ 0 & \alpha_1\tau_1^2 & \alpha_2\tau_2^2 & \dots & \alpha_N\tau_N^2 \\ \cdot & \cdot & \cdot & \dots & \\ \cdot & \cdot & \cdot & \dots & \\ \cdot & \cdot & \cdot & \dots & \\ 0 & \alpha_1\tau_1^N & \alpha_2\tau_2^N & \dots & \alpha_N\tau_N^N \end{bmatrix} \begin{bmatrix} 1 & x & x^2 & \dots & x^N \\ 1 & \tau_1 & \tau_1^2 & \dots & \tau_1^N \\ 1 & \tau_2 & \tau_2^2 & \dots & \tau_2^N \\ \cdot & \cdot & \cdot & \dots & \cdot \\ \cdot & \cdot & \cdot & \dots & \cdot \\ \cdot & \cdot & \cdot & \dots & \cdot \\ 1 & \tau_N & \tau_N^2 & \dots & \tau_N^N \end{bmatrix} \tag{9}$$

When $x = \tau_1, \tau_2, \dots, \tau_n$, $P_N(X)$ becomes zero. Therefore, the $N$ roots of polynominal $P_N(X) = 0$ are the solutions of lifetimes.

Once the lifetimes are determined, the amplitudes can be determined from the following set of linear equations derived from equation (8):

$$G_1 = \sum_i^N \alpha_i \tau_i \tag{10a}$$

$$G_2 = \sum_i^N \alpha_i \tau_i^2 \tag{10b}$$

$$G_{2N} = \sum_i^N \alpha_i \tau_i^{2N} \tag{10c}$$

*2.1.1.1. Correction for Cutoff Error.* The most prominent problem peculiar to the method of moments is the so-called cutoff error. The cutoff error arises from moment integrals in equations (3) and (4) which must be carried to infinity for data which are known only to finite time $T$. To correct this problem, an iteration procedure is used. We first calculate the following moment integrals from 0 to $T$ (time span of the MCA):

$$\mu_k = \int_0^T t^k F(t)\,dt \qquad k = 1, 2, \dots \tag{4c}$$

$$m_k = \int_0^T t^k E(t)\,dt \qquad k = 1, 2, \dots \tag{4d}$$

and determine an initial set of $\alpha_i$ and $\tau_i$ using these moments [substituting equations (4a) and (4b)]. The difference in moment integrals between $T$ and infinity is calculated using the equation

$$\Delta\mu_k = \int_T^\infty t^k \sum_i^N \alpha_i \exp(-t/\tau_i)\,dt \tag{11}$$

and the above initial estimates of $\alpha_i$ and $\tau_i$. Corrections to the set of moments for excitation [equation (4d)] are not necessary because $E(t)$ reaches 0 before $T$ in most cases. The moment-integral differences for successive values of $k$ are then added to the moments in equation (4c) and a new set of $\alpha_i$ and $\tau_i$ are determined which in turn are used to calculate $\Delta\mu_k$ in equation (11). This iteration procedure is carried on until internal self-consistency in moments and in the $\alpha_i$ and $\tau_i$ is reached. This condition is called convergence.

*2.2.1.2. Exponential Depression Parameter.* The iteration procedure above works well for one- or two-component decay data. It can usually converge in less than 20 loops. However, when the decay data contains three components, the looping becomes excessive. This problem can be remedied by applying a weighting function to the moments calculation that permits reaching convergence more rapidly. In the MOM method, $F(t)$ and $E(t)$ are replaced by $F_\lambda(t) = \exp(-\lambda t)F(t)$ and $E_\lambda(t) = \exp(-\lambda t)E(t)$, respectively, in all the above equations, where $\lambda$ is the exponential depression parameter (EDP). The MOM computer program results in

$$I_\lambda(t) = \exp(-\lambda t)I(t) = \sum_i^N \exp(-\lambda t/\tau_i)$$

Therefore the real $\tau$ is the calculated $\tau$ divided by $\lambda$. In addition to the benefit of reducing computer CPU time in the calculations, use of EDP also reduces the cutoff correction error because EDP weighs against data toward the tail of the decay curve where data are the noisiest. Therefore, use of EDP improves the analyses by reducing the contribution from the noisy data points.

*2.2.1.3. Determination of Number of Decay Components.* In many experiments, the number of decay components contained in the sample may not be known beforehand. The determination of the number of components in a decay curve is a difficult problem for any analysis method. A decay curve may be well fitted by a monoexponential function. However, the same curve may also be fitted equally well or better by two or three exponentials. Since more data counts are collected at shorter time intervals, a long-lived decay component with a small amplitude contributes little to the counts in the late channels and may look like a tail of the short-lived decay component. Unless the number of decay components can be experimentally determined, we must rely largely on the data analysis method to resolve them.

Several approaches have been suggested for the estimation of $N$. One idea is to analyze for $N = 1, 2, 3, \ldots$ components until an analysis for $N + 1$ yields the same lifetime parameters as an analysis for $N$ components, plus an additional component with a negligibly small amplitude. Another method is to treat $N$ as a parameter along with $\alpha$ and $\tau$ and judge the curve fitting by least-squares criteria. Sometimes these methods may not work at all because two sets of data may both satisfy these criteria.

We approached this problem in two ways. We adapted the Eisenfeld and Cheng[8] method originally introduced by Isenberg et al.[6] which computes a sequence of determinants $D(k)$ defined as

$$D(k) = \det \begin{bmatrix} G_1 & G_2 & G_3 & \ldots & G_k \\ G_2 & G_3 & G_4 & \ldots & G_{k+1} \\ G_3 & G_4 & G_5 & \ldots & G_{k+2} \\ \cdot & \cdot & \cdot & \ldots & \cdot \\ \cdot & \cdot & \cdot & \ldots & \cdot \\ \cdot & \cdot & \cdot & \ldots & \cdot \\ G_k & G_{k+1} & G_{k+2} & \ldots & G_{2k+1} \end{bmatrix} \qquad (12)$$

until a zero value or a negative value of the determinant is reached. Theory reveals that $D(k) \geqslant 0$ for $k < C$, when $D(C + 1)$ first reaches zero or becomes negative, where $C$ is the number of decay components.

In addition, we define a matrix conditioning number, COIN($C$), which is equal to $1/(\text{norm } G)(\text{norm } G^{-1})$, where $G$ is the matrix of moments of the estimated $I(t)$ curve. As $G$ become singular, $G^{-1}$ approaches infinity and COIN($C$) goes to zero. This additional parameter can therefore be used to complement the $D(k)$ value in judging $N$. In our experience, in more than 90% of cases, these two parameters agree with each other (sometimes, a zero value of COIN cannot be found).

*2.2.1.4. Nonrandom System Errors Correction – The Use of Moment Index Displacement.* In any fluorescence decay experiment, inevitably some non-random system errors (other than counting error and background noises) may exist. Most serious errors are scatter light and zero-time shift. Scatter is due to inperfect isolation of light from the excitation in the collection of $F(t)$ either from the sample itself or instrument. A zero-time shift may occur when $E(t)$ and $F(t)$ are collected at very different wavelengths since the instrument response time is wavelength dependent. As a result, time zero for $E(t)$ and $F(t)$ are not coincident, leading to errors in the deconvolution of $I(t)$. These errors enter the description of $F(t)$ as

$$F(t) = \int_0^T I(s)E(t - s + u)ds + \gamma E(t + u) \tag{13}$$

where $\gamma$ is the scatter coefficient and $u$ is the zero-time shift.

In the MOM, these errors can be largely eliminated by moment-index displacement. Recall equation (9), from which it can be generalized that the decay times are the $N$ roots of the polynominal

$$P_{MD}^N(x) = \begin{bmatrix} 1 & \tau & \tau^2 & \cdots & \tau^N \\ G_{MD+1} & G_{MD+2} & G_{MD+3} & \cdots & G_{MD+N} \\ G_{MD+2} & G_{MD+3} & G_{MD+4} & \cdots & G_{MD+1+N} \\ \cdot & \cdot & \cdot & \cdots & \cdot \\ \cdot & \cdot & \cdot & \cdots & \cdot \\ \cdot & \cdot & \cdot & \cdots & \cdot \\ G_{MD+N} & G_{MD+1+N} & G_{MD+2+N} & \cdots & G_{MD+2N} \end{bmatrix} = 0 \tag{14}$$

where MD is called the moment displacement index. We can easily see that equation (9) is a special case of equation (14) of MD $= 0$.

Mathematical derivations by Isenberg[7] lead to the following important conclusions about the use of MD in correcting nonrandom errors in decay-data analysis. First, light-scatter errors are completely eliminated automatically by the use of MD ≥ 1. Second, use of MD ≥ 1 will reduce a zero-time-shift error from a large one to a small one. However, it corrects this error more in the lifetime parameters than in the amplitude factors. It should be emphasized that nonrandom errors exist in almost any instrument. What appear to be small errors in measurement may result in large errors in the determination of the parameters of a multiexponential decay. The use of MD can detect these errors and correct them.

## 2.2.2. Real Calculation

The $F(t)$ and $E(t)$ profiles collected in the MCA are rearranged in appropriate formats after being dumped into the computer for access by the MOM program. This program offers a series of options available interactively on a computer terminal. After specifying the $F(t)$ and $E(t)$ files to be analyzed, the program begins with the correction of background counts which are collected in the 30 to 40 empty channels to the left of $F(t)$. The counts in these channels are averaged and subtracted from each channel in $F(t)$. When long collection times are needed for weakly fluorescent samples, subtraction of background counts usually improves the analysis significantly.

Options are chosen for selecting an EDP ($>0$) and MD (0, 1, or 2), and the number of moments to be calculated. We usually use ten moments although in

*Table 2. Error Analysis of Lifetimes of 1,5-IAEDANS-Actin[a]*

| EDP | MD | N | R (%) | τ | $E_{2N}$ (%) | $E_{all}$ (%) | RRMS × $10^2$ | SWSR × $10^3$ |
|-----|----|----|-------|-------|----------|----------|-----------|-----------|
| 0.20 | 0 | 2 | 90.8 | 20.48 | 1.041 | 4.253 | 2.600 | 2.910 |
|      |   |   | 9.2  | 2.38  |       |       |       |       |
| 0.27 | 0 | 3 | 78.4 | 21.79 | 0.152 | 0.264 | 0.787 | 0.420 |
|      |   |   | 16.6 | 11.15 |       |       |       |       |
|      |   |   | 5.0  | 0.72  |       |       |       |       |
| 0.40 | 1 | 3 | 76.4 | 22.05 | 1.052 | 1.223 | 1.244 | 0.719 |
|      |   |   | 17.9 | 12.05 |       |       |       |       |
|      |   |   | 5.7  | 1.07  |       |       |       |       |
| 0.67 | 1 | 3 | 81.0 | 21.58 | 0.188 | 0.261 | 0.885 | 0.489 |
|      |   |   | 13.9 | 10.61 |       |       |       |       |
|      |   |   | 5.1  | 0.78  |       |       |       |       |
| 1.00 | 2 | 3 | 88.9 | 21.51 | 9.64  | 9.78  | 1.322 | 1.518 |
|      |   |   | 4.5  | 10.63 |       |       |       |       |
|      |   |   | 6.7  | 1.20  |       |       |       |       |
| 1.31 | 2 | 3 | 90.0 | 20.74 | 1.324 | 1.548 | 0.945 | 0.712 |
|      |   |   | 4.5  | 8.28  |       |       |       |       |
|      |   |   | 5.5  | 0.84  |       |       |       |       |

[a] 7 μM 1,5-IAEDANS-labeled actin, 1 μM each of troponin and tropomyosin in 0.15 M KCl, 10 mM phosphate (pH 7.0), 1 mM $MgSO_4$, and 2.5 mM EGTA.

theory one needs only $2N$ moments to determine $N$ components. The program calculates the moments, determinants $D(k)$, and matrix conditioning number COIN (see Section 2.2.1.3) and displays them on the CRT terminal. The operator judges on the values of $D(k)$ and COIN and determines the number of decay components for the analysis entering $N$ (1, 2, or 3). The program calculates an initial set of $\alpha$ and $\tau$ for $i = 1, 2, \ldots, N$ and proceeds with the iterations (for correction of cutoff errors) until values of $\alpha$ and $\tau$ converge. The program displays the results and performs an error analysis (see the next section). After its completion, the main program calls for entry of a new set of EDP and MD. The above processes are repeated until the error analyses indicate that a best-fit data set has been obtained. We routinely analyze the data with MD fixed to 0, 1, or 2 and varying EDP ($>0$) until error is minimized. At the end of the calculation, a hard-copy printout is obtained containing the moments $G_k$ at convergence, the determinants, the roots $\alpha$ and $\tau$, and a variety of statistical information related to error analyses. In the end the parameters MD, EDP, $N$, $\alpha$, $\tau$, number of loops used for convergence, and the four critical error parameters for each run is tabulated in a summary page (similar to Table 2). We then select the best-fit data set based on the four error parameters.

## 2.2.3. Error Analyses

After $\alpha$ and $\tau$ are calculated, the error analysis subroutine reconvolutes $I(t)$ and $E(t)$ to $F_c(t)$, the reconvoluted decay curve, and compares $F_c(t)$ with the observed $F(t)$ channel by channel. The program first searches the peak channel and calculates the normalized residual root-mean-square deviation (RRMS) and the sum of the weighted squares of residuals (SWSR) of the reconvoluted curve from the observed curve at each channel according to the following equations:

$$\Delta_i^2 = (F_{ci}/F_{cmax} - F_i/F_{max})^2 \tag{15}$$

$$\text{RRMS} = \left(1/S \sum_i^S \Delta_i^2\right)^{1/2} \tag{16}$$

$$\text{SWSR} = \sum_i^S (\Delta_i^2/F_i)\bigg/\left(1/S \sum_i^S 1/F_i\right) \tag{17}$$

where $F_{ci}$, $F_i$, $F_{cmax}$, and $F_{max}$ are fluorescence counts at channel $i$ and at peak channel (max.), respectively, for the reconvoluted and observed curves, and $S$ is the number of channels used in the error analyses. Note that RRMS and SWSR are analogous to the parameters $\chi$ and $\Phi$ used in the nonlinear least-squares analysis by Grinvald and Steinberg[5] but differ in that the fluorescence counts in each channel of the reconvoluted curve and observed curve is normalized, with respect to their own peak channel. The normalization of the reconvoluted curve with respect to its own maximum deemphasizes the weight of the amplitude which depends on the calculated lifetime [equation (8)] in MOM calculations. The program also provides options to compute $\chi$ and plot the residuals $\chi$ distribution and autocorrelation function of the residuals.

We find that parameters RRMS and SWSR are not very sensitive to changes in MD and EDP. A more reliable method is to calculate the error in moments between the observed and reconvoluted data. A set of good-fit data usually has less than 5% error in the first $2N$ moments (errors in higher moments generally are larger since moment is a power function). The rms of moment errors are calculated according to the equation

$$E = \left[\left(\sum_{i}^{M}(G_i - G_{ci})/G_i\right)^2 \Big/ M\right]^{1/2} 100 \qquad (18)$$

where $G_i$ and $G_{ci}$ are the $i$th moments of the observed and reconvoluted curves, respectively, and $M$ is the number of moments used in the analyses. Two parameters $E_{2N}$ and $E_{all}$ are calculated; the former measures the error in the first $2N$ moments and the latter measures the error in all moments.

For each MD tried, the best-fit EDP which yields the smallest $E_{all}$ and/or $E_{2N}$ is determined. We noted that the best-fit data do not always give the minimal values of RRMS or SWSR. In our experience, judging data fit based on the error analysis in moments is a much more satisfactory and sensitive way to select the best sets of answers from those that differ very little in goodness of fit by RRMS or SWSR. In experiments where nonrandom system errors are expected to be small, e.g., nonturbid samples, we find that the calculated lifetimes using different MD's and the corresponding best-fit EDP are in remarkably good agreement.

In Table 2, we show a typical error analysis for 1,5-IAEDANS-labeled actin. In this analysis, the best-fit data are found for three components using MD = 0 and EDP = 0.27 or MP = 1 and EDP = 0.67. Note that even in the data set using MD = 2, the longest-lifetime component (20.74 ns) agrees with the data set using MD = 0 or MD = 1.

# 3. Fluorescence Lifetime Studies

## 3.1. Lifetime of Tyrosine and Tryptophan Residues in Proteins

The intrinsic fluorescence in proteins is due mainly to Trp residues and to a lesser extent Tyr residues. In proteins where Trp and Tyr residues are both present, the protein fluorescence is usually dominated by the contribution of the

Trp residues. Not only are the emission lifetimes for Tyr and Trp residues different, but there are different lifetimes for either Trp and Tyr residues located at different sites in proteins where the microenvironment surrounding the fluorophore may vary from site to site. It is difficult, if not impossible, to quantify the contribution of each residue to the total emission by the quantum yield method. On the other hand, small differences in lifetimes can sometimes still be discerned from multiple lifetime decay analysis. In practice, up to three components can be easily analyzed.

The lifetimes of aromatic residues are sensitive to their microenvironment, including such parameters as local pH, ionic charges, and hydrophobicity. Therefore, lifetime studies provide a useful way to probe conformational changes of proteins which may cause environmental changes around the aromatic residues. Environmental sensitivity of Trp lifetime in proteins has been fully demonstrated in the case of lysozymes and homologous proteins.[9,10] The amino acid sequences of lysozymes have been fully characterized. Although Tyr residues are present, the emission is dominated by the contribution from Trp residues. Lysozyme contains six Trp residues with Trp-62, Trp-63, and Trp-108 located in the active site. Most if not all six Trp residues are preserved in the same position in the sequence of other homologous lysozymes. Steady-state fluorescence measurements have shown that most of the lysozyme emission is due to Trp-62 and Trp-108.[11] Modified proteins have also been made, in which one or two Trp residues become nonfluorescent, e.g., oxidation of Trp-62 or Trp-108 in the oxidized lysozyme derivatives.

Lifetime analyses have shown that Trp fluorescence falls into three classes, Trp-62, Trp-108, and the composite emission from the remaining Trp residues (R-Trp). By measuring the lifetimes of various derivatives in which one or both of the two dominant Trp residues, i.e., Trp-62 and Trp-108, are oxidized and their fluorescence quenched, individual lifetimes for each type of Trp residue can be obtained. R-Trp has a very short decay lifetime of 0.5 ns, and both Trp-62 and Trp-108 have somewhat longer lifetimes, about 2–2.8 ns. But the lifetime of Trp-62 is not pH dependent (about 2.6–2.8 ns), in contrast to the lifetime of Trp-108 (1.5–2.0 ns). The indole fluorescence lifetimes appear to increase with the exposure to solvent but are shortened considerably by a quencher such as $I^-$. Indole quenching experiments with various derivatives showed that Trp-108 is not quenched by $I^-$, and Trp-62 is the only Trp quenched by iodide. This is consistent with the X-ray crystallographic data which showed that Trp-62 and Trp-63 are the two Trp residues most exposed to solvent. The cause for the short lifetime of R-Trp residues is attributed to the collisional quenching by large intramolecular contacts with sulfur in cystine residues. Both Trp-62 and Trp-108 have very little or no contact with cystine. The lifetime of Trp-108 is shorter than in the oxidized derivative of Trp-62; this can be ascribed to the resonance energy transfer from Trp-108 to Trp-62 which is absent in the oxidized derivative. From decay analyses, the relative intensities emitted by individual Trp can also be estimated, as well as the number of emitting species. Interestingly, binding of lysozyme with substrate tri-(N-acetyl-glucosamine) results in changes of both the number of emitting Trp residues and their lifetimes, indicating conformational changes upon complex formation.

Tryptophan lifetime studies have also been made in several dehydrogenases to gain structural information on the NADH binding site of proteins. Two dehydrogenases, cytoplasmic malate dehydrogenase (s-MDH) and lactate dehydrogenase (LDH), both from pig hearts, have been studied by Forster and his colleagues.[12,13] Structural homology between dehydrogenases has been established; the similarities between s-MDH and LDH are particularly evident. Each subunit of the tetrameric isozyme of pig heart LDH contains six Trp residues. On the other hand, in s-MDH, there are five Trp residues in each subunit of the dimeric enzyme. Binding of NADH results in quenching of Trp fluorescence of both LDH and s-MDH. The quenching is due to resonance energy transfer from Trp to NADH. NADH quenching in s-MDH is linearly proportional to NADH concentration, while in LDH it is not. The nonlinear NADH quenching is the result of intersubunit energy transfer with a competition between two or more NADH acceptors for excitation energy localized on a single Trp donor (Trp-248). Nanosecond pulse fluorimetry shows that there are three classes of Trp residues with distinctly different lifetimes. In LDH, three Trp residues which have short decay lifetimes (about 1 ns) are not quenched by bound NADH; the other three Trp residues have considerably longer lifetimes in the range 4–8 ns and are quenched by energy transfer from Trp to NADH. In the free enzyme one of these chromophores has an 8-ns lifetime, and two have 4–5-ns lifetimes. Quenching by NADH reduces the lifetime to 2 ns or less. In s-MDH there are also three or more decay components, but there is no 8-ns component as in LDH. The fluorescence is quenched by $I^-$ and acrylamide, but the quenching is unaffected by pH. It appears that the conformation of dehydrogenases is stable over the range of pH 6–8. In summary, nanosecond pulse fluorimetry has often been used to distinguish different classes of Trp residues in proteins. The lifetimes of Trp residues are different, so are their accessibility to quenchers and response to changes in environment such as changes in pH or temperature.

## 3.2. Fluorescence Decay of Coenzymes NADH and NADPH

Several dehydrogenases have also been studied utilizing the naturally occurring fluorescent coenzymes NADH and NADPH. Beef liver glutamate dehydrogenase, GDH, the enzyme which utilizes both NAD(H) and NADP(H) as coenzymes, catalyzes the oxidative deamination of L-glutamate into 2-oxoglutarate. The active GDH consists of six polypeptide chains with identical amino acid sequence. It has been postulated that the reduced coenzymes have two classes of binding sites on each protomer enzyme–substrate complex, the catalytic active sites and the allosteric regulatory sites. Supporting evidence for the two types of coenzyme binding sites has been obtained from nanosecond fluorescence decay studies. Brochon *et al.*[14] showed that the fluorescence decay curves of the enzyme-bound complexes, GTP–GDH–NADPH, L-glutamate–GDH–NADPH, and GTP–GDH–L-glutamate–NADPH, are not monoexponential. The decay curves of NADPH in these complexes can be best fitted with two decay components with lifetimes of 1.4–2.6 ns and 2.9–4.2 ns. The relative amplitudes of the two depend significantly on the complex studied but do not depend on the coenzyme concentration. These results are taken as evidence for the existence of two types of binding sites having different micro-environments in the NADPH binding sites. The protein fluorescence decay of the enzyme–coenzyme–substrate complexes also differ from one complex to another and are quenched by the binding of NADPH, suggesting an energy transfer between Trp residues and dihydronicotinamides.

In another NADPH-related study, Brochon *et al.*[15] examined the decay behavior of NADH bound to octopine dehydrogenase isolated from muscle of *Pecten maximus*. Octopine dehydrogenase (ODH) catalyzes reversibly the dehydrogenation of D-octopine to yield L-arginine and pyruvate. Unlike GDH, ODH is monomeric with a single polypeptide chain and only a single active site. It is, therefore, of great interest to compare the fluorescent decay curves of the binding of NADH to the two enzyme systems in various enzyme–substrate complexes. Like the binding of NADH to GDH, the fluorescent intensity of the reduced coenzyme is strongly increased after it binds with ODH. Both the adenine and pyridine parts of the coenzymes are involved in the binding of the enzyme. Also like the NADH–GDH complexes, the fluorescence decays of the reduced coenzyme bound to ODH are not a single exponential function, but rather better fitted by two exponentials with lifetimes of 1.3–2 ns and 7–7.3 ns. Since ODH has only a single binding site, the biexponential decays are presumably attributed to the heterogeneity of the environment surrounding the dihydronicotinamide. Biexponential decay has also been found in the lactate dehydrogenase-bound NADH. Gafni and Brand[16] interpreted the biexponential decay behavior as indicating the existence of a reversible excited state reaction which would transform the fluorescent chromophore to a nonfluorescent product. However, such a mechanism has been regarded by Brochon *et al.*[14] as unlikely. The possible causes of the heterogeneity in the environment which affect the NADH fluorescence lifetime are attributed by Brochon *et al.* to the presence of a mixed population of several conformations of the amino acid side chains surrounding the coenzyme binding sites.

# 4. Fluorescence Quenching Studies

## 4.1. Background

Fluorescence quenching experiments yield important information about the microenvironment of specific fluorophores in proteins, either intrinsic or extrinsic probes, and their accessibility to solvent and external quenchers. Thus, whether a specific fluorophore is located on the surface or buried in the interior of the protein not accessible to quencher can be inferred from fluorescence quenching experiments. Furthermore, dynamic structural fluctuations in proteins can be revealed if the "buried" fluorophore is dynamically quenched.

In dynamic (collisional) quenching, the decrease in emitter fluorescence is described by the Stern–Volmer equation

$$F_0/F = 1 + K_{SV}(Q) \tag{19}$$

If only the dynamic quenching process is present, then

$$F_0/F = \tau_0/\tau \tag{20}$$

In the above equations, $F_0$ is the fluorescence yield (or intensity) and $\tau_0$ is the lifetime in the absence of quencher, where $F$ and $\tau$ are the corresponding values at some quencher concentration $Q$. The quantity $K_{SV}$ is the Stern–Volmer quenching constant. If dynamic quenching is the only cause of quenching, then $K_{SV} = K_q$, where $K_q$ is the dynamic quenching rate constant.

For the dynamic bimolecular quenching processes, $K_q$ can be used as a quantitative measure for the accessibility of the emitter to solvent and quenchers. When the emitter is shielded and inaccessible, $K_q$ is small. When the fluorophore has high accessibility, $K_q$ approaches the upper limit of the free diffusion rate constant of the quencher or emitter.

In contrast to dynamic quenching which occurs during the lifetime of the excited state, static quenching, which is dependent upon the formation of a nonfluorescent complex between fluorophore and quencher prior to excitation, can also contribute to the decrease in the observed fluorescence intensity and follows the modified Stern–Volmer equation

$$F_0/F = 1 + K_{SV}(Q) \exp (V/Q) \tag{21}$$

where $V$ can be regarded as the volume of a sphere within which the emitter is quenched.

Fluorescence quenching can be measured either by the decrease in fluorescence intensity or by the decrease in lifetime. However, only collisional quenching contributes to lifetime shortening since it competes with the emission for depopulation of the excited state. Therefore, collisional quenching is usually studied by the lifetime technique. Static quenching reduces only fluorescence yield, because the lifetime of the complexed emitters (in the ground state) remains constant. If the quenching measurement is done by intensity measurement, the presence of static quenching will yield a concave plot of $1/F$ vs. $Q$. On the other hand, the presence of multiple components causes the plot to become convex. If plots of $F_0/F$ vs. $Q$ and $\tau_0/\tau$ vs. $Q$ both yield straight lines and also if $F_0/F = \tau_0/\tau$, then it is almost certain that the quenching process involves only a dynamic collision type of mechanism.

## 4.2. Fluorescence Lifetime Reference Standards

A good example of the application of lifetime quenching is the study of Chen[17] for lifetime reference standards. The linearity of $Q$ vs. $\tau$ was utilized to establish the quinine/NaCl system as lifetime reference standards in the range 0.189–18.9 ns and the pyrene butyrate/KI system as the lifetime standards in the range 18–115 ns. The first system uses $10^{-5}$ M quinine bisulfate solutions plus varying amounts of NaCl in 0.1 N $H_2SO_4$. The concentration of NaCl needed for a particular lifetime can be calculated by the equation

$$[\text{NaCl}] = 6.21 \times 10^{-3}(\tau_0/\tau - 1) \qquad (22)$$

where $\tau_0 = 18.9$ ns is the lifetime of quinine in the absence of quencher, and $\tau$ is the desired lifetime standard.

Similarly, reference standards in the range 18–115 ns can be prepared using $\gamma$-pyrenebutyric acid (one-tenth saturated solution in 1 mM KOH, 10 mM Tris buffer, pH 8.0) plus varying amounts of KI which can be calculated according to the relation

$$[\text{KI}] = 1.927(\tau_0/\tau - 1) \qquad (23)$$

where $\tau_0 = 115$ ns is the lifetime of pyrene butyrate in the absence of KI.

## 4.3. Long-Lived Pyrene Probes in Quenching Studies

Many protein studies utilized intrinsic Trp and Tyr fluorescence (see Section 3). However, both Trp and Tyr residues have rather short lifetimes (1–4 ns), and in many proteins there is usually more than one aromatic residue present and their distribution may not be known precisely. The use of extrinsic probes, particularly the long-lived pyrene derivatives, provides many advantages in lifetime quenching studies. Pyrene derivatives can be specifically attached to certain functional sites of proteins. More than a dozen such pyrene derivatives are available commercially. Derivatives with fatty acid side chains are frequently used in membrane studies. Compared to the typical shorter-lived 10–20 ns fluorophores, pyrene probes have lifetimes in the range 100–300 ns. This long lifetime increases by about 5- to 10-fold the time quenchers can diffuse and effectively collide with probes. This extra time gain is equivalent to an increase of about 100–1000 times for the effective volume of the sphere in which the probe can interact with the quencher.

Pyrene is sparsely soluble in water, but derivatives which contain long hydrocarbon chains such as pyrenebutyric acid (PBA), the ester of PBA and 10-hydroxyldecanoic acid (PBDA) can be solubilized in aqueous solutions of BSA presumably by adsorption to the protein. The usefulness of these probes is illustrated in a classical paper by Vaughan and Weber[18] and more recently in a paper by Cooper and Thomas.[19] Free PBA has a fluorescence lifetime of about 135 ns in deoxygenated water, and its lifetime is quenched by oxygen to about 100 ns in air-saturated water. The quenching process follows the Stern–Volmer equation and is shown to be a diffusion-controlled process in which virtually every collision is effective. The quenching rate constant is temperature dependent; at $35°C$, it is $1.5 \times 10^{10} \text{ M}^{-1} \text{ s}^{-1}$. When PBA is adsorbed or covalently attached to BSA, it becomes inaccessible to oxygen, and the quenching rate constant is greatly reduced. The excited state of the probe is also stabilized by the surrounding protein environment, as suggested by an increase in $\tau_0$ (the lifetime in the absence of quencher). Lowering the pH of the BSA solution causes the protein to expand; PBA becomes more accessible to oxygen with a five- to tenfold increase in the quenching rate constant without appreciable changes in $\tau_0$. The accessibility increases further upon urea denaturation, to the extent that the rate constant approximates that of free PBA in urea.

An interesting complementary study was done by Vaughan and Weber on the quenching behavior of PBA-polylysine conjugate by oxygen. The secondary structure of polylysine is pH dependent. At neutral pH and below (6.85), poly-L-lysine exists as a loosely open structure, and $\tau_0$ of the conjugate with PBA is about 100 ns. The shorter lifetime is probably caused by the charges carried by the $\text{NH}_4^+$ groups. The quantity $K_q$ is about $0.68 \times 10^{10} \text{ M}^{-1} \text{ s}^{-1}$, approaching the rate constant of PBA in 8 M urea. Once the positive charges are eliminated by raising the pH to 10.9, polylysine increases its $\alpha$-helical content and becomes more compact. Consequently, $\tau_0$ increases to 154 ns and $K_q$ is reduced by about 25%. Conjugated PBA with thyroglobulin has a $K_q$ value intermediate between that for BSA and polylysine. In BSA the hydrophobic residues of PBA are virtually buried, but in polylysine with an open structure, PBA is almost freely accessible to solvent. The experiments discussed above thus demonstrate that PBA may be used as a lifetime probe in determining the accessibility of oxygen in a microenvironment of biological interest. Because oxygen is small, does not carry a charge, and is itself a potent quencher, it is ideal for probing small "holes" and "clefts" in the structure of proteins.

In the study by Cooper and Thomas,[19] the quenchers iodine, nitromethane, and thallous ion ($\text{Tl}^+$) were employed. The quenching effects on several pyrene probes, pure pyrene sulfonic acid (PSA), PBA, and PBDA were compared. In these studies, the probes were all incorporated into the proteins BSA or lysozyme by absorption. The lifetime in the absence of quencher was longest for pure pyrene (344 ns at 25°C) and shortest for PSA (103 ns). Thus, in BSA solution $\tau_0$ approached the theoretical limit of 400 ns (at 0°C). Increasing temperature shortened the lifetime as predicted, but there was no sharp breaking point in the temperature profile using BSA solution. In RNase solution $\tau_0$ was smaller than its theoretical limit, indicating that less stabilizing effect was exerted by the protein on the hydrophobic probes. However, there was a sharp break in the temperature profile in RNase solution at 60°C, coinciding with the transition point when RNase starts to unfold. The rate constants for the quenching of PSA by $O_2$ and $CH_3NO_2$ are a function of pH; it gradually increases at pH 4 and rises to a maximum at pH 2.4 followed by a sharp drop below pH 2.4. The latter decrease in $K_q$ may be due to the increasing mobility of both probe and quencher molecules in the protein. Among the three quenchers mentioned above, $I^-$ yielded the lowest rate constant for probes embedded inside a protein matrix under ordinary circumstance, owing to the fact that $I^-$ is bulky in size and therefore diffuses more slowly before colliding with a fluorophore.

When $I^-$ was used as the quencher, the rate constant was the highest for pyrene ($1.2 \times 10^9$ $M^{-1}$ $s^{-1}$), being only a factor of 2 or 3 smaller than that observed in free solution, i.e., in the absence of BSA. This $I^-$ rate constant was considerably larger than those of other small molecules, i.e., $I_2$ and $CH_3NO_2$. Quenching of other pyrene probes, e.g., PSA, PBA, and PBDA, by $I^-$ was considerably less, at least by one or two orders of magnitude, than for pure pyrene. Furthermore, quenching reactions of the pyrene derivatives by $O_2$ and $CH_3NO_2$ were substantially higher than by $I^-$. These results could be explained by assuming that $I^-$ binds to the pyrene binding site in BSA and that the pyrene binding site is different from binding sites for other probes. The binding site for pure pyrene presumably carries cationic residues and is more easily accessible to a bulky quencher. This interpretation was consistent with the results obtained using thallous cation as quencher. Quenching by $Tl^+$ in probes other than pyrene was at least one to two orders of magnitude lower than by all three other quenchers; even in the case of pyrene, the quenching rate constant by $Tl^+$ was ten times lower than that by $I^-$, presumably due to the repulsive charges carried by the residues in the pyrene binding site with $Tl^+$.

The effects of inert additives on the quenching behavior can be predicted from the quenching mechanism of the individual quencher and the nature and effect which the inert additives exert on the proteins. Thus, the addition of $Cl^-$, which competes for the $I^-$ binding site of protein, will considerably reduce the rate constant of quenching of pure pyrene by $I^-$ but has little effect on quenching by $O_2$ and $CH_3NO_2$. The addition of the long hydrocarbon chain amphiphile, cetyltrimethylammonium bromide (CTAB), causes conformational changes in proteins which allow small quenchers, $O_2$ and $CH_3NO_2$, to penetrate further into the protein.

In summary, several generalizations can be made. There are basically six factors that affect fluorescence quenching: (1) *Size factor*: small quencher molecules are, in general, more permeable in proteins and, therefore, yield higher quenching efficiency; (2) *Charge factor*: for quenchers carrying a charge, the quenching reaction may be facilitated by opposite net charges on amino acid side chains in the surrounding environment at the probe binding site. Counterions which neutralize the charge effect reduce the quenching rate; (3) *pH factor*: the pH factor is often related to the charge effect and structural factor. By raising or lowering the pH of protein solutions, ionization of amino or carboxyl residues at the probe binding site can facilitate or repress the quenching reaction depending on the kind of charge carried by the quencher molecules; (4) *Temperature factor*: since dynamic collisional quenching is diffusion controlled, increasing temperature increases the frequency of encounters between fluorophore and quencher molecules and, therefore, increases the rate constant. At higher temperature, unfolding of polypeptide may occur and the protein matrix becomes more permeable to quencher molecules. In the quenching efficiency versus temperature profile, the transition temperature for protein unfolding may be revealed by a sharp break; (5) *Structure factor*: protein denaturants which cause protein to unfold increase the accessibility of fluoropores to quenchers; therefore, the addition of chaotropic agents may significantly increase the quenching rate constant to the point where the rate constant approaches that in free solution. Thus, quenching studies can be used to monitor protein denaturation or unfolding by pH, temperature, or denaturants; (6) *Viscosity factor*: a viscous medium slows down the diffusion rate of quencher and protein. Therefore, increasing the viscosity of solutions suppresses the quenching reaction. However, if the probe is deeply embedded inside the protein matrix, the macroviscosity of the bulk solvent becomes irrelevant. The quenching behavior is predominantly dependent on the microvicosity of the surrounding environment near the fluorophore binding site. A case in point will be discussed in more detail in the following section.

Evidently, all of the above six factors play a part in determining the overall quenching behavior. Information obtained from quenching experiments regarding a specific structural feature of a given protein must be carefully evaluated from several parallel studies in which each of these factors has been examined individually.

## 4.4. Dynamical Structural Fluctuations of Proteins in Solution

Proteins in solution often undergo rapid, reversible conformational changes. Such structural fluctuations can be monitored by quenching of protein fluorescence by a small molecular quencher such as oxygen. Nanosecond pulse fluorimetry has been used to probe the internal segmental polypeptide chain movement within the protein matrix because the relaxation time of these movements is in the same time domain, namely a few nanoseconds.

Lakowicz and Weber[20,21] were the first to show that the fluorescence of Trp residues buried within proteins can be collisionally quenched by oxygen. This quenching phenomenon was utilized to demonstrate that rapid structural fluctuations occurred in proteins which facilitated the diffusion of $O_2$ through the polypeptide matrix. The structural fluctuations in liver alcohol dehydrogenase (LADH) were demonstrated by Barboy and Feitelson[22] using a similar approach. LADH is composed of two subunits. Each subunit contains a Trp residue (Trp-314) which is buried within the hydrophobic area of binding interface between the two subunits. A second Trp residue in each subunit (Trp-15) lies in the periphery of the protein and is probably exposed to the solvent. When the protein is excited at 280 nm, nanosecond decay analysis showed two lifetimes of 2.2 and 5.7 ns. However, when excited above 300 nm, only the Trp-314 residue buried in the interior fluoresced and yielded a single decay curve with a 5-ns lifetime. This was verified by quenching the Trp fluorescence with iodide which quenched the Trp-15 fluorescence (in the periphery of the protein) and showed only a single lifetime of 4.3 ns when excited at 280 or 300 nm. The quencher effects on the buried residue (Trp-314) by KI and CsCl were monitored by following the lifetime changes with excitation above 300 nm. The quenching rate constant by quencher KI and CsCl in the indole model compound, N-acetyltryptophanamide (NATA) was practically independent of temperature. In contrast, in LADH, the quenching rate decreased drastically with decreasing temperature. These data were interpreted by Barboy and Feitelson as indicative of conformational fluctuations in LADH, whereby the temporal movement of the polypeptide chains, which is facilitated by increasing temperature, opens up channels through which the quencher molecules can diffuse into the protein interior and collisionally quench the fluorescence of the buried residue. The likely diffusional channel for the ion quenchers has also been identified on the three-dimensional crystalline structure of LADH.

In a similar study in RNase T, Eftink and Ghiron[23] used a much larger quencher, acrylamide, and found that the fluorescence of the supposedly "buried" Trp residue in RNase T was collisionally quenched by acrylamide with a high reaction rate constant of $3 \times 10^8$ M$^{-1}$ s$^{-1}$. This Trp residue has been shown to be "buried" and chemically very unreactive.[24] The quenching mechanism was found to be of the dynamic collisional type, because there was a parallel drop in both the fluorescence lifetime (34% decrease) and quantum yield (30% reduction) as quencher was added. If the quenching mechanism were of a static quenching type, there should have been no change in lifetime. The Trp residue was not fully exposed to solvent, since if it were, a rate constant of about $3-5 \times 10^9$ M$^{-1}$ s$^{-1}$ would be expected. When the viscosity of the solution was increased by approximately fivefold, in theory a fivefold decrease in the quenching rate would be anticipated, but instead only a 20% reduction in rate constant was found. The failure of macroviscosity changes to significantly affect the rate of collisional quenching indicated that the encounters between the Trp residues and acrylamide molecules must be limited by the microviscosity in the region of the Trp residues. Thus, the diffusion of acrylamide leading to encounters with the "buried" Trp residues was not through the bulk solvent, but rather through the protein matrix. These results, taken together, led Eftink and Ghirton to suggest that for a large molecule such as acrylamide to penetrate into a protein matrix and collide with a Trp residue that is "buried," there must be openings or pores in the tertiary structure of the enzyme. These holes were formed during rapid fluctuations of the protein matrix. Since this rapid fluctuation occurred on a time scale of a fraction of a nanosecond, detecting such motion can only be accomplished by nanosecond pulse fluorimetry and other compatible techniques such as NMR, EPR, and dielectric relaxation studies.

## 5. Fluorescence Energy Transfer for Distance Measurements in Proteins

A detailed knowledge of the three-dimensional structure of protein is very desirable in order to understand the structure–function relationships in proteins. At present, x-ray crystallographic determinations remain the only means for ultimately resolving the three-dimensional structure of proteins. The use of x-ray diffraction is limited by many technical problems inherent in the crystallographic method as well as by the tremendous investment of time needed to compute the data and build models. For proteins in solution whose structures are in dynamic equilibrium between several conformations, it is also not possible to use x-ray diffraction to determine the exact three-dimensional structure of proteins.

Fortunately, often even a limited knowledge of the structure of a part of a protein molecule is very valuable in understanding how the protein or enzyme functions, e.g., information regarding spatial arrangement of substrate in the active site of an enzyme or spatial relationship of ligand binding sites in a multiple-subunit protein. Distance measurement by fluorescence energy transfer provides a unique way of measuring the distance between pairs of functional sites in proteins. In fact, this technique has been used as a spectroscopic ruler for measuring distance between two physiologically interesting loci in a wide variety of biological macromolecules and assemblies. Even though distance measurement by fluorescence energy transfer is only one-dimensional, if several pairs of loci are measured, the spatial relationship between several sites can be reconstructed.

For energy transfer to occur, several conditions must be met. There must exist a pair of donor and acceptor chromophores where the absorption spectrum of the latter overlaps with the fluorescence emission spectrum of the former. Since energy transfer efficiency is inversely proportional to the sixth power of the separation distance between the donor and acceptor, the distance between the two cannot be too great. Experimentally, 70 Å is about the greatest measurable distance.

Of the fluorophores used for donor or acceptor, one has a wide choice. The intrinsic Trp fluorophore in proteins was widely used as a donor in early studies; however, its application is mostly restricted to qualitative proximity measurements. This limitation largely stems from the low quantum yield and short lifetime of tryptophan residues. Flavin, nicotinamide, and their related nucleotides have also been employed in a number of studies. The fluorescent analogs of ATP and coenzyme A, their 1,$N^6$-ethenoadenosine derivatives, have also been used as donors in a number of studies of proteins containing ATP or enzymes which utilize ATP as substrate or CoA as cofactor.

Even more versatile in actual practice are a group of extrinsic fluorescent probes which have been "tailor-made" to optimize their usefulness in distance measurements. Some of these donors have long lifetimes, such as pyrene and its derivatives, making energy transfer measurements more accurate. Also pairs of donors and acceptors which match to yield optimum spectral overlap and suitable critical distance can be selected from available extrinsic probes. Another advantage in using extrinsic probes is that reactive derivatives of these probes can be made which react specifically with certain amino acid residues thereby permitting labeling of specific sites. Also, a given pair of donor–acceptor fluorophores can be used repeatedly, attaching them to different loci within the same protein or to different subunits of a protein complex, or even simply interchanging the labeling of donor and acceptor at two different loci. This can be achieved by using derivatives that contain different chemically reactive groups. Donor and acceptor can often be labeled separately to individual components of a complex system and then reconstituted. Thus, by appropriate choice of donor–acceptor pair and labeling of important sites selectively, a variety of distance measurements can be made, e.g., distance between enzyme-active site (catalytic site) and regulatory site, or center-to-center distance between two globular subunits in a protein assembly.

In this section, a brief overview of the theory of fluorescence energy transfer will be given first, followed by a brief discussion of some problems frequently encountered in the experiments, and, lastly, a brief review of some distance measurements that have been made.

## 5.1. Evaluation of Energy Transfer Efficiency and Calculation of Separation Distance

The quantum theory of fluorescence energy transfer was first developed by Forster.[25] The theory has been experimentally proven using small organic model compounds.[26] Several good accounts of the original theory have appeared in review papers. The present section will only summarize some of the equations necessary in the distance calculation. For more detailed and complete discussions of the subject, the reader should consult Stryer.[27]

A donor fluorophore may dissipate part of its excitation energy to a nearby acceptor chromophore capable of absorbing emission energy of the donor via fluorescence energy transfer. The transfer efficiency is inversely proportional to the sixth power of the distance between the donor and acceptor according to the Forster equation:

$$E = R^{-6}/(R^{-6} + R_0^{-6}) \tag{24}$$

$$R = (E^{-1} - 1)^{-1/6} R_0 \tag{25}$$

where $R_0$ is the critical distance at which $E$ is 50% and can be calculated from the relation

$$R_0 = (J\kappa^2 Q_0 n^{-4})^{1/6} \, 9.79 \times 10^3 (\text{in angstroms}) \tag{26}$$

where $Q_0$ is the quantum yield of the donor in the absence of acceptor, $n$ is the refractive index of the solution, and $\kappa$ is the orientation factor which measures the angular dependence of the dipole interaction between donor emission and absorption of acceptor. Although $\kappa$ can have values between 0 and 4, it has been shown[27] that, in practice, the $k$ factor introduces relatively small errors in estimating distances, largely owing to the fact that $R_0$ is proportional only to the cube root of $\kappa$, and because of geometric constraints imposed by the specific environment on donor and acceptor orientation. The spectral overlap $J$ is given by

$$J = \int F(\lambda)\epsilon(\lambda) \cdot \lambda^4 \, d\lambda / \int F(\lambda) \, d\lambda \tag{27}$$

where $F(\lambda)$ is the fluorescence intensity of the donor and $\epsilon(\lambda)$ is the extinction coefficient of the acceptor at wavelength $\lambda$. The integral is calculated over the entire region where the donor emission spectrum overlaps with the acceptor absorption spectrum. The greater the overlap of the two spectra and the higher the efficiency of acceptor absorbance, the greater the value of the $J$ integral and the larger the distance of the separation that can be measured. The transfer efficiency is determined experimentally by donor fluorescence

$$E = 1 - (Q_{d-a}/Q_d) \tag{28a}$$

or

$$E = 1 - (\tau_{d-a}/\tau_d) \tag{28b}$$

where $Q_{d-a}$ is the quantum yield of donor in the presence of acceptor and $Q_d$ is that in the absence of acceptor, and $\tau_{d-a}$ is the lifetime of donor in the presence of acceptor and $\tau_d$ is that in the absence of acceptor. Similarly, $E$ can also be determined by measuring acceptor fluorescence:

or

$$E = (Q_{a-d}/Q_a) - 1 \qquad (29a)$$

$$E = (\tau_{a-d}/\tau_a) - 1 \qquad (29b)$$

where $Q_{a-d}$, $Q_a$ and $\tau_{a-d}$, $\tau_a$ are the quantum yield and lifetime of acceptor in the presence or absence of donor.

If the fluorescence emission spectra of two solutions are measured with the same instrumental geometry and at the same excitation light intensity, the following relations are generally valid:

$$Q_1/Q_2 = \text{Area}_1/\text{Area}_2 = (\Gamma_1 OD_1)/(\Gamma_2 OD_2) \qquad (30)$$

where $Q_1$, $Q_2$, $\text{Area}_1$, and $\text{Area}_2$ are, respectively, the quantum yields and areas under emission spectra 1 and 2, $\Gamma$ is the absolute fluorescence efficiency, and OD is the optical density of the samples.

### 5.1.1. Uncertainty in Distance Calculation

The only parameter in the energy transfer measurement that cannot be determined by a simple experiment is the orientation factor $\kappa^2$ which is the measure of the relative orientation of donor molecules with respect to acceptor molecules. If the donor and acceptor probes rotate rapidly and in all directions in the protein adducts, dynamical averaging of angular dependence in all possible orientations of donor to acceptor gives rise to a $\kappa^2$ value of 2/3. It is customarily assumed that $\kappa^2 = 2/3$ in most distance measurements by energy transfer. The calculated distance is therefore subjected to the uncertainty in the assumed value of $\kappa$.

This problem has been discussed thoroughly by Dale and Eisinger.[28,29] Hillel and Wu[30] presented a statistical method to circumvent the uncertainty. Their method enables calculation of the "probability" that a specified calculated distance corresponds to the actual distance to be measured. A partial resolution of the orientation ambiguity has also been attempted by Dale and Eisenger[29,31] using depolarization measurements to define upper and lower bounds of the actual distance to be measured.

The calculated separation distance based on assumption of $\kappa^2 = 2/3$ is called the apparent distance $R'$. The actual distance $R$ is related to $R'$ by

$$R = \alpha R' = (1.5\kappa^2)^{1/6} R' \qquad (31)$$

where $\alpha$ is a parameter which sets the range of the actual distance in relation to the calculated distance and can be estimated reasonably well from the donor emission anisotropy data. A donor not rigidly embedded in the protein attachment site undergoes rapid depolarization independent of the protein motion. The donor can be treated as though it were wobbling rapidly and confined to a cone of semiangle $\theta$. The semiangle $\theta$ can be calculated according to Kawato et al.[32] by the equation

$$A/A_0 = [0.5(1 + \cos \theta) \cos \theta]^2 \tag{32}$$

where $A_0$ is the limiting anisotropy value measured in a frozen state and $A$ is the donor emission anisotropy in the steady state after the decay due to rapid wobbling is virtually completed. The range of $\alpha$ is related to $\theta$ by

$$[0.75(1 - \cos^2 \theta)]^{1/6} \leqslant \alpha \leqslant (6 \cos^2 \theta)^{1/6} \tag{33}$$

It is evident from equation (33) that the range of $\alpha$ is quite narrow because $\alpha$ is proportional to the cube root of $\cos \theta$. For example, for a donor with a limiting cone semiangle of $\theta = 30$, $0.76 \leqslant \alpha \leqslant 1.29$. For a donor with a large rotating cone semiangle of $\theta = 60$, $0.91 \leqslant \alpha \leqslant 1.07$, i.e., the uncertainty in the calculated apparent distance is only less than 10%. It should be noted that the range of $\alpha$ could be even narrower if the acceptor also has rotational freedom, and in most cases they do. Alternatively, several pairs of donor/acceptor probes can be employed to set limits for the separation distance.

## 5.2. Some Practical Considerations and Problems Involved in Energy Transfer Experiments

### 5.2.1. Correction for Quantum Yield Measurements

As expressed in equations (28) and (29), energy transfer efficiency can be determined by measuring either the quantum yield or the lifetime, with either donor or acceptor fluorescence measured. In the quantum yield method, intensity measurements are complicated by overlap of donor emission with the absorption and/or emission band of the acceptor. In a typical quantum yield experiment, the excitation wavelength is chosen at the absorption maximum of the donor or slightly lower than the maximum to reduce possible scattered light interference. The emission wavelength is chosen either for donor or for acceptor. The transfer efficiency is calculated from either the increase in acceptor fluorescence or the decrease in donor fluorescence. The transfer efficiencies calculated by the two methods should be identical.

In the quantum yield method, subtraction of appropriate background is very important. If the emission band of donor is well separated from the acceptor emission, which is rarely the case, the intensity at a fixed emission wavelength can be substituted for the quantum yield used in equation (30). It is more correct to use the area under each emission band in which the optical density difference has also been corrected.

If donor and acceptor are both located in the same protein (or subunit), the following equations which correct for the acceptor emission measured at the donor emission wavelength or the donor emission contribution in emission measurement should be used:

and

$$E_d = 1 - [Q_d(P^{d-a}) - Q_d(P^a)]/Q_d(P^d) \qquad (34)$$

$$E_a = [Q_a(P^{d-a}) - Q_a(P^d)]/Q_a(P^a) - 1 \qquad (35)$$

where $E_d$ and $E_a$ are energy transfer efficiencies for donor and acceptor, respectively. The subscripts d and a in $Q$ and $E$ refer to fluorescence measured at donor and acceptor emission wavelength, respectively. The superscripts in protein P refer to protein labeled with donor d and/or acceptor a. For example, $Q_d(P^{d-a})$ refers to fluorescence of donor and acceptor double labeled protein measured at donor emission wavelength.

If donor and acceptor are placed on separate proteins or subunits, the following equations should be used:

$$E_d = 1 - [Q_d(P^d + P^a) - Q_d(P_1 + P^a)]/Q_d(P^d + P_2) \qquad (36)$$

$$E_a = [Q_a(P^d + P^a) - Q_a(P^d + P_2)]/Q_a(P_1 + P^a) - 1 \qquad (37)$$

where $P_1$ and $P_2$ inside the parentheses refer to proteins 1 and 2 which are labeled with donor and acceptor, respectively.

## 5.2.2. Problems in Energy Transfer Measurements

One common problem associated with the quantum yield method in determining the transfer efficiency is that frequently donor emission overlaps with acceptor emission. Furthermore, the acceptor may absorb at the wavelength used to excite the donor and emit fluorescence that cannot be segregated from its own donor emission.

The example given below illustrates this problem. In an energy transfer experiment done in our laboratory, 1,5-IAEDANS attached to tropomyosin was used as donor (the fluorescence moiety is the dansyl group) while 5-IAF attached to troponin was used as an acceptor (the chromophore is the fluorescein group). Labeled tropomyosin and troponin bind in a 1:1 molar ratio with a concentration of $1 \mu M$ each. The excitation wavelength was 340 nm, approximately the absorption maximum of the donor. The emission spectra of donor-labeled tropomyosin, acceptor-labeled troponin, and the combined system are shown in Figure 4 as curves 1, 2, and 4, respectively. Even though the acceptor absorbed only weakly at the donor excitation wavelength, because of its high quantum yield, its emission was substantial. In fact, the area under the acceptor emission band was almost three times as large as the area under the donor emission band. If one calculates the transfer efficiency by the intensity decrease at the donor emission maximum (at about 480 nm) or the intensity increase at the acceptor emission maximum (at about 518 nm), 39.1 and 7.8% efficiencies are obtained for donor and acceptor respectively. Because of the overlap between the donor emission band and the acceptor emission band in the 500–580 nm region, the area method cannot be used to directly calculate the efficiency. Instead, one first synthesizes a curve corresponding to the complex of donor-labeled protein with acceptor-labeled protein assuming there is no

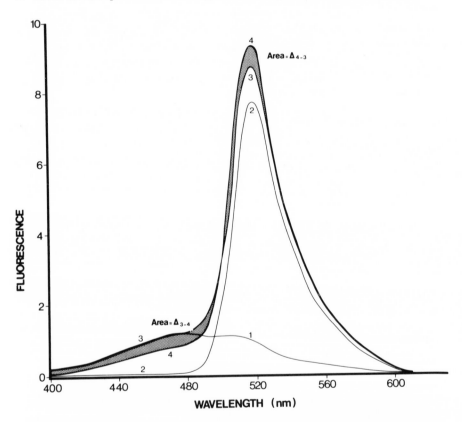

FIGURE 4. A typical problem in energy transfer efficiency determination by the quantum yield method. Curves: (1) 1,5-IAEDANS-labeled tropomyosin (donor); (2) IAF-labeled troponin (acceptor); (3) calculated sum of curves 1 and 2; (4) measured combined system of 1 and 2. Shaded area is the difference in area between curves 3 and 4.

energy transfer between them. This is done by simply adding curve 1 and curve 2 to obtain curve 3. Comparing curve 3 with curve 4, one sees immediately a substantial decrease of donor fluorescence in the 420–490 nm region and a concomitant increase of fluorescence in the acceptor peak emission. The area differences between the two curves are then used to calculate the transfer efficiency. Transfer efficiencies of 16.8 and 4.8% are obtained for donor and acceptor respectively. The large discrepancy in transfer efficiency exists not only between the donor and acceptor but also between the intensity method at a single wavelength and the area-under-the-curve method. Two factors may cause this large discrepancy. First, the optical density of acceptor-labeled protein was much higher than that of donor-labeled protein. The high extinction of acceptor-labeled protein reabsorbed the acceptor emission. Second, the number of donor molecules in the system was not matched with the number of acceptor molecules; in the above case, because of the differences in labeling ratio, there were possibly more acceptor molecules than the donor molecules. The latter problem will be discussed in more detail in Section 5.2.3.

In the lifetime method, the separation of donor emission from acceptor emission is less of a problem. The emission can be measured using a very narrow bandwidth filter so that there will be little cross-over between donor emission and acceptor emission. Since the sensitivity of nanosecond pulse fluorimetry with single-photon counting is so great, one can afford to trade off some intensity for highly selective wavelength. The donor emission can be measured at lower than peak wavelength if necessary when acceptor emission is almost zero (e.g., at 460 nm in the experiment depicted in Figure 4). Similarly acceptor emission can be measured at a longer than peak wavelength, if necessary, to avoid the donor emission band (e.g., 530 nm, Figure 4). If the donor lifetime differs substantially from that of the acceptor, complete separation of the two lifetimes can be achieved during deconvolution of the profile. One may even choose as acceptor a nonfluorescent chromophore that is highly absorptive at the donor emission wavelength to obtain good overlap, or a fluorescent acceptor with very short lifetime. However, in doing so the benefit of using acceptor emission for calculating transfer efficiency is lost. Two examples of such acceptors are eosin and stilbene derivatives.

## 5.2.3. Correction for Labeling Ratio Less Than Stoichiometric Ratio

Another problem sometimes encountered in energy transfer experiments is that the number of donor and acceptor molecules is not 1 : 1. Theoretically, equations (28) and (29) can only be used directly for calculating transfer efficiency if there is one acceptor related to each donor; if not, a normalization factor must be applied. Thus, experimental accuracy in the determination of the concentrations of probes and protein adducts, and also of the stoichiometric ratio of probe to protein, is quite important.

Taking muscle contractile proteins as an example, the stoichiometric binding ratio of thin filament proteins is tropomyosin : troponin : actin = 1 : 1 : 7. If donor tropomyosin is 80% labeled and acceptor troponin is 75% labeled, the probability that a donor "sees" an acceptor is 0.8 × 0.75 = 0.6. The observed lifetime can be related to the lifetimes of donor alone or of donor with an acceptor present by taking a number average:

$$\tau_{obs} = (n_d \cdot \tau_d^2 + n_{d-a} \cdot \tau_{d-a}^2)/(n_d \cdot \tau_d + n_{d-a} \cdot \tau_{d-a}) \qquad (38)$$

or

$$\tau_{d-a} = 1/2\tau_{obs} + [1/4\tau_{obs}^2 - (n_d/n_{d-a})\tau_d(\tau_d - \tau_{obs})]^{1/2} \qquad (39)$$

where $n_{d-a}$ and $n_d$ are the percentages of donor having an acceptor and of donor not having an acceptor, respectively. In the above example, $n_d = 0.4$ and $n_{d-a} = 0.6$. On the other hand, if actin is labeled with donor, seven actin molecules bind to only one tropomyosin molecule, and there is only one out of seven donors that can have an acceptor. If each tropomyosin molecule is labeled with one acceptor, $n_d = 0.857$ and $n_{d-a} = 0.143$. In this case, experimentally, $\tau_{obs}$ is not very much different from $\tau_d$, and one will not observe much decrease in donor lifetime.

When there is more than one acceptor for each donor, i.e., when acceptor protein is overly labeled, problems may arise because a donor may transfer energy via two pathways to two different nearby acceptor molecules. The additional normalization or correction factor will be required but difficult to obtain without some *a priori* assumptions. In addition, two lifetime components corresponding to $\tau_{d-a}$ which may differ from each other will appear in equation (38). This is likely to occur in the above example when actin molecules are labeled with acceptors and the donor–acceptor pair has a large critical distance. It is probable that more than one actin molecule may accept excitation energy transfer from a donor placed on tropomyosin. To avoid this problem, it may be advisable at times to choose a donor–acceptor pair with a smaller $R_0$.

## 5.3. Applications in Protein Systems

### 5.3.1. Distance Measurements in a Single Protein System

The distance relationships between the visual photoreceptor, retinol, and three distinct fluorescently labeled sites on the receptor protein rhodopsin have been closely examined in an early work by Wu and Stryer[33] using the energy transfer technique. Three types of fluorescent probes were attached to rhodopsin. Site A (the designation of these sites by the original authors was arbitrary) was labeled with one of the three iodacetylamide derivatives: *N*-iodoacetylaminoethyl)-5-naphthylamine-1-sulfonic acid (1,5-IAEDANS), 1,8-IAEDANS, and 5-iodoacetamidosalicylic acid. The reaction was alkylation of free sulfhydryls of the protein. Site B was labeled with one of the two disulfide derivatives as donor: didansyl-L-cystine or difluorescein isothiocarbamido-cystamine. The reaction was a disulfide–sulfhydryl reaction. Site C was labeled noncovalently with one of the two acridine derivatives as donors: 9-hydrazino-acridine or proflavin. The labeled proteins were checked for their photosensitivity and regenerability after bleaching and were found to be functional. A probe at one of these sites served as an energy donor, while 11-*cis*-retinol served as the energy acceptor. The critical distance of these seven donors which pair with the acceptor 11-*cis*-retinol was calculated to be in the range 33 Å (with 9-hydrazinoacridine as donor) to 52 Å (with didansyl-L-cystine as donor). The transfer efficiency was determined from the donor lifetime in the presence and absence of acceptor. The energy-transfer efficiencies of these seven donors to the 11-*cis*-retinol acceptor varied widely as measured by the lifetime method, being lowest in site A with efficiencies of 3, 4, and 9%, respectively, for the three probes listed above; higher in site C with efficiencies of 12 and 23% for the two probes; and highest for the two probes in site B with efficiencies of 22 and 36%, respectively. Despite the wide range in their transfer efficiencies, the apparent distances calculated from the lifetime data are quite consistent for the distances between these three sites to the retinol bound to the light-receptor site in the rhodopsin. The apparent distances from these sites to the retinol binding site were estimated to be about 75, 55, and 48 Å, respectively, for sites A, B, and C. Energy transfer experiments were also performed by attaching a pair of donor and acceptor to two of these sites. The apparent distances were calculated to be 35 Å for site A to site B, 32 Å for site A to site C, and 30 Å for site B to site C. Since the molecular weight of rhodopsin is about 40,000, it can be shown that if rhodopsin were spherical in shape, its diameter would be about 45 Å. The distance measurement between site A to retinol of 75 Å thus suggests that the rhodopsin molecule is not spherical but rather elongated. It is also clear that the three fluorescent probe binding sites are clustered in one region of the rhodopsin molecule and are at least 40 Å away from 11-*cis*-retinol. The 75 Å long axis of the molecule is sufficiently long to traverse the flat membrane of the rod cell visual pigments which is considered to be physiologically essential for the function. A model of the rhodopsin molecule based on the above distance calculations is depicted in Figure 5.

In another earlier study using the energy transfer technique, the distance between the two hapten binding sites on rabbit immunoglobulin G (IgG) was

FIGURE 5. A model of the rhodopsin molecule based on distance measurement by fluorescence energy transfer. A, B, and C are fluorescent probes labeling sites (see text). From Wu and Stryer.[33]

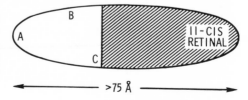

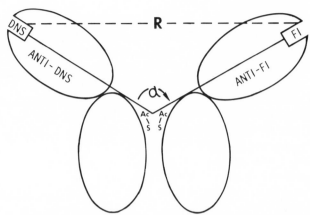

FIGURE 6. A schematic representation of the structure of the anti-DNS–anti-fluorescein hybrid antibody–hapten complex. $R$, the distance between hapten-binding sites, is about 55–70 Å; $\alpha$, the angle between the two Fab arms, is about 40 to 50°. From Werner *et al.*[34]

studied.[34] As depicted in Figure 6, IgG molecule is a Y-shaped molecule with the two hapten binding sites located at the two ends of the Fab fragments. A hybrid antibody molecule was prepared in which one active site bound an energy-donor probe, dansyl-L-lysine, and the other site bound an acceptor, fluorescein. The distance between the donor and the acceptor was measured by energy transfer using the lifetime method. For this donor–acceptor pair, the critical distance $R_0$ was calculated to be about 48 Å. By comparing the lifetime of the donor in the presence or absence of acceptor, it was found that no energy transfer occurred between the pair. Since the maximal distance over which transfer is measurable for this pair is about 80 Å (at 5% transfer efficiency), it was estimated that the average distance between the two immunoglobulin binding sites was greater than 80 Å. Assuming the length of the Fab fragment as 70 Å, the minimal angle between the two Fab arms was estimated to be about 80 to 90°.

The work of Werner *et al.* was criticized as overestimating the distance between the two Fab segments because an important interheavy-chain disulfide linkage had been eliminated irreversibly during their preparation of the hybrid antibody. In a recent study, Luedtke *et al.*[35] reexamined the proximity of the two antigen binding sites using a hybrid rabbit IgG antibody with antilactose on one site and antidansyl on the other site. The new hybrid preparation was claimed to have retained 80% of the single disulfide bond between the two heavy chains. The dansyl probe incorporated in one Fab arm in the hybrid antibody served as energy donor, while one of the three fluorescent derivatives of lactose, p-aminophenyllactoside (PAPL), dimethylaminobenzeneazo-PAPL, or N-fluorescyl-PAPL bound to the antilactose segment, served as acceptor. As in the earlier study, it was found that the dansyl donor was unaffected by the presence of acceptor, i.e., there was no energy transfer between the probes on two antigen binding sites. Taking into consideration the segmental flexibility of the two Fab arms which affects the distance distribution as well as the time-dependent variation of the angle between the two arms, these authors performed several model calculations and estimated the minimal distance between the two arms to be in the range 55–70 Å, considerably less than that calculated by Werner et al. The minimal angle between the two Fab arms was estimated to be about 40 to 50°.

Intrinsic Trp fluorescence energy transfer has been studied on the active site of papain.[36] The Trp-69 residue was reported to be capable of exchanging its excitation energy with the other four Trp residues in papain. The fluorescence of Trp-69 at the active site was enhanced by the phenyl group of Phe-glycinal inhibitor when the latter was bound to the active site. Efficient energy transfer occurred between Trp-69 and Trp-26 with an estimated distance of 11 Å between the two residues. However, when an amino-terminal mansyl [6-(N-methylanilino)-2-napthalene sulfonyl] derivative of Phe-glycinal was bound at the active site of papain, the Trp fluorescence is quenched, indicating energy transfer from papain Trp-26 to the mansyl group of Phe-glycinal inhibitor. From the transfer efficiency calculated, it was estimated that the distance between Trp-69 and Trp-26 was about 11 Å, and about 17 Å between Trp-26 and the mansyl moiety of the inhibitor.

The energy transfer from adenine to dihydronicotinamide in coenzyme NADH is a very interesting system. In neutral aqueous solution, NADH exists in two conformations: an open form, in which the ribose–diphosphate–ribose backbone is extended, and a closed form, in which the backbone is folded and the two heterocyclic rings are "stacked" in close proximity. Energy transfer can occur only when NADH is in the stacked form. However, when the coenzyme is bound to dehydrogenases, it is assumed to exist in only one of the two forms. Simultaneous determination of the quantum yields and lifetimes of $1,N^6$-ethenoadenine dinucleotide ($\epsilon$-NAD$^+$) and of the "half molecule" of $\epsilon$-AMP allows one to calculate the proportion of stacked and open conformations of the dinucleotide in solution. At 25°C, in neutral aqueous solution, it was determined by Gruber and Leonard,[37] that about 45% of the coenzyme was in the stacked conformation. Furthermore, in order to determine to what extent quenching of $\epsilon$-adenosine fluorescence can be caused by tryptophan if the latter was located near adenosine in a protein, energy transfer measurements were made using a fluorescent model compound, $1\text{-}N^6$-etheno-9-[3-(indol-3-yl)propyl]adenine ($\epsilon$-Ade$^9$-C$_3$-Ind$^3$) in which indole was used as a neutral substitute for tryptophan. By comparisons of the lifetime and the quantum yield of

$\epsilon$-Ade$^9$-C$_3$-Ind$^3$ with those of the half molecules $\epsilon$-Ade$^9$-C$_3$ and Ind$^3$-C$_3$, it was predicted that positioning the $\epsilon$-adenosine moiety close to a Trp residue in proteins will result in total quenching of the $\epsilon$-adenosine fluorescence.

### 5.3.2. Distance Measurement in Multisubunit and Multiprotein Systems

A big advantage that many researchers enjoy in using extrinsic fluorescent probes in protein studies is the additional information obtained by the combined use of several probes. This is best illustrated in protein studies involving more than one subunit or in a multiprotein system. Selected probes can be attached to different subunits in a complex system and a series of distance measurements made. A pair of donor and acceptor probes can be used twice in the same system by exchanging the sites of labeling. This technique is especially useful for checking the validity of the assumption of random orientation; the two distance measurements obtained should be essentially the same for random distribution.

*5.3.2.1. Pyruvate Dehydrogenase Complex.* The pyruvate dehydrogenase complex of *E. coli* is a multienzyme system consisting of three enzymes which catalyze the decarboxylation of pyruvate and formation of acetyl-CoA through a multistep reaction mechanism with involvement of the cofactors thiamine pyrophosphate (TPP), FAD, and lipoic acid (Lip-S$_2$). The three enzymes are pyruvate dehydrogenase (Enz$_1$), dihydrolipoyl transacetylase (Enz$_2$), and dihydrolipoyl dehydrogenase (Enz$_3$). The sequence of reactions has been postulated as follows:

$$\text{Pyruvate} + \text{Enz}_1(\text{-TPP}) \overset{\text{Mg}^{2+}}{\rightleftharpoons} \text{CO}_2 + \text{Enz}_1 \text{ (hydroxyethyl-TPP)}$$

$$\text{Enz}_1(\text{hydroxyethyl-TPP}) + \text{Enz}_1(\text{Lip-S}_2) \rightleftharpoons \text{Enz}_1(\text{-TPP}) + \text{Enz}_2(\text{HS-Lip-S-acetyl})$$

$$\text{Enz}_2(\text{HS-Lip-S-acetyl}) + \text{CoA} \rightleftharpoons \text{Enz}_2(\text{Lip-(SH)}_2) + \text{acetyl-CoA}$$

$$\text{Enz}_2(\text{Lip-(SH)}_2) + \text{Enz}_3(\text{FAD}) \rightleftharpoons \text{Enz}_2(\text{Lip-S}_2) + \text{Enz}_3(\text{FAD}_{red})$$

$$\text{Enz}_3(\text{FAD}_{red}) + \text{NAD}^+ \rightleftharpoons \text{Enz}_3(\text{FAD}) + \text{NADH} + \text{H}^+ \tag{40}$$

The reaction mechanism postulated above requires Lip-S$_2$ binding to all three enzymes, two at a time. Koike and Reed[38] proposed a mechanism in which a single lipoic acid rotates between the binding sites of the three enzymes. The pyruvate dehydrogenase (PDH) contains regulatory binding sites for acetyl-CoA and GTP and the catalytic binding sites for pyruvate and thiamine pyrophosphate. The lipoyl transacetylase (LT) contains the binding site for CoA and lipoic acid which links to an amino group of lysine. The dihydrolipoyl dehydrogenase (DHLDH) contains FAD at the active site. Because the fully extended lipoic acid is about 14 Å in length, it follows that the three active sites must be situated within 28 Å from each other if a single Lip-S$_2$ can rotate between any two enzymes at a time. Furthermore, to transfer the acetyl group between the two lipoic arms, the distance between the two catalytic sites must not be greater than 56 Å.

Several energy transfer experiments have been performed by Hammes and his colleagues in an effort to test the mechanism postulated above. In an earlier experiment using thiochrome diphosphate, a TPP analogue which binds to PDH, as the donor, and the tightly bound FAD on the DHLDH as the energy acceptor, Moe et al.[39] concluded that the distance between the active sites on the two enzymes was 45 Å, with a possible range of 30–60 Å. There were two sources of uncertainty in these energy transfer experiments, a theoretical one due to the uncertainty in the relative orientation factor, and the other due to thiochrome diphosphate being a short-lived probe (about 1.7 ns), whose lifetime could only be measured with a large error.

The issue was reexamined by Shepherd and Hammes[40] who used 8-anilino-1-naphthalene sulfonic acid (ANS) as the donor and FAD as the acceptor. The former binds tightly and specifically to the acetyl-CoA site on PDH, and the latter binds on DHLDH. The advantage of using ANS as the energy donor in this experiment was its high quantum yield and longer lifetime. However, in either set of experiments, the energy transfer efficiencies were found to be very small. The critical distance was calculated to be about 40 Å. The maximal efficiency of energy transfer between the bound ANS and FAD which would be consistent with both the observed quantum yield and the lifetime changes was found to be 10%. The minimal separation distance between the two sites was estimated to be 58 Å. In another related study, Shepherd et al.[41] measured the distance between CoA bound on the dihydrolipoyl transacetylase enzyme and the FAD on the DHLDH enzyme to be at least 50 Å.

In a recent study, Angelides and Hammes[42] preferentially labeled the lipoic acids on the PDH multienzyme complex using a variety of fluorescent probes, e.g., N-ethylmaleimide (NEM), N-(3-pyrene)maleimide (PM), and N-[4-(dimethylamino)-3,5-dinitrophenyl]maleimide (DDPM). As the extent of labeling of the enzyme with PM increased, the quantum yield decreased, the polarization increased, and excimer fluorescence increased. Thus the pyrene moieties on different lipoic acids interacted strongly, suggesting that some lipoic acids were located close to each other. The average distance between donor PM and acceptor DDPM on lipoic acids increased from 24 to 33 Å as the ratio of donor to acceptor increased. The intermolecular distance between PM and acceptor FAD at the catalytic site of the PDH and DHLDH was measured to be from 23 to more than 47 Å, depending on the lipoic acids labeled. Little energy transfer was found between the donor thiochrome diphosphate located at the catalytic site of the PDH and a variety of lipoid acids labeled with acceptor DDPM; the intermolecular distances calculated range from 38 to greater than 45 Å.

In summary the energy transfer measurements by Hammes and his colleagues are a vivid example of the usefulness of distance measurements in elucidating the complex molecular mechanism of a multienzyme system.

*5.3.2.2. Bacterial Luciferase.* Distance measurements have been performed on a naturally bioluminescent protein, bacterial luciferase from cells of *Beneckea harveyi* strain MAV 392.[43] Bacterial luciferase is a monooxygenase which catalyzes the oxidation of reduced riboflavin 5'-phosphate and a long-chain aldehyde by molecular oxygen. In the process, an excited state of a flavin bound to the enzyme is generated as an intermediate which in turn falls to the ground state and emits light. Bacterial luciferase is a heterodimer with a catalytic subunit of molecular weight 42,000 and a second subunit which must be present for the enzyme to be active. Luciferase has a single flavin site for both the substrate $FMNH_2$ and the oxidized product FMN. The enzyme is inhibited by ANS which competes with $RMNH_2$ but does not share the same binding site. Luciferase can be labeled at reactive sulfhydryls with $N$-(1-pyrene)malemide or $N$-[$p$-(2-benzoxazolyl)phenyl] malemide. The modified enzymes showed similar affinities for ANS as the unmodified enzyme. The energy transfer efficiencies between the two fluorescent probes as donors and bound ANS as acceptor based on the enhancement of emission by acceptor was found to be about 37% per ANS acceptor, corresponding to a separation distance in the range 21–37 Å. On the other hand, with bound ANS as donor and bound FMN as acceptor, the distance was 30–58 Å from ANS to FMN. The rotational correlation time of the probe–enzyme conjugate was also determined by the time-resolved anisotropy technique to be about 47 ns. The observed rotational correlation time was much longer than that calculated for the enzyme if it were assumed to be spherical, thus suggesting that luciferase is an elongated molecule.

*5.3.2.3. Heme Proteins.* The spatial arrangement of the heme group in cytochrome $cd_1$ oxidase from *Pseudomonas aeruginosa* was studied by Mitra and Bersohn[44] using fluorescence energy transfer. The oxidase consists of two identical subunits and has a two-fold symmetry axis. Each subunit (molecular weight about 60,000) has a covantly linked heme $c$ as well as a noncovalently linked heme $d_1$. The enzyme converts $O_2$ to water by a four-electron process or nitrite to NO by a one-electron process. The electrons are accepted by heme groups from reduced azurin or ferrocytochrome C-551. It has been suggested that the two heme $d_1$ (one in each subunit) in the reduced enzyme bind CO and $CN^-$ cooperatively.[45] It has also been found that while reducing agents initially reduce the heme $c$, the electron is transferred from heme $c$ to heme $d_1$ at a rate of 8 per second.[46] To investigate the cooperativity among the heme groups, the spatial relation of these heme groups in the protein must be evaluated.

The disposition of heme groups within the oxidase can be inferred qualitatively from nanosecond pulse fluorescence decay studies of its Trp residue and a dansyl probe attached to the protein (two probes per mole of the oxidase). In myoglogin, cytochrome $c$, and hemoglobin, Trp residues do not fluoresce because the heme groups are centrally located in these relatively small proteins and are capable of quenching the Trp fluorescence by electron-energy transfer. On the other hand, horseradish peroxidase and hemoglobin, which have two of their heme groups absent, do fluoresce, presumably because the heme groups are asymmetrically located in a relatively large protein; some of the Trp residues are thus too far away from the heme to be quenched.

Lifetime measurements of cytochrome $cd_1$ oxidase show that the Trp residues and the two dansyl probes, which attach to the $\epsilon$-amino groups of lysine residues in the enzyme all exhibit quite normal decay lifetimes. The Trp residues have an average lifetime of about 4 ns, and the dansyl probe exhibits two lifetimes of 4.2 and 13.2 ns. There is no indication of energy transfer from these donors to the heme groups. The critical distance $R_0$ for Trp to heme was calculated to be 34.5 Å and for the dansyl–ferriheme pair to be 58 Å. It has been estimated that for Trp and dansyl groups not to transfer excitation energy to the heme groups, they must be located at least 80 Å (about 1.5 $R_0$, corresponding to 8% transfer efficiency) away from the nearest heme. Since the dimeric oxidase molecules are oblong in shape with a length of 90 to 100 Å, the results suggest that all four hemes presumably are located at one end of the protein. The spatial relation of the heme groups based on these measurements is shown in Figure 7.

*5.3.2.4. Glutamate Dehydrogenase Complexes.* The problem of determining the rates of energy transfer for a finite number of donor–acceptor pairs has been investigated by Brochon *et al.*[47] in beef liver glutamate dehydrogenase complex, utilizing the naturally fluorescent NADH or NADPH coenzyme. The enzyme contains several Trp residues, their intrinsic emission properties differing from each other. It has been shown that excitation-energy transfer occurs from

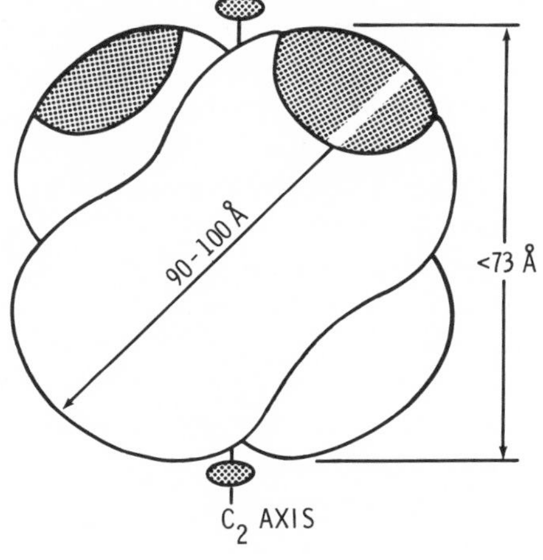

FIGURE 7. The shape, dimension, and summetry of *P. aeruginosa* cytochrome $cd_1$ oxidase as obtained from electron-microscope and preliminary x-ray studies. The shaded area at the end of each molecule is the approximate location of the hemes. From Mitra and Bersohn.[44]

the enzyme Trp residues to the coenzyme bound to the protein. The distance and the relative orientation between each Trp residue and the coenzyme ligand, in general, are different. The situation is further complicated by the fact that the enzyme binds several NADH or NADPH ligands.

Brochon *et al.* developed a theory and derived mathematical expressions to determine the transfer rate constants for each individual pair by resolving individual decay constants in the overall decay curves which represent a sum of multiple exponential functions. In general, the observed decay curve can be written as

$$I(t) = \sum_i A_i \exp\left(-t/\tau_i\right) \tag{41}$$

where $\tau_i$ and $A_i$ are the lifetimes and amplitude of donor or acceptor in pair $i$.

Fluorescence decay curves were measured under the following conditions: (1) direct excitation and emission of donor; (2) direct excitation and emission of acceptor; and (3) excitation of donor and emission of acceptor. For glutamate dehydrogenase–NADPH complexes, under condition (1), the excitation wavelength was fixed at 295 nm to excite the Trp residues in the enzyme. The Trp residues emit at 350 nm, and there is also a second emission band at 460 nm from the NADPH acceptor via energy transfer. On the other hand, under condition (2), the system is illuminated at 336 nm which excites only the bound NADPH; it emits at 460 nm only. To simplify the decay analyses, it was assumed that all the NADPH binding sites were identical and each Trp residue transferred its energy to only one nearest neighbor of the bound NADPH molecules. Also it was assumed that there were only three classes of Trp donors (the maximal number of decay components that their data analyses could handle). Since the environment of each of the Trp residues and its distance from the nicotinamide site were different from the others, it is likely that there were at least three classes of Trp decay constants, each shortening upon transfer of energy to NADPH.

The first class consists of residues having the highest rate of energy transfer, which contribute only to the 460 nm emission. Their distance from NADPH is smaller than the critical distance. The second class contains residues with transfer rates comparable to the rate of emission, contributing emission both at 350 and 460 nm. The 460 nm emission was expected to have longer decay lifetimes than NADPH itself. The distance of the second class of Trp residues from NADPH was estimated to approximate the critical distance. The third class of Trp residues do not contribute to energy transfer. Their lifetime was expected to be the same as in the absence of acceptor. They are separated from NADPH by a distance at least twice as great as the critical distance.

Using the theory they developed and measuring the fluorescence decay curves under the above conditions, Brochon *et al.* determined the individual energy transfer rate constants in a ternary complex, enzyme–NADPH–L-glutamate, and a quaternary complex, enzyme–NADPH–L-glutamate–GTP. They found that all Trp residues behave as would be expected of the first class, which can be divided into two subclasses having different transfer rates. One subclass had a decay lifetime of about 1.5 ns and the other had a lifetime of about 6 ns. The distances between these residues and the NADPH site were of the order of 25 Å. In addition, ligand binding induced a protein conformational change leading to quenching of Trp fluorescence.

*5.3.2.5. Aspartate Transcarbamylase.* Aspartate transcarbamylase catalyzes the formation of carbamyl-L-aspartate in the biosynthetic pathway, leading to the production of pyrimidine nucleotides. The reaction is subject to feedback regulation by the accumulation of nucleotides. The enzyme has six catalytic sites located on two trimeric subunits as well as six regulatory sites for the nucleotide effectors located on three dimer subunits. The intact enzyme has a molecular weight of 300,000, containing two catalytic trimers (100,000 daltons each) and three regulatory dimers (34,000 daltons each).

Fluorescence energy transfer has been employed by Hammes and his colleagues to measure the distance between several sites on this enzyme.[48,49] In one experiment, the Trp groups on the catalytic subunit were chosen as donors while pyridoxamine phosphate, covalently attached to an amino group at the active site, or ANS, which bound noncovalently at the active site, was used as acceptor. The distance between Trp residues and the active site was determined to be about 23–27 Å. In this case, energy transfers involved multiple donors and acceptors; the determination of energy transfer rate was more complicated than in simple cases. Expressions have been derived to determine the average distance between the donor and the acceptor in different pairs. In another experiment, the pyridoxamine phosphate (PDP) label was employed as donor (bound to the active site), while a mercurinitrophenol derivative was attached to the sulfhydryl group of the catalytic subunit and served as an acceptor. The sulfhydryl group was also adjacent to the active site. By determining the energy transfer between donor and acceptor on the nonadjacent sites, the distance between the active sites of subunits in the catalytic trimer was estimated to be about 26 Å. This value was found to be consistent with that obtained from the experiment in which PDP at the active site was the donor while ANS bound to another active site of the subunit of the catalytic trimer was the acceptor. The distance between PDP bound at the active site in the catalytic trimer and a fluorescent probe, a fluorescamine derivative of cytidine 5′-triphosphate attached at an amino group of the regulatory subunit as acceptor, was found to be greater than the measurable distance of 42 Å for this donor–acceptor pair. Using a different donor–acceptor pair, Hahn and Hammes[49] performed a similar distance measurement and reached a similar conclusion. The spatial relationship between

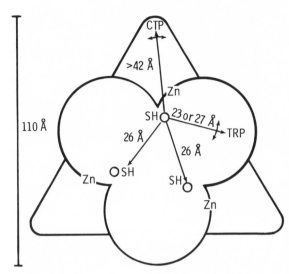

FIGURE 8. A schematic cross-sectional diagram of aspartate transcarbamylase showing the central catalytic trimer and peripheral regulatory subunits. Small circles are the proposed active site which may overlap with the adjacent sulfhydryls. From Angelides and Hammes.[42]

various Trp and sulfhydryl residues in this multisubunit system is illustrated in Figure 8.

# 6. Nanosecond Pulse Fluorimetry Studies of Muscle Contractile Proteins

Contractile proteins of muscle and other cells play an essential role in a wide variety of important physiological functions associated with movement. Contractile proteins from skeletal muscle have been studied most extensively. Their ever-increasing importance has been well recognized since proteins analogous to muscle contractile proteins have been found in many nonmuscle cells. They are involved in many seemingly diverse cellular activities, such as cytoplasmic streaming, hemostasis by blood platelets, cell division, spindle formation during mitosis, movement of spermatoza, etc. Because of their important role in many of these physiological and cellular activities, contractile proteins and similar proteins are among the most widely studied proteins.

Besides their important functional role, the organization of muscle proteins into a macromolecular assembly is of special interest. These proteins are held together tightly without covalent bonding and yet the assembly can withstand tremendous mechanical stresses during tension development. The assembly of contractile proteins has a periodic structure. As a system, there are several divalent cation binding sites ($Mg^{2+}$ and $Ca^{2+}$), as well as two types of nucleotide binding sites, one enzymatic (ATPase of myosin) and one nonenzymatic (on actin). When separated, each protein possesses its own distinct biological function independent of the others. In assembly, their interactions are harmonious and highly cooperative. Muscle as an energy-transducing machine has a very high efficiency in converting chemical energy into mechanical work.

Muscle contractile proteins have been studied by biophysical and biochemical experimental techniques. All of the three types of nanosecond pulse fluorimetry techniques discussed in the previous three sections have been applied to muscle proteins. We consider the muscle system separately as an example of how various kinds of fluorimetry measurements can be used in concert to study a given complex protein system.

## 6.1. Pyrene Lifetime and Excimer Fluorescence in Labeled Actin and Tropomyosin

Long-lived pyrene derivatives are very useful probes for protein conformation. They offer many advantages in time-resolved anisotropy decay studies as well as in energy transfer experiments because small changes can be easily detected. Two sulfhydryl-directed pyrene probes, 1-(N-pyrenyl)maleimide and 1-(N-pyrenyl)iodoacetamide, have been utilized in two recent nanosecond pulse fluorimetry studies using actin and tropomyosin.[50,51] Actin and tropomyosin are both located in the thin filaments of muscle together with another regulatory protein, troponin. Thin filaments, about 80 Å in diameter, are formed around a core of two strands of actin monomers wound helically with a repeat of about 740 Å. Each turn has about 13 subunits. Tropomyosin has two highly helical subunit chains arranged in a stranded coil–coil helix. The tropomyosin molecule is rod shaped, about 400 Å long, and wound loosely around the actin filament. The molar ratio of the three proteins in the thin filament is troponin : tropomyosin : actin = 1 : 1 : 7.

The two cys-190 residues in the two tropomyosin subunits are oriented in parallel and in register such that when two pyrene molecules are linked to the two cysteine residues, they lie in such close proximity that excimer fluorescence can occur. An excimer is a dimer formed briefly by translational diffusional collision of an excited fluorophore molecule with a molecule in the ground state. For an intramolecular excimer to occur, the two pyrene molecules in tropomyosin must be located within 3–8 Å to each other. In PIA-labeled tropomyosin, excimer fluorescence occurs immediately upon attaching two pyrene molecules to the cys-190 SH of each chain. In the case of PM, two probes are also attached to the cys-190 SH of each chain, but excimer fluorescence does not occur until the subsequent reaction of the succinimido ring of the probe with the ε-amino group of the nearby lys-189 (intramolecular aminolysis).

In actin, Lin[52] suggested that the two reactive SH, cys-373 and cys-10, were in close proximity. Double labeled actin at cys-10 and cys-373 with either PM or PIA exhibited weak pyrene excimer fluorescence. In aged PM-labeled F-actin, intramolecular aminolysis results in greater excimer fluorescence. Thus, pyrene excimer fluorescence can be used as a sensitive proximity probe. The probing range for pyrene excimer is about 3–8 Å. Excimer fluorescence differs from monomer fluorescence in both excitation and emission spectra. Monomer fluorescence exhibited characteristic multiple peak emission of pyrene in the 360–400 nm region, while excimer fluorescence exhibited a broad emission band with maximum at 480 nm and a shoulder at 505 nm. Excimer fluorescence was sensitive to changes in ionic strength and pH of the medium, while monomer fluorescence was not. The protein denaturants Gu-HCl and urea, which cause dissociation of protein polypeptide chains, also caused partial disappearance of excimer fluorescence, but not as effectively as the hydrophobic surfactant sodium dodecyl sulfate.

There has been inconsistency in the literature regarding fluorescence lifetimes of pyrene compounds and their adducts with proteins. The question of whether pyrene derivatives possess multiple fluorescence lifetimes as an intrinsic property is of particular concern. The fluorescence lifetimes of several pyrene adducts measured by nanosecond pulse fluorimetry in our laboratory are shown in Table 3. All adducts exhibited three lifetimes, ranging from 1.3–17 ns for the shortest component, from 6.1–48 ns for the medium-lifetime component, and from 27–111 ns for the longest-lifetime component.

The fraction each component contributes to the total fluorescence varies considerably in different proteins, as does the total fluorescence, which is a measure of the quantum efficiency. The fractional contribution of each lifetime component and the total fluorescence intensity reflect microenvironment effects on the fluorophore which can vary from protein to protein and even among different binding sites on a single protein. The general observation of three lifetime components in the decay of PM and PIA-labeled proteins suggests that multiple decay components of pyrene–protein adducts is an intrinsic property of the probe. In protein adducts having more than one binding site for pyrene derivative, the three measured lifetimes presumably are the weighted averages of the different lifetimes within each class, i.e., short, medium, and long, at each reactive site.

The strongest evidence for the existence of three decay components as intrinsic, nontrivial properties of pyrene–protein adducts comes from lifetime measurements of excimer fluorescence (Table 3). In all three adducts, PM–TM, aminolyzed PM–TM, and PIA–TM, the excimer lifetimes are almost identical, about 9, 35, and 65 ns, and fall into the same ranges as the three components of monomer fluorescence. Because only one kind of excimer can exist in each of these three adducts, it is difficult to explain more than one lifetime component unless all three decay components are derived from monomer. The excimer of pure pyrene exhibits only one lifetime (45–65 ns), as does the monomer. These

*Table 3. Fluorescence Lifetimes of PM and PIA Adducts[a]*

| Compound | $R_1$ | $\tau_1$ | $R_2$ | $\tau_2$ | $R_3$ | $\tau_3$ | EDP | MD |
|---|---|---|---|---|---|---|---|---|
| A. Small conjugates | | | | | | | | |
| PM in DMF | 0.21 | 1.41 | 0.16 | 11.0 | 0.63 | 47.3 | 0.4 | 0 |
| PM—mercaptoethanol | 0.02 | 2.64 | 0.04 | 20.9 | 0.94 | 64.8 | 0.38 | 0 |
| PM—cysteine (pH 8.4) | 0.39 | 1.35 | 0.44 | 7.16 | 0.18 | 26.6 | 0.35 | 0 |
| PM—cysteine (pH 11) | 0.24 | 1.34 | 0.60 | 6.17 | 0.17 | 30.2 | 0.37 | 0 |
| PM—lysine (pH 7.2) | 0.35 | 1.64 | 0.53 | 6.27 | 0.12 | 37.7 | 0.3 | 0 |
| PM—lysine (pH 10.4) | 0.40 | 1.27 | 0.48 | 6.12 | 0.12 | 29.5 | 0.35 | 0 |
| B. Protein adducts | | | | | | | | |
| PM—actin (fresh) | 0.19 | 16.94 | 0.50 | 48.3 | 0.31 | 110.8 | 0.70 | 2 |
| PM—TM (monomer) | 0.44 | 2.98 | 0.34 | 22.1 | 0.22 | 86.7 | 0.23 | 1 |
| PIA—actin | 0.51 | 2.92 | 0.44 | 13.5 | 0.05 | 59.8 | 0.25 | 0 |
| PIA—TM | 0.67 | 2.51 | 0.27 | 15.2 | 0.06 | 73.8 | 0.11 | 0 |
| Aminolyzed PM—actin | 0.23 | 7.52 | 0.38 | 23.3 | 0.39 | 66.6 | 0.21 | 0 |
| Aminolyzed PM—TM | 0.36 | 2.41 | 0.52 | 11.1 | 0.12 | 50.9 | 0.15 | 0 |
| PM—actin in Gu-HCl | 0.51 | 4.08 | 0.36 | 15.2 | 0.13 | 40.6 | 0.21 | 0 |
| PM—TM in Gu-HCl | 0.38 | 3.14 | 0.37 | 12.2 | 0.25 | 37.7 | 0.46 | 1 |
| Aminolyzed PM—actin in Gu-HCl | 0.34 | 4.20 | 0.39 | 9.89 | 0.27 | 53.5 | 0.18 | 0 |
| C. Excimers | | | | | | | | |
| PM—TM | 0.14 | 8.85 | 0.54 | 35.13 | 0.32 | 65.85 | 0.23 | 1 |
| Aminolyzed PM—TM | 0.12 | 8.80 | 0.53 | 35.57 | 0.36 | 65.97 | 0.32 | 1 |
| PIA—TM | 0.14 | 7.11 | 0.43 | 33.87 | 0.43 | 63.57 | 0.26 | 1 |

[a] Excitation at 340 nm, emission at 400 nm (A and B) and 480 nm (C).

observations of excimer fluorescence lifetimes also militate against heterogeneity in the pyrene probes as a cause for multiple lifetimes, because an impurity is unlikely to form an excimer with three identical lifetimes.

Kawasaki et al.[53] also found three lifetimes for PM-labeled actin, but two of the lifetimes were shorter, probably because these investigators studied actin adducts after intermolecular aminolysis by Tris buffer. On the other hand, Kouyama and Mihashi[54] recently reported three lifetimes of 170 and 80 ns, and a minor component of very short lifetime in PM-labeled actin. However, lifetimes as long as 170 ns have never been reported before. In measurements involving multiple-lifetime components that differ by an order of magnitude in both lifetime and amplitude, it is essential to choose an appropriate timebase for the experiment. In general, a timebase which can accommodate five decays of the longest-lifetime component should be used. A difficulty inherent in using a longer timebase is the fact that the late channels accumulate very few counts as compared to the early channels, a problem which is worsened if the longest-lifetime component is of weak intensity; the use of too long a timebase results in an abnormally high value calculated for the longest-lifetime component.

## 6.2. 1,N⁶-Ethenoadenosine Triphosphate (ε-ATP)

The fluorescent ATP analog 1-$N^6$-ethenoadenosine triphosphate (ε-ATP) is a very useful probe for several kinds of studies, e.g., the binding interaction of adenosine nucleotides with proteins, conformational changes associated with the nucleotide binding in proteins, and interaction of the nucleotide-containing proteins with other proteins. It has also been used as donor in energy transfer experiments for determining the proximity of the nucleotide binding site to other acceptor sites in proteins. Use of ε-ATP is particularly popular among muscle protein chemists. It has been used extensively in studies of two major contractile proteins, actin and myosin, both of which bind ATP. Actin is a major contractile protein found in the thin filaments of muscle cells. Myosin is the chief protein constituent of the muscle thick filaments. Both proteins are also found in many other nonmuscle cells exhibiting motility. The fundamental mechanism of muscle contraction involves the interaction of actin with myosin, which permits the chemical energy stored in ATP to perform mechanical work. The enzymatic ATPase activity lies in two globular subfragments of myosin (commonly called myosin head or subfragment-1, S-1) and is greatly enhanced by the interaction with actin. Actin contains one mole of nucleotide per monomer. In the presence of physiological concentration of salts and the divalent cations $Ca^{2+}$ or $Mg^{2+}$, the bound ATP is hydrolyzed and G-actin monomers are polymerized into a double-stranded helically arranged F-actin filament. ε-ATP has been used to replace ATP as a structural probe in many actin studies, including binding-constant determination,[55] studies of Brownian rotational motion of G-actin and flexibility of F-actin filament using time-resolved anisotropy measurement,[56] and several lifetime decay studies.

G-actin containing ε-ATP possesses essentially full polymerizability.[57] In the absence of EDTA (presence of either $Ca^{2+}$ or $Mg^{2+}$), the binding constant of G-actin for ε-ATP (about $5 \times 10^6$ $M^{-1}$) is about half that for ATP, while in the presence of EGTA (absence of $Ca^{2+}$), the binding constants are comparable for the two nucleotides (about $2-3 \times 10^5$ $M^{-1}$).[55] The fluorescence lifetime of ε-ATP is about 36 ns, quite long compared to Trp. When ATP is free in solution, the probe is vulnerable to collisional quenching by a number of ionic quenchers, e.g., iodide ion, methionine, tryptophan, and cysteine. In the presence of quenchers, the lifetime of ε-ATP is shortened drastically to between 8.7 to 16 ns, depending on the quencher. Water molecules can also shorten the lifetime quite substantially, to 27 ns. Harvey and Cheung[58] have utilized the lifetime quenching properties of ε-ATP to study the accessibility of various quenchers, including water, to the nucleotide binding site on actin. It was found that none of the above-mentioned collisional quenchers had any effect on the lifetime of ε-ATP when the nucleotide was bound to actin, either in G-actin or in F-actin. These results were interpreted as indicating that the ethenoadenosine base was tightly bound in a region of the protein inaccessible to even small water molecules.

This technique has been further extended to study the binding interaction between actin and myosin.[59] Subfragments of myosin can be obtained by limited digestion with various proteolytic enzymes to yield the so-called S-1 and heavy meromyosin (HMM). Both ATPase and actin binding activities are preserved in either subfragments. The fluorescence lifetime of $\epsilon$-ATP bound to actin is sensitive to the binding of S-1 and HMM. Presumably conformational change at the myosin binding site on actin induces environmental changes around the nucleotide binding site, to which the fluorescent probe is sensitive. Fully saturated binding of HMM or S-1 to actin reduces the fluorescence lifetime of $\epsilon$-ATP by 4 ns. The titration curve of lifetime versus protein concentration shows complex binding of actin with S-1. At full saturation, each actin molecule binds one mole of S-1. However, binding of only one-sixth of the actin monomers produces a 50% reduction of the total lifetime. This result was interpreted as evidence for cooperative interaction among actin monomers in binding with myosin S-1. Furthermore, HMM and S-1 exhibit an identical titration curve when their concentrations were expressed in molarity. Since HMM contains twice as many globular heads as S-1 per mole of protein, this result was taken as evidence that only one of the heads of each HMM molecule was necessary for producing cooperative binding.

Tao and Cho[60] studied the lifetime quenching of 1,5-IAEDANS-labeled actin by acrylamide. They reported two lifetime components for labeled G-actin, a major lifetime component of 17.3 ns that accounted for 96% of the fluorescence, and a minor component lifetime of 33.3 ns. The lifetime of the major component increased to 19.5 ns upon polymerization to F-actin. On the other hand, lifetime measurement of IAEDANS-actin labeled under polymerization conditions (Lin, unpublished result; see Table 2) yields three lifetime components. The major component of lifetime 21.8 ns accounts for 78–83% of the total fluorescence. A medium-lifetime component of about 11 ns accounts for 13–18% of the fluorescence, and the shortest-lifetime component of about 0.9 ns, which contributes less than 5% of the total intensity, may be attributed to a light scatter artifact. Deconvolution analysis of the nanosecond decay curve can sometimes yield more than one set of equally well-fitting data.

Tao and Cho identified the major-lifetime component with a major labeling site at cys-373. Quenching of IAEDANS-labeled actin was studied by both lifetime and intensity methods, using acrylamide as the quencher. The quenching rate constant was found to be $2.31 \times 10^8$ M$^{-1}$ s$^{-1}$ for the major and minor components, respectively. The rate constant decreased substantially in F-actin and also in the presence of tropomyosin for both major and minor labeling sites. It was concluded that the major labeling site is more exposed than the minor labeling site and polymerization shielded both sites so that they were less accessible to the quencher. Tao and Cho also suggest that tropomyosin binds to actin at a site near cys-373 of actin. However, this suggestion is in contrast to the result of energy transfer experiments (see Section 6.3) which seems to suggest that tropomyosin attaches to actin at the opposite side and about 65 Å away from cys-373. It is also interesting to mention that with all the fluorescent-labeled actins studied by Lin et al.[68,69] including the labels dansyl aziridine, 1,5-IAEDANS, DMAMS, MIANS, PM, and PIA, no significant changes were found in fluorescence intensity in labeled actin caused by the addition of tropomyosin. On the other hand, addition of myosin subfragment-1 always causes fluorescence changes (either enhancement or quenching) in all of the above labeled actins. This observation seems to suggest that cys-373 of actin is more likely to be involved in myosin binding than in tropomyosin binding.

## 6.3. Distance Measurement Studies of Muscle Contractile Proteins by Fluorescence Energy Transfer

Information concerning the spatial arrangement of the various protein constituents in the highly organized muscle filaments is essential for the understanding of the mechanisms of muscle contraction. Of special interest are the changes in the spatial relationship of thin-filament proteins in response to calcium stimulation and the cyclic movement of myosin heads in the thick filaments during the muscle contraction—relaxation cycles. In the past, such information came largely from x-ray diffraction studies and three-dimensional image reconstruction of electron micrographs.[61] While these studies have provided the basis for a general model for the structural organization of proteins in the filaments, the techniques lack the fine resolution necessary to distinguish individual protein constituents in the filament assembly. Consequently, disagreements exist in the interpretation of data.[62,63] These problems can be partly resolved by performing fluorescence energy transfer experiments. Since fluorescent probes can be specifically labeled on selected (preferably functionally important) sites on individual filament proteins before assembly, their locations are known precisely. The separation distance between these labeled sites can be measured. When several distance measurements are made by varying the locations of probes in different protein components or by using several different pairs of probes, the spatial relationship of the protein constituents in the filament can be reconstructed.

There are a few energy transfer studies made recently on muscle contractile proteins. Miki and Mihashi[57] measured a distance of about 30 Å between the donor (a fluorescent ATP analogue ε-ATP attached to the nucleotide binding site of actin) and the acceptor, N-(4-dimethylamino-3,5-dinitrophenyl)-maleimide attached to cys-373 of actin. Takashi[64] labeled specifically and covalently the fast-reacting thiol (SH₁) of myosin subfragment-1 (S-1) with 5-(iodoacetamide)fluorescein (5-IAF), and cys-373 of actin with N-(iodoacetyl)-N′-(1-sulfo-5-naphthyl)ethylenediamine (1,5-IAEDANS). He observed about

30% fluorescence energy transfer from the donor 1,5-IAEDANS on actin to the acceptor 5-IAF on S-1 in the rigor complex of acto-S-1 and calculated the distance between the donor and acceptor to be about 60 Å assuming an orientation factor of 2/3. Marsh and Lowey[65] employed the same fluorescence donor/acceptor pair to label the fast-reacting thiol of chymotrypsin-digested myosin S-1 and the single cysteine on the alkali light chain-1 of myosin. The two fluorophores were calculated to be about 40 Å apart, again using an orientation factor of 2/3. The assumption of 2/3 as the orientation factor was partially justified because the same value was obtained upon reversal of the donor and acceptor attachment sites. This distance was not changed when Mg-ATP and F-actin were added to the doubly labeled S-1.

It is noted that the majority of the above energy transfer studies employed the quantum yield method of donor quenching. As discussed in Section 5.2.2, there are some shortcomings to this approach. Foremost is the problem that the presence of acceptor chromophore sometimes introduces additional quenching of the donor fluorescence in addition to the decrease in fluorescence from energy transfer by the dipole—dipole interaction mechanism. Also, both donor and acceptor concentrations must be sufficiently diluted to avoid absorption by the inner filter effect. In general, transfer efficiency determinations based on quantum yield measurements involve more correction procedures. Also, the accurate determination of probe-to-protein labeling ratio is crucial to the calculation. Transfer efficiency determinations based on careful lifetime measurements are more reliable. Not only can more accurate data be obtained with the lifetime method, but also the determination of labeling ratio of probe to protein is less critical than in the case of the quantum yield method so long as the acceptor protein is 100% labeled (or even labeled with greater than a 1:1 labeling ratio). When the acceptor is labeled 1:1 or more, irrespective of the labeling ratio of the donor protein, the only lifetimes observed for the donor-labeled protein are either the donor itself when acceptor is not present or the donor in the presence of the acceptor.

The donor-lifetime-decrease method of energy transfer has been used in experiments to measure the distance between a pair of fluorophores attached to the cys-36 of β-tropomyosin and the cys-373 of actin in reconstituted muscle thin filaments.[66] Two pairs of donor/acceptor fluorophores, 1,5-IAEDANS/5-IAF and PM/DMAMS (dimethylamino-4-maleimidostilbene), were covalently attached to tropomyosin and actin. The critical distances were 45 and 38 Å, respectively, for the IAEDANS/IAF and PM/DMAMS pairs, assuming $\kappa = 2/3$. For the PM/DMAMS pair there was no change in donor lifetime when acceptor protein was present under various conditions in which myosin S-1, calcium, or EGTA was present. The separation distance between the two labeled sites was estimated to be greater than 62 Å (distance for 5% transfer efficiency). For the IAEDANS/IAF pair labeled, respectively, on tropomyosin and actin, there was an 8 to 12% energy transfer when acceptor protein was present in the reconstituted filaments. A separation distance of 65 Å between the donor and acceptor was calculated.

Since cys-373 of actin is thought to be involved in the myosin binding site, the distance measured is an approximate measure of the distance between the tropomyosin and myosin binding site on actin. Taking into account the distances from the centers of the donor and the acceptor to their respective protein labeling sites (maximally about 10 and 8.5 Å for donor and acceptor, respectively) and assuming 12 Å for the diameter of the tropomyosin rod (Figure 9), the minimal distance between the tropomyosin and myosin binding site is estimated to be $65 - 10 - 8.5 - 12 = 34.5$ Å. Since actin molecules are spaced about 55 Å apart along the actin thin filament and each actin binds one myosin head, the above results suggest that myosin binds to actin on the side opposite to tropomyosin attachment (Figure 9).

As mentioned in Section 5.1.2, the calculated separation distance is subject to the uncertainty in the assumed value of the orientation factor, 2/3 in this case. From independent time-resolved anisotropy measurements in the literature, it was concluded that the donor was rapidly rotating and confined to a cone with a semiangle of about 32 to 46° according to equation (32). The calculated distance using $\kappa = 2/3$ is estimated to be within 80% of the corrected value based on equations (31) and (33).

The effect of calcium ions (the initiator of muscle contraction) on the separation distance between donor IAEDANS-labeled tropomyosin and acceptor

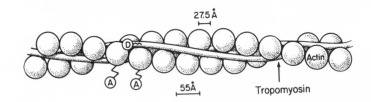

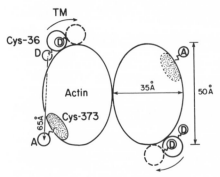

FIGURE 9. A simplified schematic drawing of a tropomyosin-actin reconstituted filament labeled with fluorescent donors and acceptors. Top: longitudinal view. Bottom: cross-sectional view. Arrow representing the distance between donor and acceptor is pointing outward to the viewer and is not scaled to the real dimension. The shaded area (bottom figure) in actin is the myosin-binding site in which cys-373 is located. The dashed circles represent tropomyosin in the contracting state.

IAF-labeled actin was also studied. A small change in the calculated distance, about 1.5 Å, was found between the "relaxed" state when $Ca^{2+}$ was absent and the "activated" state when $Ca^{2+}$ ions were bound to troponin. This small distance change is perhaps not very significant compared to the possible error in calculating the distance $R'$, which is about 3 Å. Ford[67] recently measured the distance between the donor and acceptor on the opposite strands of tropomyosin across the actin filament using three donor/acceptor pairs, including PM/DMAMS, IAEDANS/IAF, and IAF/eosin. His results suggest that the distance between the tropomyosin in the reconstituted filament in the presence of TN and Ca is about the same as in the tropomyosin–F-actin filament without troponin, about 77 to 79 Å. The minimal center-to-center distance between the two tropomyosins across the actin filament was about 47 Å (taking into account the sum of the distances between the center of the fluorophores to the proteins and the diameter of the tropomyosin rod). In the presence of EGTA, the separation distance was found to be even greater, suggesting that the two tropomyosin strands across the actin filaments are moving away toward the periphery. However, the exact distance cannot be measured with certainty because of the low energy transfer efficiency observed.

In conclusion, nanosecond pulse fluorimetry has provided protein chemists, cell biologists, and others with a powerful means for exploring the properties of proteins in solution, and of the interactions of proteins with low-molecular-weight substances. It provides a method for studying complex protein systems, giving information that is complementary to data obtained from other biophysical techniques. It will undoubtedly become more commonly used as time goes on and as more investigators become familiar with the methodology.

## References

1. W. R. Ware, in *Creation and Detection of the Excited State,* Vol. 1, Part A, Marcel Dekker, New York (1971).
2. V. J. Koester, and R. M. Dowben, *Rev. Sci. Instrum.* 49, 1186–1191 (1978).
3. I. Isenberg, Time-decay fluorimetry by photon counting, in *Biochemical Fluorescence: Concepts* (R. F. Chen and H. Edelhoch, eds.), Vol. 1, Marcel Dekker, New York (1975), pp. 43–77.
4. M. G. Badea, and L. Brand, Time-resolved fluorescence measurements, in *Methods in Enzymology,* Vol. 61, *Enzyme Structure,* Part H (C. H. W. Hirs and S. N. Timasheff, eds.), Academic Press, New York (1979), pp. 378–425.
5. A. Grinvald and I. Z. Steinberg, *Anal. Biochem.* 59, 538–598 (1974).
6. I. Isenberg, R. D. Dyson, and R. Hanson, *Biophys. J.* 13, 1090–1115 (1973).
7. I. Isenberg, *J. Chem. Phys.* 59, 5696–5713 (1973).
8. J. Eisenfeld and S. W. Cheng, *Appl. Math. Comput.* 6, 335–357 (1980).
9. C. Formoso and L. S. Forster, *J. Biol. Chem.* 250, 3738–3745 (1975).
10. C. Formoso and L. S. Forster, *Biochim. Biophys. Acta* 427, 377–386 (1976).
11. T. Imoto, L. S. Foster, J. A. Rupley, and F. Tanaka, *Proc. Natl. Acad. Sci. USA* 69, 1151–1155 (1972).

12. T. Torikata, L. S. Forster, R. E. Johnson, and J. A. Rupley, *J. Biol. Chem.* **254**, 3516–3520 (1979).
13. T. Torikata, L. S. Forster, C. C. O'Neal, Jr., and J. A. Rupley, *Biochemistry* **18**, 385–390 (1979).
14. J. C. Brochon, P. Wahl, J. M. Jallon, and M. Iwatsubo, *Biochemistry* **15**, 3259–65 (1976).
15. J. C. Brochon, P. Wahl, M. O. Monnense-Doublet, and A. Olomucki, *Biochemistry* **16**, 4594–4599 (1977).
16. A. Gafni, and L. Brand, *Biochemistry* **15**, 3165–3170 (1976).
17. R. F. Chen, *Anal. Biochem.* **57**, 593–604 (1974).
18. W. M. Vaughn and G. Weber, *Biochemistry* **9**, 464–473 (1979).
19. M. Cooper and J. K. Thomas, *Radiat. Res.* **70**, 312–314 (1977).
20. J. R. Lakowicz and G. Weber, *Biochemistry* **12**, 4161–4170 (1973).
21. J. R. Lakowicz and G. Weber, *Biochemistry* **12**, 4171–4179 (1973).
22. N. Barboy and J. Feitelson, *Biochemistry* **17**, 4923–4926 (1978).
23. M. R. Eftink and C. A. Ghiron, *Proc. Natl. Acad. Sci. USA* **72**, 3290–3294 (1975).
24. M. Irie, *J. Biochem.* **68**, 31–37 (1970).
25. T. Forster, *Ann. Phys.* **2**, 55–75 (1948).
26. L. Stryer and R. P. Haugland, *Proc. Natl. Acad. Sci. USA* **58**, 719–726 (1967).
27. L. Stryer, *Annu. Rev. Biochem.* **47**, 819–846 (1978).
28. R. E. Dale and J. Eisenger, *Biopolymers* **13**, 1573–1605 (1974).
29. R. E. Dale and J. Eisenger, in *Biochemical Fluorescence: Concepts* (R. F. Chen and H. Edelhoch, eds.), Vol. 1, Marcel Dekker, New York (1975), pp. 115–284.
30. Z. Hillel and C.-W. Wu, *Biochemistry* **15**, 2105–2112 (1976).
31. R. E. Dale and J. Eisenger, *Biophys. J.* **26**, 161–193 (1979).
32. A. Kawato, K. Kinosita, Jr., and A. Ikegami, *Biochemistry* **16**, 2319–2324 (1977).
33. C.-W. Wu and L. Stryer, *Proc. Natl. Acad. Sci. USA* **69**, 1104–1108 (1972).
34. T. C. Werner, J. R. Bunting, and R. Cathou, *Proc. Natl. Acad. Sci. USA* **69**, 795–799 (1972).
35. R. Luedtke, C. S. Owen, and F. Karush, *Biochemistry* **19**, 1182–92 (1980).
36. J. B. Henes, M. S. Briggs, S. G. Sligar, and J. S. Fruton, *Proc. Natl. Acad. Sci. USA* **77**, 277–288 (1980).
37. B. A. Gruber and N. J. Leonard, *Proc. Natl. Acad. Sci. USA* **72**, 3966–3969 (1975).
38. M. Koike and L. J. Reed, *J. Biol. Chem.* **235**, 1931–1938 (1960).
39. O. A. Moe, Jr., D. A. Lerner, and G. G. Hammes, *Biochemistry* **13**, 2552–2557 (1974).
40. G. B. Shepherd and G. G. Hammes, *Biochemistry* **15**, 2953–2958 (1976).
41. G. B. Shepherd, N. Papadakis, and G. G. Hammes, *Biochemistry* **15**, 2888–2893 (1976).
42. K. J. Angelides and G. G. Hammes, *Biochemistry* **18**, 5531–5537 (1979).
43. S.-C. Tu, C.-W. Wu, and J. W. Hastings, *Biochemistry* **17**, 987–93 (1978).
44. S. Mitra and R. Bersohn, *Biochemistry* **19**, 3200–3203 (1980).
45. D. Barber, S. R. Parr, and C. Greenwood, *Biochem. J.* **175**, 239–249 (1978).
46. D. C. Wharton and Q. H. Gibson, *Biochim. Biophys. Acta* **430**, 445–453 (1976).
47. J. C. Brochon, P. Wahl, J. M. Jallon, and M. Iwatsubo, *Biochim. Biophys. Acta* **462**, 759–769 (1977).
48. S. Matsumoto and G. G. Hammes, *Biochemistry* **14**, 214–224 (1975).
49. L.-H. Hahn and G. G. Hammes, *Biochemistry* **17**, 2423–2429 (1978).
50. T.-I. Lin, *Biophys. Chem.* **15**, 277–288 (1982).
51. T.-I. Lin and R. M. Dowben, *Biophys. Chem.* **15**, 289–298 (1982).
52. T.-I. Lin, *Arch. Biochem. Biophys.* **185**, 285–299 (1978).

53. Y. Kawasaki, K. Mihashi, H. Tanaka, and H. Ohnuma, *Biochim. Biophys. Acta* **446**, 166 (1976).

54. T. Kouyama and K. Mihashi, *Eur. J. Biochem.* **105**, 279–287 (1980).

55. K. E. Thames, H. C. Cheung, and S. C. Harvey, *Biochem. Biophys. Res. Commun.* **60**, 1252–61 (1974).

56. K. Mihashi and P. Wahl, *FEBS Lett.* **52**, 8–12 (1975).

57. M. Miki and K. Hihashi, *Biochim. Biophys. Acta* **533**, 163–172 (1978).

58. S. C. Harvey and H. C. Cheung, *Biochem. Biophys. Res. Commun.* **73**, 865–868 (1976).

59. S. C. Harvey, H. C. Cheung, and K. E. Thames, *Arch. Biochem. Biophys.* **179**, 391–396 (1977).

60. T. Tao and J. Cho, *Biochemistry* **18**, 2759–2765 (1979).

61. H. E. Huxley, *Cold Spring Harbor Symp. Quant. Biol.* **37**, 361–376 (1972).

62. J. Seymour and E. J. O'Brien, *Nature (London)* **283**, 680–682 (1980).

63. R. Mendelson, *Nature (London)* **298**, 665–667 (1982).

64. R. Takashi, *Biochemistry* **18**, 5164–5169 (1979).

65. D. J. Marsh and S. Lowey, *Biochemistry* **19**, 774–784 (1980).

66. R. M. Dowben and T.-I. Lin, The spatial arrangement of muscle thin filament proteins as determined by fluorescence energy transfer, in *Cell and Muscle Mobility* (R. M. Dowben and J. W. Shay, eds.), Vol. 4 Plenum Press, New York (1982).

67. C. Ford, Fluorescent and ionic probe studies of tropomyosin with consideration of multiexponential decay analysis, Ph.D. Thesis, University of Texas Health Science Center, Dallas, Texas (1981).

68. T.-I. Lin, C. Ford, and R. M. Dowben (1978a) Fluorimetric studies of tropomyosin and F-actin labeled with 3-pyrenemaleimide. *Biophys. J.* **21**, 16a.

69. T.-I. Lin, S. Carlilie, and R. M. Dowben, (1978b) Fluorescent probe studies of binding of actin with myosin subfragment-one. *Biophys. J.* **21**, 16a.

# 4

# The Use of Fluorescence Anisotropy Decay in the Study of Biological Macromolecules

## ROBERT F. STEINER

## 1. Introduction

### 1.1. Application of Fluorescence Anisotropy

In recent years fluorescence measurements have evolved into one of the most widely used techniques in biochemistry. We shall be concerned here with one type of fluorescence measurement, fluorescence anisotropy decay, which has found extensive application in studies of the properties of biological macromolecules.

Fluorescence anisotropy decay belongs basically to the category of relaxation techniques in which the time-dependent transition of the system from a biased to a random arrangement is monitored. In this case the observed transition corresponds to the change from a specific to a random orientation in space, occurring via Brownian rotational diffusion. In contrast to other physical techniques, such as electrical birefringence, in which orientation is achieved by the application of an external force field, fluorescence anisotropy relies upon the initial selection of a population of fluorophors of specific orientation from a large collection of randomly oriented fluorophors. From the change in anisotropy with time, information as to the rotational mobility of the fluorophor may be obtained. When the fluorophor is a fluorescent label attached to a biological macromolecule, its rotation will be modified or dominated by the properties of the macromolecule and will, in general, reflect both the overall

ROBERT F. STEINER ● Department of Chemistry, University of Maryland Baltimore County, Catonsville, Maryland 21228.

rotation of the macromolecule and any internal rotational modes which may be present.

The basic kinds of information which may be derived from fluorescence anisotropy decay experiments are of two kinds. To the extent that the fluorescent label and the macromolecule to which it is attached rotate as a rigid unit, anisotropy decay provides a potential source of information as to the size and shape factor of the macromolecule. In addition, it is also perhaps the most convenient means of studying the internal rotations of macromolecules and of assessing the nature of the molecular flexibility present.

In what follows we shall first summarize the theory of fluorescence anisotropy decay and then describe some representative applications to the study of biopolymers.

# 2. Theory

## 2.1. Basic Principles

The polarization of the radiation emitted by a fluorophor is normally characterized with reference to a system of laboratory coordinates. Fluorescence is usually observed at an angle of 90° to the exciting beam. If the origin, $O$, is taken as the center of the volume irradiated and the $x$ and $y$ axes are chosen along the direction of observation and along the direction of the exciting beam, respectively, then the plane defined by the $Ox$ and $Oy$ directions contains all the instrumental elements. The $Oz$ axis is perpendicular to this plane.

The polarization of the fluorescent radiation emitted by a fluorophor within the irradiated volume may be described in terms of the components of the intensity along the three coordinate axes: $I_x$, $I_y$, and $I_z$. The sum of these, $S$, is proportional to the total radiant energy emitted in all directions:

$$S = I_x + I_y + I_z \qquad (1)$$

The exciting beam usually consists of either unpolarized (natural) light or linearly polarized light whose direction of polarization is along the $z$ or $x$ axis. If the exciting beam is unpolarized, so that its electric vector may have any orientation within the $xz$ plane, then, from symmetry considerations

$$I_x = I_z \qquad (2)$$

and the total intensity is given by

$$S = 2I_z + I_y \qquad (3)$$

If the exciting beam is polarized along the $Oz$ axis, then, again from symmetry,

$$I_x = I_y \tag{4}$$

and

$$S = I_z + 2I_y \tag{5}$$

If the exciting beam is linearly polarized along the $Ox$ axis, then

$$I_y = I_z \tag{6}$$

and

$$S = 2I_z + I_x \tag{7}$$

Fluorescence anisotropy measurements are usually made with exciting light which is polarized in the $z$ direction or, less commonly, with unpolarized light. In either case, the quantities which are observed are the components of fluorescence intensity polarized in the $Oz$ and $Oy$ directions. It is conventional to define

$$I_{\parallel} = I_v = I_z \tag{8}$$

and

$$I_{\perp} = I_H = I_y$$

The quantities $I_{\parallel}$ and $I_{\perp}$, which designate the components of fluorescence intensity polarized parallel and perpendicular to the $Oz$ axis, respectively, are often referred to as the vertically and horizontally polarized components.

The emission anisotropy, $A$, is defined by

$$A = (I_{\parallel} - I_{\perp})/S = D/S \tag{9}$$

where

$$D = I_{\parallel} - I_{\perp}$$

From equations (7) and (9) we have, for the case of vertically polarized exciting radiation

$$A_v = \frac{I_{\parallel} - I_{\perp}}{I_{\parallel} + 2I_{\perp}} = \frac{I_v - I_H}{I_v + 2I_H} \tag{10}$$

and, for unpolarized exciting radiation,

$$A_u = \frac{I_{\parallel} - I_{\perp}}{2I_{\parallel} + I_{\perp}} \tag{11}$$

The emission anisotropies for the two cases are related by

$$A_v = 2A_u \tag{12}$$

If the exciting light is horizontally polarized, we have

$$I_\parallel = I_\perp \tag{13}$$

and

$$A = 0 \tag{14}$$

If more than one fluorescent species is present, the observed emission anisotropy is given by[1,2]

$$A = \sum_i A_i f_i \tag{15}$$

where $A_i$ and $f_i$ are the emission anisotropy of species $i$ and its fractional contribution to the total intensity, respectively.

In the older literature, data were generally reported in terms of polarization ($P$) rather than anisotropy.[1,3-6] The emission polarization is defined by

$$P = \frac{I_\parallel - I_\perp}{I_\parallel + I_\perp} \tag{16}$$

However, since the use of anisotropy leads to simpler formulations, it has largely replaced polarization in the recent literature and will be exclusively employed in this chapter.

## 2.2. The Effect of the Shape of the Excitation Pulse

Measurements of the time decay of fluorescence intensity or anisotropy generally involve determinations of the summed response to a repetitive series of exciting light flashes. If the exciting pulses were infinitely sharp, corresponding to a $\delta$-function, and if only a single fluorescent species were present, then the fluorescence intensity would decay with time according to a simple exponential law:

$$I = I_0 \exp\left(-t/\tau\right) \tag{17}$$

where $I_0$ is the intensity at zero time (which corresponds to the time of the excitation pulse), $t$ is the time after excitation, and $\tau$ is the average decay time of fluorescence, or the average excited lifetime.

It is possible to obtain excitation pulses of picosecond width, which approximates the infinitely sharp case, by employing pulsed laser sources. However, most anisotropy decay studies have employed repetitive spark discharges, whose widths are of the order of nanoseconds. In this case excitation and decay occur simultaneously over a significant period, and it is no longer permissible to ignore the finite duration of the excitation pulse.

It is convenient to relate the observed time decay profile $i(t)$ to the profile $I(t)$ which would be observed if the excitation pulse were infinitely sharp by the convolution integral:

$$i(t) = \int_0^t I(t-u)E(u)du \qquad (18)$$

Here, $t$ is the time and $E(u)$ is the time profile of excitation. (Here and elsewhere the convention will be followed of designating the experimentally observed quantities $i$, $s$, $d$, etc., by lower case letters, while capital letters will be used for the corresponding quantities $I$, $S$, $D$, etc. corrected for convolution effects so as to correspond to the behavior expected if the excitation pulse were infinitely sharp.) The mathematical techniques which have been developed for analyzing the time decay of fluorescence intensity in terms of discrete decay times provide for the deconvolution of the experimental decay curve so as to remove the distorting effects of the finite duration of the excitation pulse.[7-10]

## 2.3. The Time Decay of Fluorescence Anisotropy

The observation of fluorescence anisotropy depends upon the selective and nonrandom excitation of fluorescent molecules or groups. The absorption of a quantum of radiant energy by a fluorophor results in its transition to some vibrational level of a higher electronic state. A very rapid (several picosecond) process of internal conversion next places it in a low vibrational level of the lowest electronic excited state. This process is normally completed prior to emission of fluorescence. As a consequence, the transition moments associated with absorption and emission are generally nonequivalent.

For the common case where absorption and emission are strongly permitted electronic transitions and the wavelength of measurement is close to the 0–0 vibronic transition, the transition moments of absorption and emission may be represented as linear with a well-defined direction. This model will be assumed in the discussion to follow.

The probability of excitation of a fluorophor will depend upon the angle $\phi$ between its absorption transition moment and the electric vector of the exciting beam, being proportional to $\cos^2 \phi$. If the exciting beam is vertically polarized, preferential excitation will occur of those molecules whose absorption moments are oriented parallel to the $z$ axis; if the exciting beam is unpolarized, preferential excitation occurs of those whose moments lie in the $xz$ plane.

At short times after excitation, before significant Brownian rotation has had time to occur, the relative magnitudes of $i_{\parallel}$ and $i_{\perp}$ reflect this biased distribution.

In the complete absence of rotational diffusion, as in a rigid medium, the anisotropy has its limiting value $A_0$, being governed entirely by the angle $\lambda$ between the linear transition moments of absorption and emission. We have, for vertically polarized exciting light,[11]

$$A_0 = (3\cos^2\lambda - 1)/5 \qquad (19)$$

The maximum and minimum values of $A_0$ are thus 0.4 and $-0.2$, corresponding to $\lambda = 0$ and $\lambda = 90°$, respectively. In practice, experimental values of $A_0$ have never attained 0.4. This may be due to the effects of torsional vibration.[12]

When the fluorophor has rotational mobility in a fluid medium, a progressive randomization occurs with increasing time of the directions of the transition moments as a consequence of Brownian rotation, so that ultimately a uniform distribution is approached with $i_\parallel = i_\perp$ and $A = 0$.

The principal kinds of information obtainable from the time profile of fluorescence anisotropy decay are concerned with the rate of rotational motion of the fluorophor, which is in turn related to its molecular characteristics and, in the case of a fluorescent conjugate, to those of the macromolecule to which it is attached. In practice, one is usually interested in the behavior of a fluorescent probe attached covalently or noncovalently to a larger macromolecule. In order to obtain a manageable expression for the time dependence of anisotropy, it is necessary to approximate the actual, normally somewhat irregular, shape of the macromolecule by a smoothed geometrical form, such as an ellipsoid of revolution. Let us further assume that the label is rigidly attached to the macromolecule, with a fixed orientation of its transition moments with respect to the coordinate axes of the latter. In this general case the time decay of anisotropy depends not only upon the characteristics of the macromolecule, but also upon the respective orientations of the transition moments of absorption and emission.

At this point it is appropriate to digress to a discussion of the rotational motion of rigid macromolecules, approximated as ellipsoids.

## 2.4. The Rotational Diffusion of Rigid Ellipsoidal Particles

Rotational diffusion is characterized by a rotational diffusion coefficient, which is defined in a manner analogous to that of the familiar translational diffusion coefficient. The rotation of the diameter of a rigid sphere with a particular orientation at zero time is described by

$$\partial W(\omega, t)/\partial t = D\nabla^2 W(\omega, t) \qquad (20)$$

Here $\omega$ is the rotational angle of the spherical diameter, $W$ is a probability density function characterizing the distribution of rotational angles, and $D$ is the rotational diffusion coefficient.

For a spherical particle, $D$ is given by

$$D_0 = kT/6\eta V \tag{21}$$

where $k$ is the Boltzmann constant, $T$ is the absolute temperature, $\eta$ is the solvent viscosity, and $V$ is the effective hydrodynamic volume. The value of $V$ is equal to the anhydrous volume plus an increment reflecting the bound water of hydration.

An ellipsoidal particle has three principal axes, each of which is associated with a rotational diffusion coefficient. These are designated as $D_1$, $D_2$, and $D_3$, where $D_i$ is the diffusion coefficient for rotation about the $i$th axis. For a symmetrical ellipsoid of revolution, if axis 1 corresponds to the axis of symmetry and axes 2 and 3 to the (equivalent) equatorial axes, then $D_2 = D_3$.

The quantities $D_1$ and $D_2$, the rotational diffusion coefficients for rotation about the axis of symmetry and the axes perpendicular to it, respectively, may be related to the axial ratio of the ellipsoid, $\gamma$, and to the rotational diffusion coefficient $D_0$ of the sphere of equivalent hydrodynamic volume. For a prolate ellipsoid of revolution, the basic equations are[13,14]

$$\frac{D_2}{D_0} = \frac{3}{2}\gamma\frac{[(2\gamma^2 - 1)\beta - \gamma]}{(\gamma^4 - 1)} \tag{22}$$

$$\frac{D_1}{D_0} = \frac{3}{2}\frac{\gamma(\gamma - \beta)}{(\gamma^2 - 1)} \tag{23}$$

where $\beta = (\gamma^2 - 1)^{-1/2} \ln [\gamma + (\gamma^2 - 1)^{1/2}]$.

The value of $D_0$ for real proteins is subject to some uncertainty because of the difficulty of determining the hydration quantitatively. The values assumed have generally ranged from 0.2 to 0.5 ml/g.

It is also useful to define a set of three rotational correlation times, $\sigma_i$, which are functions of the rotational diffusion coefficients. For an ellipsoid of revolution[1,13,15]

$$\sigma_1 = 1/6D_2$$
$$\sigma_2 = 1/(5D_2 + D_1) \tag{24}$$
$$\sigma_3 = 1/(2D_2 + 4D_1)$$

For the case of a spherical particle, $\sigma_1 = \sigma_2 = \sigma_3 = \sigma_0 = 1/6D_0$. From equation

(21), we have

$$\sigma_0 = \eta V/kT \tag{21'}$$

## 2.5. The Time Decay of Anisotropy for Ellipsoidal Particles

Belford et al.[16] have examined the general case of a rigid ellipsoidal particle which is not necessarily an ellipsoid of revolution. Here and elsewhere in this section the fluorescent label is assumed to be rigidly attached to the particle so that its transition moments have well-defined orientations with respect to the axes of the particle. In this case the time decay of anisotropy is governed by the three rotational diffusion coefficients ($D_1$, $D_2$, and $D_3$) corresponding to the three axes and by the directions of the linear transition moments of absorption and emission. The equation derived by Belford et al. is as follows:

$$A(t) = \frac{6}{5} \left\{ \sum_{i=1}^{3} c_i \exp\left(-t/\sigma_i'\right) + [(F+G)/4] \exp\left[-(6D-2\Delta)t\right] \right.$$

$$\left. + [(F-G)/4] \exp\left[-(6D+2\Delta)t\right] \right\} \tag{25}$$

Here, $D = (D_1 + D_2 + D_3)/3$, the mean rotational diffusion coefficient; $\Delta = (D_1^2 + D_2^2 + D_3^2 - D_1D_2 - D_1D_3 - D_2D_3)^{1/2}$, and $c_i = \alpha_j \alpha_k \epsilon_j \epsilon_k (ijk = 123, 231,$ or $312)$, where $\alpha_1$, $\alpha_2$, $\alpha_3$ are the cosines of the angles formed by the transition moment of absorption with the three axes; $\epsilon_1$, $\epsilon_2$, $\epsilon_3$ are the corresponding direction cosines of the transition moment of emission,

$$\sigma_i' = 1/(3D + 3D_i)$$

$$F = \sum_{i=1}^{3} \alpha_i^2 \epsilon_i^2 - \frac{1}{3}$$

$$G\Delta = \sum_{i=1}^{3} D_i(\alpha_i^2 \epsilon_i^2 + \alpha_j^2 \epsilon_k^2 + \alpha_k^2 \epsilon_j^2) - D \qquad i \neq j \neq k$$

A number of special cases of this general solution are of interest. If the transition moments are randomly oriented with respect to the axes of the particle, so that

$$\alpha_i = \alpha_j = \alpha_k = \epsilon_i = \epsilon_j = \epsilon_k = 1/\sqrt{3} \tag{26}$$

then, for low values of $t/\sigma_i$, corresponding to the initial slope of the anisotropy decay curve, we have[17,18]

$$A(t) = A_0 \exp(-t/\sigma_h) \tag{27}$$

where $\sigma_h$ is the harmonic mean of the three correlation times and is defined by

$$\frac{1}{\sigma_h} = \frac{1}{5}\left(\frac{1}{\sigma_1} + \frac{2}{\sigma_2} + \frac{2}{\sigma_3}\right) \tag{28}$$

In practice, this case is likely to be realized when the label is distributed among a large number of possible sites of attachment.

For the limiting case of a spherical particle, for which $D_1 = D_2 = D_3 = D_0$, equation (25) reduces to

where
$$A(t) = A_0 \exp(-t/\sigma_0) \tag{29}$$
$$\sigma_0 = 1/6D_0 \tag{30}$$

Here $D_0$ is the rotary diffusion coefficient of the spherical particle and $A_0$ is given by equation (19).

In general, for a symmetrical ellipsoid of revolution, the time-dependent anisotropy is given by the sum of three exponentials:[19]

$$A(t) = \alpha_1 \exp(-t/\sigma_1) + \alpha_2 \exp(-t/\sigma_2) + \alpha_3 \exp(-t/\sigma_3) \tag{31}$$

where

$$\sigma_1 = 1/6D_2$$
$$\sigma_2 = 1/(5D_2 + D_1)$$
and
$$\sigma_3 = 1/(2D_2 + 4D_1)$$
$$\alpha_1 = 0.1(3\cos^2\theta_1 - 1)(3\cos^2\theta_2 - 1)$$
$$\alpha_2 = 0.3\sin 2\theta_1 \sin 2\theta_2 \cos\phi$$
$$\alpha_3 = 0.3\sin^2\theta_1 \sin^2\theta_2(\cos^2\phi - \sin^2\phi)$$

Here $\theta_1$ and $\theta_2$ are the angles formed by the absorption and emission transition moments, respectively, with the axis of symmetry of the ellipsoid and $\phi$ is the angle formed by the projections of the two moments in the plane perpendicular to the axis of symmetry.

If the two moments coincide, then $\phi = 0$ and $\theta_1 = \theta_2 = \theta$, and we have

$$\alpha_1 = 0.1(3 \cos^2 \theta - 1)^2$$
$$\alpha_2 = 0.3 \sin^2 2\theta \tag{32}$$
$$\alpha_3 = 0.3 \sin^4 \theta$$

For an ellipsoid of revolution for which the direction of the absorption moment is that of the axis of symmetry, the time decay of anisotropy is given by equation (29), but with $\sigma_0$ replaced by[20]

$$\sigma_1 = 1/6D_2 \tag{33}$$

where $D_2$ is the rotational diffusion coefficient corresponding to the rotary motion of the particle about either of the equatorial axes.

In the case of an ellipsoid of revolution for which the transition moments of absorption and emission are perpendicular to the axis of symmetry, a more complex expression is obtained:

$$A(t) = 0.1 \exp(-t/\sigma_1) + 0.3(2 \cos^2 \lambda - 1) \exp(-t/\sigma_3) \tag{34}$$

where
$$\sigma_1 = 1/6D_2$$
$$\sigma_3 = 1/(2D_2 + 4D_1)$$
and
$$\lambda = \phi$$

Here $D_1$ is the rotational diffusion coefficient of the particle about the axis of symmetry and $\lambda$ is the angle between the transition moments of absorbtion and emission.

The predicted time decay of anisotropy is here clearly nonexponential. Moreover, if $\cos^2 \lambda < 0.5$, the second term on the right-hand side of equation (34) will be negative, leading to the possibility of negative anisotropies at short times after excitation. Since $\sigma_1$ is always greater than $\sigma_3$, a transition to positive anisotropies occurs, followed by a gradual decay to zero.

In principle, equations (25)–(34) would provide a means of estimating the shape of a rigid asymmetric protein, to the extent that its shape can be approximated by an ellipsoid. In practice, instrumental limitations, including, especially, lamp instability, limit the time of observation so that it is difficult to monitor anisotropy decay over more than one decade before scatter becomes unacceptably large. This is insufficient to permit the accurate detection of multiple correlation times arising from molecular asymmetry. A fruitful exploitation of these relations will probably have to await improvements in instrumentation.

## 2.6. Computer Simulation of Anisotropy Decay

Harvey and Cheung[21] have developed a computer-simulation approach to the time decay of fluorescence anisotropy. A collection of a large number of identical fluorescent molecules, each with its own set of internal coordinates and characteristic orientation of the linear transition moments of absorption and emission, is generated by means of a random-number generator, which is also used to yield the random rotational steps of Brownian movement. The laboratory frame of reference is selected with the $z$ axis parallel to the polarization vector of the exciting light and the $x$ and $y$ axes in the directions of excitation and observation, respectively. The distribution of orientations in the initial excited state is accounted for by utilizing the fact that the probability of excitation is proportional to $\cos^2 \Omega_1$, where $\Omega_1$ is the angle formed by the absorption transition moment with the $z$ axis.

It may be shown that the normalized components of fluorescence intensity polarized parallel $(i_\parallel)$ and perpendicular $(i_\perp)$ to the direction of polarization of the exciting light are given by[21]

$$i_\parallel = 0.75 \cos^2 \Omega_2 \left(1 + \cos^2 \Omega_2\right)$$
$$i_\perp = 0.75 \sin^2 \Omega_2 \left(1 - 0.25 \sin^2 \Omega_2\right) \tag{35}$$

where $\Omega_2$ is the angle formed by the emission moment with the $z$ axis.

The Brownian rotation was simulated by increasing the time by a series of small increments $\delta t$ and rotating the molecule at each step about each of its axes by angular increments $\pm \delta \omega_i (i = 1, 2, 3)$ with the sign of $\delta \omega_i$ assigned randomly. The magnitude of each step was governed by the three diffusion coefficients, according to the classical theory of Brownian rotation:

$$\delta \omega_i = (2D_i \delta t)^{1/2} \tag{36}$$

After a set of several time steps, the anisotropy was computed using equations (35) and (36). This was done for a series of such sets, and the resultant anisotropies plotted as a function of time. Computations of this kind were performed for prolate and oblate ellipsoids of revolution with a range of axial ratios and of orientations of the absorption and emission moments. In each case the simulated decay (Figure 1) matched closely the behavior predicted by the equation of Belford et al.[16]

An interesting prediction of the theory of Belford et al., which was confirmed by the simulations of Harvey and Cheung, is that, for certain orientations of the transition moments of absorption and emission, the anisotropy increases with time for short times after excitation, passing through a maximum before declining to zero at longer times.

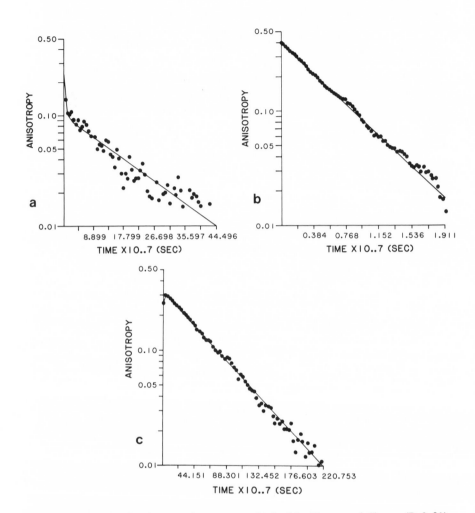

FIGURE 1. Simulated anisotropy decay curves obtained by Harvey and Cheung (Ref. 21) for ellipsoids of revolution. The smooth curves are calculated from the equation of Belford *et al.* (Ref. 16) for an ellipsoid of revolution of molecular weight 120,000, molecular density 1.30 g/ml, and hydration 0.20 ml/g. The viscosity of the solvent is 1.0 centipoise and the temperature is 27°. (a) A prolate ellipsoid of axial ratio 20; $\theta_1 = \theta_2 = 90°; \phi = 25°$; (b) an oblate ellipsoid of axial ratio 0.40; $\theta_1 = \theta_2 = 30°; \phi = 0°$; (c) a prolate ellipsoid of axial ratio 40; $\theta_1 = \theta_2 = 15°; \phi = 180°$. Note the initial increase in anisotropy at low times for case (c). $\theta_1, \theta_2$, and $\phi$ are as defined by equation (31).

## 2.7. Internal Rotation

In practice, the case of a fluorescent label which is rigidly oriented with respect to the coordinate axes of a rigid macromolecule is often not realized. For many fluorescent conjugates of proteins and nucleic acids, internal rotation is present in some degree. The forms of internal rotation which may occur may be loosely grouped into the following categories (Figure 2):

(1) rotation of the chromophore about the bond linking it to the protein;
(2) rotational wobble of that portion of the polypeptide adjacent to the chromophore;
(3) rotation of a molecular domain as a unit about a flexible hinge point.

Unfortunately, the complexity of the problem has thus far precluded the development of a comprehensive theoretical treatment which would adequately encompass all of these cases. All that is available at present is a theoretical description of a few limiting cases.

Gottlieb and Wahl[22] have treated the case of a fluorophor assumed to be spherical, which rotates freely about the bond linking it to a rigid macromolecule. If the rotation of the fluorophor is not hindered by interaction with the structural features of the macromolecule, or by other factors, then the

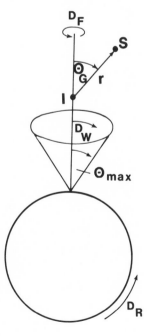

FIGURE 2. Model for rotational motion of a covalently attached label. The rotation of the probe, S, is confined to the surface of a cone of semiapex angle $\theta_G$ and corresponds to a rotational diffusion coefficient $D_F$. This is superimposed upon the rotational wobble of the polypeptide within a cone of semiapex angle $\theta_{max}$ and with rotational diffusion coefficient $D_W$. Rotation of the entire protein also occurs, with rotational diffusion coefficient $D_R$.

rotation of the label is governed solely by its own rotational diffusion coefficient.

For a spherical macromolecule, Gottlieb and Wahl derived the equation

$$A(t) = \exp\left(-t/\sigma\right)\left[\alpha_1 + \alpha_2 \exp\left(-t/6\sigma_i\right) + \alpha_3 \exp\left(-2t/\sigma_i\right)\right] \tag{37}$$

Here $\sigma$ is the rotational correlation time associated with the motion of the entire macromolecule as a unit, and $\sigma_i$ is the correlation time of the freely rotating label. The quantities $\alpha_1$, $\alpha_2$, and $\alpha_3$ have the values predicted by equations (31); however, $\theta_1$ and $\theta_2$ are now the angles formed by the two transition moments with a reference axis, chosen as that coinciding with the bond linking the label to the macromolecule, and $\phi$ is the angle between the projections of the two transition moments in the plane perpendicular to this axis. If $\sigma$ is very large, the predicted time decay of anisotropy is governed entirely by the expression within the parentheses.

In this limiting case, $\sigma$ and $\sigma_i$ will both depend upon solvent viscosity, as predicted by equation (21'), provided that the microviscosity sensed by the label is equivalent to the bulk viscosity. This is in fact a test for the validity of the model.

If $\sigma_i \ll \sigma$, equation (37) may be rewritten in the form

$$A(t) \cong a_1 \exp\left(-t/\sigma_1\right) + a_2 \exp\left(-t/\sigma_2\right) \tag{38}$$

where $\sigma_1$ depends entirely upon the rotation of the label and $\sigma_2$ reflects the rotation of the entire macromolecule. The sum of $a_1$ and $a_2$ is equal to the anisotropy at zero time, $A_0$. The relative magnitude of $a_1$ depends upon the value of the angle formed by the emission oscillator with the axis of rotation [$\theta_2$ in equations (31) and (37)]. In this case the emission oscillator effectively rotates about the surface of a cone of semiapex angle $\theta_2$. If the directions of the transition moments of absorption and emission coincide $(\theta_1 = \theta_2 = \theta)$, then[23,24] we have

$$\frac{a_2}{a_1 + a_2} = [(3\cos^2\theta - 1)/2]^2 \tag{39}$$

If the bond joining the fluorophor to the macromolecule has sufficient wobble so that the label may rotate freely within the cone, equation (38) still applies and equation (39) is replaced by

$$\frac{a_2}{a_1 + a_2} = \cos^2\psi\,(1 + \cos\psi)^2/4 \tag{40}$$

where $\psi$ is the semiapex angle of the cone.

A second limiting case of interest is that where the label undergoes strongly hindered rotation about the bond joining it to the macromolecule. In this case the label may exist in any of several orientations corresponding to potential energy minima. If the time spent in the specific orientations is long in comparison with the time spent in the transit between positions, then the factor which limits the rate of rotation is the probability of a jump between orientations. For this case, the following equation has been derived:[22]

$$A(t) = \exp(-t/\sigma)[(\alpha_2 + \alpha_3) \exp(-Kw't) + \alpha_1]$$
$$= \exp(-t/\sigma)[(\alpha_2 + \alpha_3) \exp(-wt) + \alpha_1] \tag{41}$$

where $w'$ is the frequency of jumps between positions and $K$ is a numerical constant.

Since $w$ depends only on the potential energy barrier opposing a transition from a particular position, it should be insensitive to solvent viscosity. This provides a potential test for the relevance of this model. If equation (41) is rewritten in the form of equation (38), then

$$1/\sigma_1 = 1/\sigma + w$$
$$1/\sigma_2 = 1/\sigma \tag{42}$$

When sucrose or some other viscosity-increasing solute is added to the solvent, the viscosity is increased by a factor $\beta$, and, from equation (21'), the new value of $\sigma_1, \sigma_1'$ is given by

$$1/\sigma_1' = 1/\beta\sigma + w \tag{43}$$

An extrapolation of $1/\sigma_1$ versus $1/\beta$ should yield $1/\sigma$ as the slope and $w$ as the intercept.[25] This assumes that the effective microviscosity sensed by the rotating label is equal to the bulk viscosity of the solvent. If the rotating unit is partially shielded from the solvent, the effective microviscosity may be less than the bulk viscosity, leading to an overestimation of $w$ and an underestimation of $\sigma$.[25]

When internal rotation which involves a significant portion of the protein is present, the available theory is not in general adequate to make unequivocal conclusions as to the size of the rotating unit without additional structural information. However, if the time decay of anisotropy detects a correlation time close to that expected for a well-defined molecular domain, to which the label is attached, this is evidence for the free rotation of the domain.

## 2.8. Analysis of Multiexponential Decay of Anisotropy

The time decay of anisotropy of a fluorescent label attached to a macromolecule is normally monitored by single-photon counting, whereby the time dependence of the intensities of the vertically and horizontally polarized components of the fluorescence are measured. The kinetics of anisotropy decay are often complex, especially if a multiplicity of both fluorescence lifetimes and rotational correlation times exists. In this case, ignoring convolution effects, we have

$$A(t) = \frac{I_{\parallel}(t) - I_{\perp}(t)}{I_{\parallel}(t) + 2I_{\perp}(t)}$$

$$= \sum \alpha_i \exp(-t/\sigma_i)$$

$$= \{I_{\parallel}(t) - I_{\perp}(t)\}/S(t)$$

$$= \{I_{\parallel}(t) - I_{\perp}(t)\} \Big/ \sum_j f_j \exp(-t/\tau_j)$$

$$= D(t)/S(t) \tag{44}$$

where $\tau_j$ is the decay time of the fluorescence intensity corresponding to component $j$ and $f_j$ is its contribution to the total intensity. Also,

$$D(t) = S(t)A(t)$$

$$= \sum_{i,j} g_{ij} \exp(-t/\alpha_{ij}) \tag{45}$$

where $\alpha_{ij} = \sigma_i\tau_j/(\sigma_i + \tau_j)$ and $g_{ij} = \alpha_i f_j$.

In actuality, convolution effects cannot be ignored and equation (45) must be replaced by

$$d(t) = s(t) A_c(t) \tag{46}$$

where $d$, $s$, and $A_c$ are the convolutions of $D$, $S$, and $A$, respectively, with the light pulse.

The time decay of $s(t)$ may be determined independently, deconvoluted, and analyzed in terms of the set of $f_j$ and $\tau_j$ by any of several available procedures, including the method of moments, developed by Isenberg et al. and the Laplace transform developed by Gafni and Brand.[7-10] Available computer programs employing these approaches may be used to determine the set of $f_j$ and $\tau_j$.

Alternatively, a nonlinear least-squares fitting procedure may be used to analyze intensity decay data.[26] In this case, the lifetime parameters are obtained by minimizing $\chi_S^2$, which is given by

$$\chi_S^2 = \sum_k v_k^2 [s_e(k) - s_c(k)]^2 \tag{47}$$

where $s_e(k)$ is the experimentally observed intensity at channel $k$, $s_c$ is the intensity predicted from the convolution of the computed values of $S(t)$ with the lamp profile, and $v_k^2$ is the variance. Since the single-photon counting error obeys a Poisson distribution, the weighting factor $v_k^2$ is equal to $s_e(k)$. When the set of $f_j$ and $\tau_j$ yielding the optimum fit is obtained, $s_c(t)$ may be computed.

The quantity $d(t)$ may be evaluated by the same procedure used for computing $s_c(t)$. The set of $\alpha_i$ and $\sigma_i$ which govern the time decay of $A(t)$ may be obtained by minimizing $\chi_A^2$:

$$\chi_A^2 = \sum_k v_A^2(k)[A_e(k) - A_c(k)]^2 \tag{48}$$

where $A_e(k)$ and $A_c(k)$ are defined analogously to $s_e(k)$ and $s_c(k)$. The variance of anisotropy, $v_A^2$, is given by

$$v_A^2(t) = v_\parallel^2(t) \left[ \frac{\partial A(t)}{\partial i_{\parallel,\,e}(t)} \right]^2 + v_\perp^2(t) \left[ \frac{\partial A(t)}{\partial i_{\perp,\,e}(t)} \right]^2 \tag{49}$$

where $v_\parallel^2(t)$ and $v_\perp^2(t)$ are the variances of $i_{\parallel,\,e}(t)$, and $i_{\perp,\,e}(t)$, respectively. The quantities $i_{\parallel,\,e}$ and $i_{\perp,\,e}$ are the experimental convoluted values of $I_\parallel$ and $I_\perp$, respectively.

Since

$$\begin{aligned} v_\parallel^2(t) &= i_{\parallel,\,e}(t) \\ v_\perp^2(t) &= i_{\perp,\,e}(t) \end{aligned} \tag{50}$$

we have

$$v_A^2(t) = 9[i_{\perp,\,e}^2(t)\, i_{\parallel,\,e}(t) + i_{\parallel,\,e}^2(t)\, i_{\perp,\,e}(t)] / [s_e(t)]^4 \tag{51}$$

The derivatives of $A_c(t)$ with respect to $\alpha_i$ and $\sigma_i$ are also needed for the minimization of $\chi_A^2$:

$$\frac{\partial A_c(t)}{\partial \alpha_i} = \left[ \sum_j f_j \frac{\partial d_c(t)}{\partial g_{ij}} \right] \Big/ s_c(t) \tag{52}$$

$$\frac{\partial A_c(t)}{\partial \sigma_i} = \left[ \sum_j \left( \frac{\tau_j}{\sigma_i + \tau_j} \right)^2 \frac{\partial d_c(t)}{\partial \alpha_{ij}} \right] \Big/ s_c(t) \tag{53}$$

Computer programs have been developed which analyze the time decay of $A(t)$ in terms of the set of $\alpha_i$ and $\sigma_i$ which yield the optimum fit.[25,26] In practice, experimental limitations preclude at present the reliable determination of more than two correlation times, provided that no more than two fluorescence lifetimes are present. The preceding approach tacitly assumes that, if more than one lifetime is present, the corresponding chromophores respond in the same way to molecular rotations.

In principle, the procedure described above does not distinguish between multiexponential decay arising from molecular asymmetry or flexibility and could be used to examine either property. However, its use has thus far been confined to studies of the internal flexibility of proteins.

## 2.9. Relation between Static and Dynamic Anisotropies

Static anisotropies are measured with a continuous light source, which is equivalent to a very large number of flashes each of which is of infinitesimally short duration. The average anisotropy is given by (for vertically polarized exciting light)

$$\bar{A} = \frac{\int_0^\infty d(t)\, dt}{\int_0^\infty s(t)\, dt} \tag{54}$$

where

$$d = i_\parallel - i_\perp$$
$$s = i_\parallel + 2i_\perp$$

Since convolution effects are absent, $d(t)$ and $s(t)$ may be replaced by $D(t)$ and $S(t)$, which are the corresponding functions for an infinitely sharp excitation pulse. If only one fluorescence decay time is present, then

$$S(t) = S_0 \exp(-t/\tau)$$
$$D(t) = A(t) S(t)$$
$$= S_0 A(t) \exp(-t/\tau) \tag{55}$$

and, from equation (54)

$$\bar{A} = \frac{1}{\tau} \int_0^\infty A(t) \exp(-t/\tau)\, dt \tag{56}$$

If $A(t)$ also decays as a single exponential, then

and
$$A(t) = A_0 \exp(-t/\sigma) \tag{57}$$

$$\bar{A} = A_0/(1 + \tau/\sigma) \tag{58}$$

which may be rewritten

$$A_0/\bar{A} = 1 + \tau/\sigma \tag{59}$$

This is the Perrin equation, which has been widely used to compute rotational correlation times.[1-6] In practice, $A_0$ is determined by making a linear extrapolation of $\bar{A}^{-1}$ versus $\tau(T/\eta)$ to $\tau(T/\eta) = 0$, according to equation (21'). In principle, either $\tau$, $\eta$, or $T$ may be varied. In the latter case, since $\eta$ is itself a function of $T$, both parameters are varied simultaneously. The quantity $\tau$ may be altered by the addition of a quencher, while $\eta$ may be increased by a neutral additive, such as sucrose. Temperature variation is not always satisfactory for this purpose, since internal rotational modes may be activated by an increase in temperature.

When the labeled macromolecule deviates from spherical symmetry, or possesses internal degrees of rotational freedom, the time decay of its anisotropy is no longer exponential. When a set of correlation times $\sigma_i$ are present, we have, if only one lifetime is present

$$\bar{A} = \sum_i \frac{\alpha_i}{1 + \tau/\sigma_i} \tag{60}$$

The limiting anisotropy at $\tau(T/\eta) = 0$, $A_0$, is now equal to $\Sigma \alpha_i$. A plot of $\bar{A}/A_0$ versus $\tau(T/\eta)$ will, in general, show significant curvature. Under these conditions, the slope at large values of $\tau(T/\eta)$ will reflect the longer correlation times, while the limiting slope at low values of $\tau(T/\eta)$ corresponds to some kind of average correlation time.

# 3. Experiment Procedures: Measurement of Anisotropy Decay

The determination of the time profile of the decay of fluorescence anisotropy and intensity may be accomplished using a single-photon counting technique.[27-31] This procedure requires a light source capable of generating a large number ($10^4$–$10^5$) of repetitive light flashes per second, as well as a system for detecting, collecting, and storing single-photon counts arising from fluorescence.

The experimental details of measurement have already been described in Chapter 3.

By monitoring the number of single-photon events as a function of time after excitation of both the vertically and horizontally polarized components, the time decay of fluorescence anisotropy may be monitored.

# 4. Applications of Fluorescence Anisotropy

## 4.1. Anisotropic Rotation

Because of the occurrence of multiple orientations of fluorescent labels with respect to the coordinates of the macromolecules to which they are attached, it is difficult to observe the effects of molecular asymmetry for fluorescent protein conjugates. However, they are readily observable for small planar aromatic fluorophors, such as anthracene and many naphthalene derivatives.[32] Such molecules may be approximated by an oblate ellipsoid of revolution, characterized by a rotational correlation time $\sigma_3$ for (in-plane) rotation about the axis of symmetry and equivalent correlation times $\sigma_1$ for (out-of-plane) rotation about the two equatorial axes, where $\sigma_1 > \sigma_3$.

For planar aromatic molecules of this class, the transition moments corresponding to $\pi \to \pi^*$ transitions are confined to the equatorial plane. In this case, equation (34) applies for the time-dependent anisotropy. Upon carrying out the integration with respect to time of equation (34), we obtain, for the static anisotropy

$$\bar{A} = 0.3 \left[\frac{(2\cos^2\lambda - 1)}{1 + \tau/\sigma_3}\right] + 0.1 \left(\frac{1}{1 + \tau/\sigma_1}\right) \tag{61}$$

Substituting from equation (19), we have

$$\bar{A} = 0.1 \left[\frac{10A_0 - 1}{1 + \tau/\sigma_3} + \frac{1}{1 + \tau/\sigma_1}\right] \tag{62}$$

It follows from equations (61) and (62) that the relative contribution of the two rotational modes is dependent upon the cosine of the angle between the transition moments of absorption and emission and is correlated with the value of $A_0$. Upon expanding in a power series and retaining only terms in $\tau$, we obtain[32]

$$(A_0/\bar{A}) - 1 = \tau \left[\frac{1}{\sigma_3} + 0.1 A_0^{-1} \left(\frac{1}{\sigma_1} - \frac{1}{\sigma_3}\right)\right] \tag{63}$$

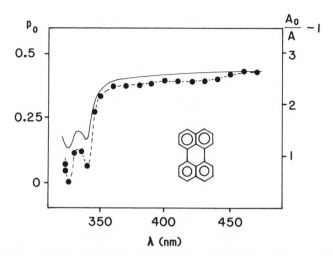

FIGURE 3. Wavelength dependence of static anisotropy for perylene in propylene glycol solution (Ref. 32). The solid curve is for limiting polarization [equation (16)] for the immobilized fluophor at $-52°$. The dashed curve is for $A_0/A = 1$.

Thus, in contrast to the case of isotropic rotation, as for a spherically symmetrical molecule, $(A_0/\bar{A}) - 1$ is dependent upon $A_0$. A dependence of $(A_0/\bar{A}) - 1$ upon $A_0$ has been observed experimentally by Valeur and Weber[32] for a series of aromatic fluorophores in the viscous solvent propylene glycol. The dependence upon excitation wavelength was measured for both $\bar{A}$ and $A_0$. (The latter was determined at $-58°$, where the molecules are effectively immobilized.)

Equation (63) predicts that the variation of $(A_0/\bar{A}) - 1$ with wavelength should qualitatively parallel that of $A_0$. This is what is observed experimentally (Figure 3) except, in some cases, for the long-wavelength end of the spectrum, where anomalous "red edge" effects arise, as will be discussed in the following section. Equation (63) also predicts that, when $A_0 = 0.1$, only $\sigma_1$ makes a contribution and the only rotation affecting the anisotropy is that of the axis of symmetry about an equatorial axis. In terms of these planar aromatic molecules, this corresponds to an out-of-plane rotation, which would be associated with a longer correlation time and with a lower value of $(A_0/\bar{A}) - 1$ than an in-plane rotation of the aromatic ring. In actuality, the minimum value of $(A_0/\bar{A}) - 1$ occurs at a wavelength which is close to, but usually not coincident with, that for which $A_0 = 0.1$.[32] This deviation is not unexpected in view of the approximations involved in the model.

## 4.2. The Red Edge Effect in Aromatic Molecules

The fluorescence of many aromatic molecules shows anomalous properties

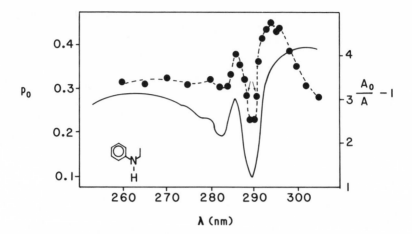

FIGURE 4. Wavelength dependence of static anisotropy for indole in propylene glycol solution (Ref. 32). The solid curve is for limiting polarization; the dashed curve is for $A_0/A = 1$.

when excitation occurs toward the long-wavelength, or red, edge of the absorption spectrum. In particular, the occurrence of radiationless energy transfer between identical molecules, as reflected by a depolarization of the fluorescence of concentrated solutions, is greatly reduced or lost altogether under these conditions.[33] Shifts in the emission spectra for red-edge excitation in rigid media have also been reported.[34]

Valeur and Weber[32] have examined the static anisotropies of a series of small aromatic molecules in a viscous solvent (propylene glycol) as a function of excitation wavelength. In each case the anisotropy was measured both at $-58°C$, where the molecules should be immobilized and the anisotropy is equal to $A_0$, and at a temperature sufficiently high so that the quantity $(A_0/\bar{A}) - 1$ was determined as a function of wavelength.

As discussed in Section 4.1, the wavelength variation of $(A_0/\bar{A}) - 1$ generally paralleled that of $A_0$. This was the case for indole, anthracene, perylene, and a series of naphthalene derivatives. However, all of the above, except for perylene, showed an abrupt drop in $(A_0/\bar{A}) - 1$ at the red edge of the spectrum (Figure 4). This is not predicted by the simple theory of anisotropic rotation summarized in Section 4.1 and must arise from some other factors. After ruling out various possible experimental artifacts, Valeur and Weber[32] proposed, as an explanation, that a transition moment for excitation and emission exists which is normal to the plane of the aromatic rings. This is only detected at the red edge of the absorption spectrum because at other wavelengths its effects are swamped by those of the more intense in-plane transition moment. It is theoretically possible for an out-of-plane transition to arise as a consequence of coupling between out-of-plane nuclear vibrations and an in-plane transition moment, although theoretical calculations are lacking.

Valeur and Weber have further suggested that the red edge effect may involve the splitting of the ground-state energy level into a set of energy levels as a result of the formation of various possible solvent–fluorophor complexes. At most absorption wavelengths, excitation preferentially involves the most stable complex, but at the red edge the excitation of complexes of higher and less stable states makes a significant contribution.

Qualitatively, an out-of-plane transition which is dominant at long wavelengths could readily explain the red edge effect upon anisotropy, since the correlation time corresponding to the rotation of such a moment would be substantially elevated over that corresponding to the in-plane transition. This would result in an increase in $\bar{A}$ and a drop in $(A_0/\bar{A}) - 1$.

Although the model of Valeur and Weber is somewhat speculative, it has no serious competitors at present and provides a useful guide to experiment.

### 4.3. A Comparative Study of Two Rigid Proteins: Lysozyme and α-Lactalbumin

Lysozyme and α-lactalbumin have extensive homology of their amino acid sequence and show similar circular dichroism spectra.[34,35] Moreover, model-building studies have demonstrated that the amino acid sequence of α-lactalbumin is compatible with a conformation similar to that of lysozyme.[36,37] The molecular weights, frictional ratios, and partial specific volumes of the two proteins are almost identical,[38,39] as are the radii of gyration, as determined by low angle x-ray diffraction.[40] The highly cross-linked nature of both proteins[1] suggests further that their structures are likely to be relatively rigid and that internal flexibility involving the polypeptide backbone is unlikely to be present to an important extent.

It is of interest therefore to examine the fluorescence anisotropy of lysozyme and α-lactalbumin in order to determine whether their rotational correlation times are indeed similar and whether the values of the latter are quantitatively predictable in terms of the known size and shape of the proteins. Tang et al.[41] have measured the fluorescence anisotropy decay of both proteins, employing three different kinds of fluorescent labels. These corresponded to dansyl conjugates for which the dansyl groups were linked to ε- or α-amino groups, dansyl conjugates for which the label was specifically linked to amino-tyrosine,[42] and fluorescamine [4-phenyl-spiro(furan-2(3H)-1'-phthalan)-3,3'-dione] conjugates for which the label was joined to ε- or α-amino groups.[43]

Two methods were employed to evaluate the anisotropy decay from raw data. In the first of these (Method 1) the excitation pulse was approximated as a delta function and the analysis was confined to times after complete decay of the excitation pulse. In the second procedure (Method 2) $s(t)$ and $d(t)$ were deconvoluted separately, using the method of moments,[7] to yield $S(t)$ and $D(t)$ as sums of exponentials. If only one rotational mode is present, corresponding to correlation time $\sigma$, then we have

$$A(t) = A_0 \exp(-t/\sigma)$$

$$S(t) = \sum_i \alpha_i' \exp(-t/\tau_i)$$

$$D(t) = A_0 \sum \alpha_i' \exp\left[-\left(\frac{1}{\tau_i} + \frac{1}{\sigma}\right)t\right] \tag{64}$$

where $\tau_i$ is the fluorescence decay time corresponding to component $i$ and $\alpha_i'$ is the corresponding amplitude.

Since the decay of anisotropy was exponential outside of the convolution zone, $\sigma$ was evaluated by Method 1 directly from the slope of a linear plot of $\ln A$ versus $t$. When Method 2 was used, $\sigma$ was computed from equation (64). The values of $\sigma$ obtained from Methods 1 and 2 were in reasonable agreement (Table 1). In each case the time decay of $S(t)$ could be fitted with either one or two decay times, depending upon the conjugate (Table 1).

Within experimental uncertainty, the computed correlation times were similar for lysozyme and $\alpha$-lactalbumin conjugates of each kind, except in the case of dansyl conjugates with aminotyrosine, for which significantly higher values were obtained for $\alpha$-lactalbumin conjugates (Table 1). However, the correlation times obtained for fluorescamine conjugates were consistently smaller than those found for dansyl conjugates (Table 1). Since it was possible to rule out aggregation as an explanation, this effect probably stems from differing preferential orientations of the labels with respect to the coordinate axes of the proteins. It is of interest to compare the observed correlation times with the average values computed for ellipsoids of revolution of the same molecular weight and varying axial ratios and degrees of hydration. This was done using the theoretical relationships summarized in Section 2.5 and assuming that the transition dipoles of excitation and emission were parallel ($\lambda = 0$). The latter was reasonable for the excitation wavelengths used, since $A_0$ approached its limiting value in each case. Calculations were done for the cases of random orientation of the transition moments (Case 1), orientation parallel to the axis of symmetry (Case 2), and orientation perpendicular to the axis of symmetry (Case 3). Tables 1 and 2 compare the observed and computed values. The computed values correspond to the mean values obtained from the initial slopes of $\ln A(t)$ versus $t$ (Table 2).

X-ray crystallographic studies have indicated that lysozyme is roughly

Table 1. Fluorescence Properties of Lysozyme and α-Lactalbumin Conjugates

| Conjugate[a] | Number of groups per molecule | Buffer[b] | $\alpha'_1$ | $\tau_1$ (ns) | $\alpha'_2$ | $\tau_2$ (ns) | $\sigma$ (ns) (method 1) | $\sigma$ (ns) (method 2) |
|---|---|---|---|---|---|---|---|---|
| FL-Lys | 1.2 | a | 0.26 | 8.5 | 0.74 | 3.4 | 6.7 | 7.7 |
| FL-Lac | 3.4 | a | 0.36 | 8.1 | 0.64 | 2.8 | 8.4 | 7.7 |
| FL-Lys | 1.0 | b | 0.17 | 10.2 | 0.84 | 4.3 | 7.0 | 7.8 |
| FL-Lac | 1.0 | b | 0.25 | 8.9 | 0.75 | 4.4 | 7.5 | 7.6 |
| DNS-Lys | 0.4 | b | 1.0 | 7.1 | — | — | 9.8 | |
| DNS-Lac | 0.4 | b | 0.38 | 20.6 | 0.62 | 6.1 | 10.6 | |
| DNS-Lys (T) | 0.95 | a | 0.45 | 19.5 | 0.55 | 5.4 | 8.8 | 9.0 |
| DNA-Lac (T) | 3.3 | a | 0.31 | 18.3 | 0.69 | 3.6 | 11.1 | 11.3 |

[a] FL-Lys and FL-Lac designate fluorescamine conjugates of lysozyme and α-lactalbumin, respectively; DNS-Lys and DNS-Lac are the corresponding dansyl conjugates with amino groups; DNS-Lys (T) and DNS-Lac (T) are dansyl conjugates with aminotyrosines.
[b] Buffer a is 50 mM acetate, pH 5.5; buffer b is 50 mM cacodylate, pH 7.4. The temperature is 25°. The protein concentration is 1 mg/ml.

Table 2. Theoretically Predicted Correlation Times for Lysozyme[a]

| Axial ratio | Hydration | Orientation of label | $\bar{\sigma}$(prolate) (ns) | $\bar{\sigma}$(oblate) (ns) |
|---|---|---|---|---|
| 1.0 | 0.22 | – | 4.7 | 4.7 |
|  | 0.52 |  | 6.2 | 6.2 |
| 1.5 | 0.22 | $\theta = 90°$ | 4.7 | 5.3 |
|  |  | Random | 4.9 | 5.1 |
|  |  | $\theta = 0°$ | 5.6 | 4.9 |
|  | 0.52 | $\theta = 90°$ | 6.3 | 7.0 |
|  |  | Random | 6.6 | 6.8 |
|  |  | $\theta = 0°$ | 7.5 | 6.4 |
| 2.0 | 0.22 | $\theta = 90°$ | 5.1 | 6.0 |
|  |  | Random | 5.5 | 5.8 |
|  |  | $\theta = 0°$ | 7.0 | 5.4 |
|  | 0.52 | $\theta = 90°$ | 6.8 | 7.9 |
|  |  | Random | 7.3 | 7.7 |
|  |  | $\theta = 0°$ | 9.3 | 7.2 |

[a] The partial specific volume of lysozyme is assumed to be 0.77 ml/g and its molecular weight to be 14,100.

ellipsoidal with an axial ratio of 1.5.[44] The mean correlation times computed on this basis for Case 1 approached those obtained experimentally for Fluorescamine conjugates if a hydration of 0.52 ml/g was assumed. The latter figure lies at the upper limit of the values estimated experimentally.[45]

The experimental values of correlation time obtained for the dansyl conjugates were uniformly higher than could be accounted for by this model with any reasonable assumption as to hydration (Tables 1 and 2). In all probability this is a consequence of the limitations of the smoothed ellipsoidal model. The actual, somewhat irregular, protein shape might well result in a significantly elevated correlation time which is accentuated by a preferred orientation of the dansyl labels.

## 4.4. Proteins with Internal Rotation Involving a Well-Defined Domain: The Immunoglobulins

In the higher vertebrates the $\gamma$-globulin antibodies occur as three main classes: IgG, IgA, and IgM. The most common of these, IgG, whose molecular weight is close to 150,000, consists of two "heavy" chains of molecular weight 52,000 and two "light" chains of molecular weight 23,000. The chains are joined by disulfide bridges (Figure 5) to form a molecule of roughly Y-shape appearance. The IgA molecules are of similar structure, while the IgM class consists of polymers of these basic units.

The biological role of the immunoglobulins, which are elicited in response to exposure to molecules (antigens) which are foreign to the circulation of the animal, is to combine with, and facilitate the removal of, these substances. Such antigens include the carbohydrates and proteins occurring on the surfaces of invading microorganisms. Antibodies may also be developed in response to small groups, or haptens, which are chemically linked to homologous plasma proteins, which are not themselves antigenic.

IgG molecules have been shown to be bivalent; each molecule contains two equivalent combining sites which can interact with the antigen molecule. If the antigen is multivalent, as is the case for protein and cellular antigens, a three-dimensional network is ultimately built up, resulting in an insoluble precipitate.

The IgG immunoglobulins are cleaved by papain (Figure 5) to form an Fc fragment consisting of the greater part of the two "heavy" chains plus two Fab

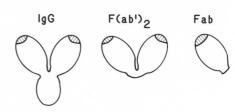

FIGURE 5. Schematic version of the shape of an IgG molecule and its fragments produced by enzymic action.

fragments, each of which contains a "light" chain plus the remainder of a "heavy" chain. Each Fab fragment contains an active site, whose combining efficiency is retained in full. The Fab fragments correspond to the arms of the Y.

From the above structural considerations, as well as electron-microscopic observation, the IgG molecule may be visualized as a multidomain protein. The fact that papain cleavage occurs near the junction of the Fab and Fc subunits raises the possibility that a randomly coiled region of the "heavy" chain may occur here, which might provide a flexible hinge point, permitting some degree of independent motion of these structural elements.

Early static anisotropy measurements employing dansyl-conjugated immunoglobulin yielded somewhat discordant results.[46] Correlation times derived from Perrin plots obtained by temperature variation yielded, in general, average correlation times which were much less than the value expected for an IgG molecule approximated by a rigid ellipsoid of revolution of the same molecular weight. In contrast, Perrin plots obtained by viscosity variation at constant temperature yielded much higher correlation times which were comparable in magnitude to the expected value.[46] Since the IgG molecule can be represented only very roughly by an ellipsoid of revolution, the predicted value is itself somewhat uncertain. However, it may probably be placed in the neighborhood of 100 ns.[47]

Both of the above experimental approaches are subject to serious ambiguities. Perrin plots obtained by temperature variation are sensitive to thermally activated rotational modes involving the probe, which tend to magnify the slope of $A^{-1}$ versus $\tau(T/\eta)$ and hence to underestimate the average correlation time for a particular temperature. On the other hand, Perrin plots obtained by viscosity variation may underestimate the effects of internal rotation because of differences between the effective viscosity sensed by the probe and the bulk viscosity.

Wahl[48] subsequently applied the more powerful technique of time-resolved measurements of anisotropy decay to a dansyl conjugate of $\gamma$-globulin. The degree of substitution (less than one dansyl group per molecule) was sufficiently low so that any depolarization via radiationless energy transfer was unlikely.

At pH 8 the convoluted curves of $d(t)$ and $s(t)$ both showed a time decay which was clearly nonexponential. Since an adequate mathematical procedure for analyzing a complex system of this kind was not available at that time, an empirical curve-fitting procedure was employed. It was assumed that the deconvoluted forms of $s(t)$ and $d(t)$, $S(t)$ and $D(t)$, could be represented as the sum of two exponentials.

$$S(t) = A_1 \exp\left(-t/\tau_1\right) + A_2 \exp\left(-t/\tau_2\right)$$
$$D(t) = A_1' \exp\left(-t/\tau_1'\right) + A_2' \exp\left(-t/\tau_2'\right)$$

(65)

By a trial-and-error curve-fitting procedure, values of the two sets of decay times were found so that the corresponding convoluted forms of $S(t)$ and $D(t)$ reproduced graphically the experimental curves of $s(t)$ and $d(t)$. The ratio $D(t)/S(t)$ then yielded the deconvoluted anisotropy $A(t)$ as a function of time.

The time decay of deconvoluted anisotropy was also nonexponential. A repetition of the graphical curve-fitting procedure was also done to reproduce $A(t)$ by an expression of the form

$$A(t) = \alpha_1 \exp(-t/\sigma_1) + \alpha_2 \exp(-t/\sigma_2) \qquad (66)$$

where $\alpha_1, \alpha_2 = 0.075, 0.14$ and $\sigma_1, \sigma_2 = 7.7, 123$ ns.

The value of $\sigma_2$ is in the range expected for the rotation of the entire $\gamma$-globulin molecule and may probably be loosely attributed to this origin. That of $\sigma_1$ can only reflect some type of internal rotation. Since the sites of attachment of the dansyl label were unknown, as well as the importance of hindered rotation confined to the probe itself, it was not possible to draw further structural conclusions.

Yguerabide et al.[49] subsequently studied the fluorescence anisotropy decay of IgG antibodies directed against the fluorescent hapten $\epsilon$-dansyl-L-lysine. The hapten itself was bound to combining sites in both the IgG molecule and the Fab fragments. The combination was accompanied by a 25-fold increase in quantum yield, making it readily feasible to analyze the fluorescence properties of the complex alone. This approach has the important advantage that the sites of attachment of the probe are known and specific.

In this study the time decay of $i_\parallel$ and $i_\perp$ was measured for the complex of $\epsilon$-dansyl-L-lysine with IgG and its Fab fragment. Curves of $s(t)$ and $A(t)$ were then constructed from these data. It was found that IgG and Fab yielded equivalent curves for the time decay of $s(t)$. These were nonexponential, indicating the presence of more than one fluorescence lifetime. This is explicable in terms of the molecular heterogeneity of antibodies.

To avoid problems of analysis arising from convolution effects, analysis of the time decay of anisotropy was confined to times after complete decay of the excitation pulse. This procedure has the disadvantage of minimizing, or losing altogether, the effects of any rotational modes of very short correlation time, such as might arise from a local motion of the probe. In contrast to the case of $s(t)$, it was found that the time decay of $A(t)$ was very different for Fab and the intact immunoglobulin. In the case of Fab the anisotropy decayed exponentially, indicating that only a single rotational mode was sensed under these conditions. The computed rotational correlation time was 33 ns. For a molecule of molecular weight 50,000 and hydration 0.3 ml/g, this would formally correspond to a prolate ellipsoid of axial ratio 2.5. However, for the reasons outlined earlier, this figure may probably be regarded as an upper limit, and the smoothed shape of the Fab fragment may not deviate greatly from a sphere.

In the case of intact IgG, the time decay of anisotropy was nonexponential and could not be fit on the assumption of a rigid prolate ellipsoidal shape. A least-squares fit of the observed $A(t)$ curve showed that it could be represented by equation 66, where $\alpha_1, \alpha_2 = 0.14, 0.18$ and $\sigma_1, \sigma_2 = 33, 168$ ns.

The value of $\sigma_1$ is much shorter than the minimum value, 47 ns, computed for a rigid spherical particle with the molecular weight of IgG and is of similar magnitude to that obtained for Fab. The longer time, $\sigma_2$, is roughly in the range expected for the rotation of IgG as a whole.

These results are consistent with a model in which the Fab domains rotate freely with respect to the balance of the molecule. However, the possible existence of any rotational modes of shorter correlation time, reflecting segmental motion within the Fab units, remains uncertain.

The independent rotation of the Fab units may have an important bearing on the antibody function, facilitating the combination with antigen.

Lovejoy et al.[50] have subsequently continued these studies, employing rabbit antibodies directed against the hapten pyrenebutyrate. Pyrenebutyrate (PBA) has a much longer lifetime than dansyl, the value for the free hapten being about 100 ns in air-saturated aqueous solution and somewhat longer in $O_2$-free solution.

The combination of PBA with the antibody was accompanied by significant red shifts of the primary excitation maxima from 326.5 to 330.5 nm and from 341.5 to 347 nm and of the emission maxima from 375 to 376 nm and from 395 to 396 nm. In contrast to the case of antibodies to dansyl-lysine, no increase in quantum yield occurred. The average binding constant was determined by fluorescence titration by taking advantage of the spectral red shift. The value obtained for an antibody sample produced by prolonged (8 month) immunization was $5 \times 10^8 \, M^{-1}$. This is high enough to permit virtually quantitative binding of hapten.

The time decay of fluorescence intensity was heterogeneous, except for antibody produced at very long (11 month) immunization times, for which a single fluorescent component of lifetime 157 ns was observed. The time decay of fluorescence anisotropy showed a similar profile for two different preparations and was fitted to equation (66) without deconvolution by employing a least-squares procedure. The values obtained were $\alpha_1, \alpha_2 = 0.042, 010$ and $\sigma_1, \sigma_2 = 24, 140$ ns.

The correlation times are of similar magnitude to those reported by Yguerabide et al.[49] Since the time decay could be measured over a sufficient time interval to permit accurate determination of the longer correlation time, these results reinforce the conclusion that mobility of the Fab unit is of general occurrence in IgG molecules.

Holowka and Cathou[51] have subsequently extended studies of this kind to the macroglobulin (IgM) class of antibodies. In animal serum IgM generally occurs as a pentamer of five disulfide-linked subunits of total molecular weight about 900,000. Each subunit is roughly similar in structure to an IgG molecule, consisting of two light (L) and two heavy ($\mu$) chains joined by disulfide links and noncovalent interactions. A third unrelated (J) chain is also present and may figure in the assembly of IgM from its subunits. The IgM pentamer has a total of ten combining sites for antigen on its ten Fab units.

IgM antibodies directed against $\epsilon$-dansyl-L-lysine were prepared by Holowka and Cathou from horse, pig, and shark antisera obtained by immunization with a dansyl-lysine streptococcal conjugate, which favors the development of antibodies of the IgM type in these species. The fluorescence characteristics of complexes of $\epsilon$-dansyl-L-lysine with purified IgM molecules were examined.

The time decay of fluorescence intensity varied with the species. For both horse and pig IgM, the dominant emitting species had a decay time close to 24 ns, with a secondary contribution from a species of decay time 8–12 ns. For shark IgM, the situation was reversed, the species of short (4 ns) decay time being dominant. However, in all three cases, the contribution of the long decay time was sufficient to permit measurements of anisotropy for times up to 200 ns.

The pattern of time decay of fluorescence anisotropy was basically similar for the porcine and equine IgM molecules and their proteolytic fragments. Pepsin or papain digestion yielded the Fab$\mu$ fragments, of molecular weight 56,000, which are analogous to the Fab fragments of IgG. Limited pepsin digestion formed the (Fab$'$)$_2\mu$ species, consisting of the two arms of the Y, plus a short segment of the stem. Its molecular weight was 105,000–120,000. Reduction and alkylation of (Fab$'$)$_2\mu$ produced Fab$'\mu$. The decay of anisotropy for the Fab$\mu$ fragments was monoexponential, corresponding to a single correlation time of 32–36 ns. A minor degree of dependence of measured correlation time upon wavelength was observed, raising the possibility of some influence of preferential orientation of the label with respect to the axes of the protein. There was no indication of any short (<10 ns) correlation time which might reflect a localized motion of the probe. The latter finding was consistent with the development of an extrinsic circular dichroism spectrum by the bound hapten.

The behavior of the porcine and equine Fab$\mu$ fragments is thus entirely similar to that of the Fab fragments from IgG. In both cases the fragments seem relatively rigid, with no indication of internal flexibility as sensed by this probe. In particular, the "switch" peptides, which connect the V and C domains within the Fab unit do not appear to provide a flexible hinge point.[51]

The time decay of anisotropy was also measured for the equine (Fab$'$)$_2\mu$

fragment obtained by pepsin-catalyzed proteolysis. The decay of $(Fab')_2\mu$ was nonexponential and could be fitted by equation (66) with the following values: $\alpha_1, \alpha_2 = 0.22, 0.06$ and $\sigma_1, \sigma_2 = 38, 211$ ns. The shorter correlation time, which makes the larger contribution to the anisotropy decay profile, is slightly larger than that of $Fab\mu$, suggesting that the rotation of the $Fab\mu$ units may be significantly hindered in the $(Fab')_2\mu$ species.

Model calculations were made in an effort to fit the decay curves of $Fab\mu$ and $(Fab')_2\mu$ to ellipsoidal models, with reasonable assumptions as to hydration. In the case of $Fab\mu$, the data could be fitted on the basis of a prolate ellipsoidal shape of axial ratio 2.0 and hydration 0.69 g $H_2O/g$ of protein, assuming random orientation of the emission dipole. While, for the reasons discussed earlier, it would be unrealistic to regard these figures as quantitatively valid, it is probably reasonable to approximate the $Fab\mu$ fragment as a rigid prolate ellipsoid of low asymmetry.

In contrast, data for the $(Fab')_2\mu$ species could not be fitted with a rigid ellipsoidal model, assuming any reasonable value for the hydration. The decay of anisotropy was more rapid than that predicted for an axial ratio of 1.0 and a hydration of 0.69. This provides further evidence for internal flexibility of the $(Fab')_2\mu$ species and for the independent rotation of the $Fab\mu$ units.

The anisotropy decay of equine and porcine IgM was nonexponential and required two correlation times for fitting by equation (66). In both cases a long ($> 500$ ns) correlation time was present which could probably be identified with the overall motion of the IgM pentamer. In addition, a shorter (61–69 ns) correlation time was detected which did not correspond closely with those of either the $Fab\mu$ or the $(Fab')_2\mu$ species. It is clear that segmental flexibility on the nanosecond time scale must exist in the IgM molecule. The nature of the internal motion appears to be somewhat different from that of IgG, perhaps reflecting a greater degree of hindrance of the rotation of the Fab units.

The behavior of shark IgM differed from that of the outer two species examined in that an additional rotational mode of quite short ($< 10$ ns) correlation time was detected. This probably reflects a localized flexibility in the vicinity of the hapten-combining site. In addition, the correlation time analogous to $\sigma_1$ had a somewhat larger value (93 ns). Clearly there must be significant structural differences between shark IgM and those from the other, phylogenetically distant species.

Some reservations must be retained as to the quantitative interpretation of the preceding results. The ill-conditioned nature of the problem and the limitations of the data do not really permit a rigorous exclusion of additional rotational modes or the possibility that the apparent correlation times may correspond in reality to complex averages. In addition, the location of the hapten probe may be such as to minimize its sensing of segmental flexibility elsewhere in the Fab unit.

In a recent study, Siegel and Cathou[52] have examined the effects of heat treatment (30 min at 60°) upon the properties of equine IgM antibodies directed against $\epsilon$-dansyl-L-lysine. This treatment resulted in a major decrease of antigen-independent complement fixation and of binding affinity for the hapten, as well as significant changes in the circular dichroism spectrum.

The thermally treated IgM, as well as its $(Fab')_2\mu$ fragment, showed qualitatively a more rapid time decay of anisotropy than the corresponding native species. Analysis of the decay curves in terms of two correlation times indicated that, while the values of $\sigma_1$ and $\sigma_2$ were almost unchanged, their relative amplitudes were substantially altered, the contribution of the shorter correlation time ($\sigma_1$) increasing markedly for both species. The implication is that the rotational mobility of the Fab$\mu$ unit is enhanced by limited thermal denaturation.

## 4.5. Internal Flexibility of Multidomain Proteins: Myosin

The muscle protein myosin is, like the IgG immunoglobulins, roughly Y-shaped, with a rodlike stem and two globular (S-1) units at the head. Current models for the mechanism whereby a myosin "cross-bridge" is linked to the thin actin filaments within a muscle fiber and causes a mechanical thrust require that at least one point of flexibility be present within the myosin molecule. Mendelson et al.[53] have adopted an approach analogous to that successfully applied to the immunoglobulins, attaching a fluorescent label to specific locations on the globular S-1 moieties, which comprise the "arms" of the Y.

Each S-1 unit contains a single highly reactive sulfhydryl group, permitting the selective attachment of iodoacetamide derivatives. Mendelson et al. employed 1,5-AEDANS ($N$-iodoacetylamino-l-naphthylamine-l-sulfonate), whose reactivity to sulfhydryl groups parallels that of iodoacetamide and whose fluorescence properties resemble those of dansyl groups. The conjugates studied by Mendelson et al. contained about two AEDANS groups per myosin molecule, substitution being confined to the S-1 units. The labeled S-1 units were also isolated by papain treatment of labeled myosin, followed by chromatographic purification.

The time decay of fluorescence anisotropy for S-1 was monoexponential. A least-squares fitting procedure yielded a correlation time of about 220 ns. From the known molecular weight ($\sim 1.15 \times 10^5$) of the S-1 fragment, assuming a hydration of 0.2 ml/g protein, it was possible to set a lower limit of 3.5 to the axial ratio of S-1, approximated as a prolate ellipsoid of revolution, by applying the equation of Belford et al.[16] in the form it assumes for ellipsoids of revolution.

The procedure adopted was to determine, for a series of assumed values of axial ratio ($\gamma$), those values of $\theta_1$ and $\theta_2$ [equation (31)] which yielded predicted anisotropy decay curves consistent, within experimental uncertainty, with the observed decay curves. In this way, a region was defined in $(\gamma, \theta_1, \theta_2)$ space, any point of which represented an acceptable description of the data. The minimal value of $\gamma$, which agreed with estimates from other approaches,[54] was attained only for very low values of $\theta_1$ and $\theta_2$, suggesting that the label was preferentially oriented with its transition moment parallel to the axis of symmetry of the ellipsoid.

The time decay of anisotropy for myosin itself was also monoexponential, with a correlation time of 400–450 ns. This value is about twice that for free S-1 and substantially less than would be expected for a protein of the molecular weight $(5 \times 10^5)$ and asymmetry of myosin, if it were a rigid unit. Since the correlation time increased to 1800 ns for an aggregated form of myosin, with no significant change in fluorescence decay time (20 ns), the possibility was ruled out that the value for monomeric myosin reflected some type of localized motion of the probe. The favored interpretation was therefore that independent rotation of the S-1 units occurs in myosin.

This conclusion was substantially strengthened by studies on heavy mero-myosin (HMM) formed by the action of trypsin on myosin.[53] This fragment (molecular weight $3.4 \times 10^5$) contains both S-1 units, plus a fraction of the rodlike stem. If HMM were a rigid molecule, its correlation time would be expected to be considerably more than twice that observed for S-1, as its molecular weight is greater by a factor of 3. The observed correlation time (400 ns) was less than twice that of S-1.

Although the interpretation is less clear-cut than in the case of the immuno-globulins, there is thus a definite indication of some degree of freedom of rotation of the S-1 units. However, in contrast to the immunoglobulin case, the rotation of the S-1 units appears to be somewhat hindered, as reflected by a correlation time which is twice that of the isolated S-1 unit. It is therefore unlikely that the S-1 units are joined to the stem by a highly flexible poly-peptide.

In a subsequent report Mendelson and Cheung[55] have examined the question of the mechanical interaction of the two S-1 units, which could be a factor in the elevation of the correlation time of labeled myosin. By limited proteolysis with papain it is possible to remove a single S-1 head. Single-headed myosin, in which the S-1 unit was labeled with AEDANS, was found to have a correlation time almost indistinguishable from that of native myosin. It is thus unlikely that mechanical interference is a dominant factor in the increase of the correlation time of myosin above that of free S-1. An appreciable stiffness of the hinge polypeptide joining S-1 to the myosin stem is a more probable explanation. The qualitative picture of the molecular dynamics of myosin which emerges from these results is that of the two rigid S-1 moieties tumbling independently about a somewhat flexible universal joint between each S-1 unit and the stem.

These observations are of importance with respect to theoretical models of muscle contraction and, in particular, to the role of the myosin heads as cross-bridges to the thin actin filaments. If the S-1 units can move as segments independently of the stem, then the translation of the myosin molecules along the actin filament by an "arm-over-arm" movement becomes plausible.

## 4.6. Internal Flexibility of Multidomain Proteins: F-Actin

The mechanical force generated in muscle contraction arises from the cyclic interaction of myosin and actin, which is coupled with ATP hydrolysis. In the course of the overall process, the myosin–actin system passes through several distinct states, one of which may be equivalent to the stable complex formed in the absence of ATP. The question of how the conformation of F-actin is influenced by complex formation and by the presence of the various modifiers involved in muscle contraction is of central importance in developing an adequate account of this process at the molecular level. Purified F-actin, the polymeric form of G-actin, exists in solution as very long fibers, whose average length is of the order of microns. If the structures of these were rigid and devoid of internal flexibility, their rotational correlation times would be immeasurably large.

Wahl and co-workers[56] have approached this problem by examining the static and dynamic fluorescence anisotropy of 1,5-AEDANS conjugates of F-actin. The time decay of fluorescence anisotropy for F-actin in 0.1 mM $Ca^{2+}$ could be fitted by equation (66) with the following values of the parameters: $\alpha_1, \alpha_2 = 0.065, 0.125$ and $\sigma_1, \sigma_2 = 13.5, 260$ ns.

The addition of HMM or S-1 to the F-actin conjugate in the presence of 0.1 mM $Ca^{2+}$ resulted in a progressive increase in $\sigma_2$ to a limiting value of 1100 ns for the case of a large excess of HMM. The fluorescence lifetime varied only slightly, being close to 19 ns for all compositions. Under dissociating conditions, in the presence of 10 mM ATP, the correlation time reverted to that of free F-actin, thereby ruling out an artifact arising from the denaturation of F-actin.

In the presence of $Mg^{2+}$ (1 mM) the decay parameters of F-actin were substantially different from their values in $Ca^{2+}$. A two-exponential fit yielded in this case $\alpha_1, \alpha_2 = 0.04, 0.25$ and $\sigma_1, \sigma_2 = 5.8, 682$ ns. The longer correlation time thus undergoes a major increase in the presence of $Mg^{2+}$.

The addition of HMM in the presence of 1 mM $Mg^{2+}$ produced a biphasic response of the magnitude of the longer correlation time, which passed through a minimum at a mole ratio [HMM]/[actin] of 0.02, followed by an increase. A similar biphasic pattern was observed upon the addition of S-1 in the presence of $Mg^{2+}$.

These results suggest that some form of segmental flexibility exists in F-actin and that the degree of flexibility is dependent upon experimental conditions. The shorter correlation time, $\sigma_1$, presumably reflects some type of localized rotational motion, while the longer time, $\sigma_2$, corresponds to a rotational mode involving a larger unit, probably a set of actin monomers. The initial decrease in $\sigma_2$ produced by complex formation with HMM or S-1 in the presence of 1 mM $Mg^{2+}$ may arise from an increase in the flexibility of the links joining actin monomers, while the increase observed at higher levels in the presence of $Mg^{2+}$ and at all levels in the presence of $Ca^{2+}$ reflects a stiffening of these contacts.

The molecular dynamics of F-actin have been examined by other physical techniques, with results which are not in quantitative agreement with those obtained by fluorescence anisotropy decay. From the quasi-elastic light scattering of F-actin solutions, Fujime and Ishiwata[57] computed a correlation time of $10^{-2}$ s. From saturation transfer electron spin resonance, Thomas et al.[58] obtained a value of $10^{-4}$ s. Both groups found that the correlation time increased substantially upon complex formation with HMM.

The reasons for the varying results obtained by the different techniques remain obscure. The molecular motions sensed by these methods are clearly different, but the existing theory is not adequate to provide an explanation.

## 4.7. Internal Flexibility of Multidomain Proteins: Fibrinogen

The plasma protein fibrinogen, the polymerization of whose activated form is the central event of blood clotting, consists of three pairs of polypeptide chains ($A\alpha$, $B\beta$, and $\gamma$) of total molecular weight 340,000. The polypeptide chains are extensively cross-linked by disulfide bonds. The overall structure of fibrinogen, as visualized by electron microscopy, corresponds to three roughly spherical nodules of diameter $5-7$ nm which are joined by thin filaments.[59] Hydrodynamic and light scattering data suggest that the shape of the fibrinogen molecule may be approximated by a prolate ellipsoid of revolution of axial ratio $\sim 10$.[60] It has been proposed that the three polypeptide chains which join the globular nodules exist as coiled coils, on the basis of amino acid sequence data and a helix prediction scheme.

Fibrinogen provides a clear-cut example of a multidomain protein. A question of particular interest is whether its conformation possesses significant flexibility, so that an individual nodule can twist or bend independently of the rotational motion of the balance of the molecule. Electron-microscopic studies have indicated a significant proportion of bent molecules which deviate from linearity.[61]

Direct evidence for molecular flexibility has been obtained from static anisotropy measurements. A rigid unhydrated spherical protein of molecular weight 340,000 and partial specific volume 0.72 ml/g would have a molecular volume of $4.0 \times 10^{-19}$ cm$^3$ and a predicted correlation time of 100 ns. For reasons discussed earlier, this value is essentially a lower limit, and the measured correlation time would be expected to be up to twice this value.

For the more realistic model of a prolate ellipsoid of revolution with major and minor axes equal to 42 and 4.2 nm, respectively, the predicted correlation time ranges from 185 ns for a randomly oriented probe to 1300 ns for a probe with the transition moments oriented parallel to the long axis (assuming coincidence of the directions of the absorption and emission moments).

In an early report, Johnson and Mihalyi[62] obtained the correlation time by static anisotropy measurements of a chemically conjugated dansyl derivative of fibrinogen containing 11–13 dansyl groups per molecule. In this case the probe is distributed among a large number of possible loci, so that the assumption of random orientation is probably a reasonable approximation.

The rotational correlation time obtained by Johnson and Mihalyi was close to 65 nm. This value is substantially less than the minimum value predicted for a spherical molecule of this molecular weight and much less than the values predicted for a prolate ellipsoidal shape. The implication is that the effective kinetic unit whose rotation is sensed by the probe is significantly smaller than the entire fibrinogen molecule.

In a recent report Hantgan[63] reexamined the question of the internal flexibility of fibrinogen, employing a more specifically located label. Dansyl cadaverine was incorporated into specific sites on fibrinogen by the action of plasma factor XIIIa. Factor XIIIa is a transglutaminase whose biological function is to catalyze the formation of peptide bonds between specific lysine donor and glutamine acceptor sites on the $\gamma$ and $\alpha$ chains of fibrinogen, thereby stabilizing the fibrin clot. Dansyl cadaverine competitively inhibits the cross-linking reaction, forming an isopeptide bond with a $\gamma$-chain glutamine group which resides in the outer D nodule, as well as a limited number of $\alpha$-chain sites. For the preparations studied, the label was distributed roughly equally between the two classes of site.

The fluorescence emission maximum of dansyl cadaverine undergoes a major blue shift from 538 to 495 nm upon conjugation with fibrinogen. This is accompanied by a fivefold increase in fluorescence intensity. This suggests that the probe is predominantly imbedded within nonpolar regions of the fibrinogen structure.

Perrin plots were constructed from static anisotropy measurements, in which $T/\eta$ was varied by altering the temperature from 5 to 40°. Interpretation of the results is therefore potentially subject to the complication that thermally activated internal rotations may be present. However, reservations of this kind are partially relieved by the agreement of the limiting anisotropies obtained by extrapolation to $T/\eta = 0$ with those obtained by direct measurement in media of high viscosity.[63]

The computed correlation times were independent of the degree of labeling and equal to $53 \pm 3$ ns, in fair agreement with the value originally obtained by Johnson and Mihalyi. Both sets of measurements assign to the effective rotational kinetic unit of fibrinogen a size much smaller than that of the intact molecule and are consistent with the idea that fibrinogen is not a completely rigid molecule, but possesses some degree of internal flexibility.

The sites of flexibility are a matter of speculation. A model proposed by Doolittle[64] suggests that coiled-coil filaments joining the nodules are the flexible regions, serving as molecular hinges permitting the twisting or bending motion of the nodules. However, in view of the uncertain location of the probes attached to the α-chains, the precise nature of the motion sensed by the probes is uncertain, although the order of magnitude of the correlation time is consistent with independent rotation of the nodules.

## 4.8. Internal Flexibility: Motion of Intrinsic Fluorophors

In a pioneering study, Munro et al.[26] have examined the time decay of fluorescence anisotropy for the tryptophan fluorescence of a series of proteins containing only a single tryptophan. A unique experimental feature of this investigation was that synchrotron radiation was employed as the exciting light source. The half-width of the excitation pulse was about 0.65 ns, which is much less than the values found typically for the usual arc discharge. This simplified the detection and measurement of correlation times in the subnanosecond range.

The time decay of anisotropy was fitted to equation (66) by a least-squares procedure. The observed mobility of the tryptophan residue varied widely for the different proteins examined. That of nuclease B of *Staphylococcus aureus* displayed only a single rotational mode of correlation time 9.8 ns. This is comparable to the correlation time, 7.6 ns, predicted for a rigid sphere of molecular weight 20,000 and hydration 0.2 ml/g. There was no indication of any rotational mode of shorter correlation time. The implication is that the tryptophan of nuclease B is effectively immobilized within the protein matrix, with which it rotates as a unit.

In contrast, the tryptophan of basic myelin protein was found to undergo quite rapid motion. The decay was biexponential, with correlation times of 90 ps and 1.3 ns. The shorter time does not greatly exceed the value expected for free tryptophan in solution, while the longer time is still much less than the value 6.9 ns computed for a rigid sphere of molecular weight 18,000 and hydration 0.2 ml/g. It is clear that the tryptophan of basic myelin protein is undergoing rotation which is virtually unhindered by the balance of the molecule. This is consistent with the conclusions of circular dichroism and hydrodynamic studies, which depict basic myelin protein as largely structureless.[65]

Holoazurin represents a somewhat intermediate case. The decay of anisotropy is biexponential, corresponding to a shorter correlation time ($\sigma_1$) of 0.51 ns and a longer time ($\sigma_2$) of 11.8 ns. The latter value somewhat exceeds that computed for a rigid hydrated sphere of molecular weight 14,000 (8.5 ns). The value of $\sigma_1$ is somewhat larger than expected for a freely rotating tryptophan, suggesting that the motion of the tryptophan is hindered to some degree, or else that a significant portion of the adjacent polypeptide is involved in its motion. From the amplitudes associated with $\sigma_1$ and $\sigma_2$, one may compute from equation (40) that the rotational movement of the tryptophan is confined within a cone of semiapex angle 34°. The most plausible model for holoazurin is that of a tryptophan undergoing restricted rotation about its link with a rigid protein molecule.

The behavior of apoazurin, which is formed by removal of $Cu^{2+}$ from the active site, suggests a major increase in flexibility. The angular range corresponding to the shorter correlation time increases to 44°, while the value of the longer correlation time decreases to 5.8 ns, which is less than that predicted for a rigid molecule. It is therefore likely that apoazurin possesses internal rotational modes not present in holoazurin.

The anisotropy decay of human serum albumin was found to depend upon the temperature. At 5°, only one rotational mode was detected, corresponding to a correlation time of 31 ns, which is comparable to the value (40 ns) predicted for a rigid hydrated sphere. At 45° the decay was biexponential, with correlation times of 0.14 and 14 ns. From equation (40) and the amplitudes corresponding to the two correlation times, a value of 26° is computed for the semiangle of the effective cone to which rotation is confined. The implication is that the tryptophan of human serum albumin, which is rigidly bound at 5°, acquires a restricted freedom of rotation at higher temperatures.

Ross et al. [66] have examined the anisotropy decay of the single tryptophan of the polypeptide hormone adrenocorticotropin. The decay was biexponential, corresponding at 3.5° to correlation times of 0.92 ns ($\sigma_1$) and 4.5 ns ($\sigma_2$). The latter figure is comparable to that predicted for the rotation of the entire molecule. The former may be attributed to a localized motion of the tryptophan. The elevation of its value above that obtained for free tryptophan may reflect either some hindrance to its rotation, or else the involvement of part of the polypeptide in the effective rotating unit. Direct evidence for the presence of hindered rotation was obtained from the viscosity dependence. While $\sigma_2$ varied linearly with viscosity, as predicted by equation (21'), $\sigma_1$ was essentially independent of viscosity, as would be predicted if the rate-controlling factor were the probability of release of the fluorophor from positions corresponding to potential energy minima. From the amplitudes of the two rotational modes, application of equation (40) leads to a value of 50° for the semiangle of the effective cone to which rotation is restricted.

It is clear that the tryptophan of adrenocorticotropin has a high degree of mobility. It is of interest that a variety of physical methods have suggested that adrenocorticotropin and several other peptide hormones are flexible molecules.

The time decay of fluorescence intensity for adrenocorticotropin is also biexponential, corresponding to lifetimes of 2.3 and 5.9 ns. Ross et al. have raised the possibility that this arises from the rotational mobility of the tryptophan, which results in its exposure to differing microenvironments, corresponding to different configurations.

An example of a protein containing two tryptophans, both of which are immobilized, is provided by horse liver alcohol dehydrogenase (ADH). This protein is of particular interest because of the extensive information as to its detailed structure and mechanism of action. One of its tryptophans (Trp-15) is exposed to solvent, while the other (Trp-314) is enclosed in a hydrophobic cage at the interface between the two identical subunits of molecular weight 40,000.[67]

Ross et al.[68] have measured the time decay of fluorescence anisotropy for ADH over a range of temperatures from 10 to 40°. A least-squares fit yielded only one correlation time at all temperatures, whose magnitude (e.g., 36 ns at 27°) was comparable to the value expected for a rigid hydrated sphere of molecular weight 80,000. Moreover, the correlation time varied linearly with viscosity, as predicted by equation (21'). It appears that, in ADH, both the exposed and buried tryptophan are devoid of mobility with respect to the protein structure.

It is of interest that the time decay of fluorescence intensity of ADH is biexponential, corresponding to lifetimes of 3.6 and 7.1 ns. Their values are essentially independent of emission wavelength, although their respective amplitudes vary. Since quenching by KI, which preferentially quenches exposed tryptophans, reduces the magnitude of the longer decay time, this may probably be attributed to Trp-15, while the shorter lifetime arises from Tyr-314.

Thus, in contrast to adrenocorticotropin, each of the immobilized tryptophans of ADH decays exponentially, reflecting their invariant microenvironment.

Lakowicz and Weber[69] have adopted an alternative approach to the detection of segmental mobilities of tryptophan residues in proteins, employing lifetime-resolved static anisotropies. In an earlier study those authors observed that $O_2$ was an efficient quencher of exposed and buried tryptophans in a variety of proteins.[70] This was interpreted as suggesting that the structural mobilities of the proteins examined facilitated the diffusion of $O_2$ through their matrices.

Lakowicz and Weber employed dynamic quenching by $O_2$ at pressures up to 100 atm to obtain Perrin plots in which lifetime, rather than viscosity, was varied. This has the advantages of avoiding ambiguities arising from possible differences between the bulk viscosity and the effective viscosity sensed by the probe and of detecting hindered rotation which may not respond to a change of viscosity. However, when more than one tryptophan is present, the correlation time observed will be a poorly defined average, which may be biased in favor of a preferentially quenched tryptophan.

Employing this approach, it was found that all of the seven proteins examined yielded correlation times which were much smaller than predicted

Table 3. Average Correlation Times of Tryptophan
Residues in Proteins Obtained by Lifetime Variation
(Ref. 70)

| Protein | $\phi^a$ (ns) | $\phi_0^b/\phi$ |
|---------|---------------|-----------------|
| Pepsin | 8.0 | 2.4 |
| Carboxypeptidase A | 1.3 | 15 |
| α-Chymotrypsin | 3.0 | 4.3 |
| Human serum albumin | 7 | 4.7 |
| Bovine serum albumin | 16 | 2.1 |
| Trypsin | 2.9 | 4.5 |
| Carbonic anhydrase | 7 | 2.4 |

$a$ Rotational correlation time obtained from equation (59)
for 25°.

$b$ Rotational correlation time predicted for a rigid sphere
of equivalent molecular weight [equation (21)].

for rigid spheres of the same molecular weight (Table 3), the ratio of predicted to observed correlation times ranging as high as 15 in the case of carboxy-peptidase A.

From these measurements on a variety of proteins of widely different structure, it is clear that mobility of tryptophans is a widespread phenomenon. It is not, however, possible from these results to distinguish between hindered rotation confined to the indole side chain and segmental rotation involving a significant portion of the polypeptide chain.

## 4.9. Librational Motion of Fluorescent Probes Linked to Hemoglobin and Its Subunits

The allosteric protein hemoglobin has been the subject of intensive physical studies for many years, and the structural origins of its properties have come to be understood in considerable detail. Adult human hemoglobin consists of two α- and two β-polypeptide chains, each of which binds a heme group, so that the overall structure may be designated as $\alpha_2\beta_2$. The α- and β-chains are of similar molecular weight (18,000), show considerable homology of sequence, and have similar conformations when present in the intact hemoglobin tetramer. The β chain contains a reactive sulfhydryl group at the β-93 position and a second sulfhydryl at the β-112 position.

Bucci et al.[71] and Oton et al.[25] have examined the static and dynamic fluorescence anisotropy of fluorescent derivatives of hemoglobin and its sub-units. The covalent conjugates studied contained 1,5-AEDANS groups linked to the β-93 or the β-112 sulfhydryl. In studies on the isolated apo-β-chains, the intrinsic fluorescence of the β-15 and β-37 tryptophans was also utilized.[25] In addition, results were also reported for the 1,8-ANS complex with the apo-$\beta$-chain.[25]

For conjugates of tetrameric CO-hemoglobin with the AEDANS group at the $\beta$-93 position, the fluorescent label is in close proximity to the heme group, so that extensive quenching of its fluorescence occurs via radiationless energy transfer to the heme. The time decay of fluorescence intensity was biexponential, with a major (99%) component of decay time $0.8 \pm 0.2$ ns and a minor (1%) component of decay time 13 ns.[25, 71] Since efforts to eliminate the minor component by further chromatographic purification proved unsuccessful for both tetrameric hemoglobin and the isolated $\beta$-chains, it was felt that it did not reflect a contaminant, but rather arose from an improbable orientation of the AEDANS label, which was such as not to favor radiationless energy transfer.

Static and dynamic anisotropy measurements upon the $\beta$-93 AEDANS conjugate of CO-hemoglobin at pH 7.5 and 23°, indicated a correlation time close to 4 ns.[71] The short fluorescence lifetime confined dynamic anisotropy measurements to times shorter than about 7 ns after decay of the excitation pulse, thereby precluding detection of any longer correlation time corresponding to the rotation of the entire molecule.

The observed correlation time, which probably corresponds to a poorly defined average, is much too small to reflect the rotation of either tetrameric hemoglobin or its $\beta$-subunit and much larger than the value ($\sim 80$ ps) expected for a freely rotating AEDANS label. It could, in principle, reflect either the hindered rotation of the label or the librational motion of a portion of the adjacent polypeptide.

In a subsequent study, Sassaroli, et al.[72] performed anisotropy decay measurements at a series of viscosities. Viscosity was varied by the addition of sucrose. The apparent correlation time was viscosity dependent. Linear extrapolation, employing equation (43), led to a value of $\sigma$ equal to $4.3 \pm 0.5$ and a value of $w$ equal to $0.028 \pm 0.017$ for CO-hemoglobin. The low value of $\sigma$ and the near-zero value of $w$ are counter to hindered rotation as the dominant factor and suggest that librational motion involving some portion of the polypeptide occurs.

The time profile of anisotropy and the values of $\sigma$ and $w$ obtained by extrapolation versus $T/\eta$, according to equation (43), were also found to be dependent upon the state of ligation of the heme groups. For deoxyhemoglobin the value of $\sigma$ increased to $9.1 \pm 0.5$ ns and that of $w$ to $0.054 \pm 0.004$. The simultaneous increase of $\sigma$ and $w$ is consistent with an enhanced mobility of the probe itself upon deoxygenation, accompanied by a general increase in rigidity of the hemoglobin subunit.

The question naturally arises as to what extent the probe itself has modified the properties of the protein. While detailed structural information is not available for AEDANS conjugates with the $\beta$-93 sulfhydryl of hemoglobin, it is possible to draw some inferences from the crystallographic studies of Moffat[73] on the analogous derivative formed with the spin label $N$-(oxyl-2,2,5,5-tetramethylpiperidinyl)iodoacetamide. These studies, together with the electron spin resonance investigations of McConnell and co-workers[74, 75] led to the conclusion that, in CO-hemoglobin, the spin label may exist in either of two orientations. One orientation is weakly immobilized and external to the tertiary structure; the other, strongly immobilized configuration is integrated into the tertiary structure and displaces the C-terminus of the $\beta$-chain, thereby causing significant perturbations in both $\beta$- and $\alpha$-chains. In deoxyhemoglobin only the weakly immobilized orientation is present.[73]

To the extent that the findings of Moffat are relevant to the case of the AEDANS label, it is of interest that they are basically compatible with the results summarized above.[72] In particular, they offer an explanation for the lower immobilization of the label for deoxyhemoglobin in comparison to CO-hemoglobin, as reflected by the larger value of $w$ in the former case.

The mobility of the AEDANS probe was also found to be strongly influenced by the allosteric effector inosine hexophosphate (IHP). In the presence of saturating levels of IHP, the values of both $w$ and $\sigma$ show a major increase, $w$ increasing to $0.31 \pm 0.01$ and $\sigma$ to $44 \pm 16$ ns. The latter figure approaches the range of correlation time expected for a rigid hemoglobin tetramer. The enhanced value of $w$ indicates that this is associated with a decreased immobilization of the probe itself. IHP thus appears to accentuate the effects of deoxygenation. This is consistent with its allosteric function. A qualitatively similar influence of IHP upon CO-hemoglobin was also observed, but the effect was much smaller, as would be expected from the low binding affinity of CO-hemoglobin for IHP.[72]

From these results it appears that a fluorescent label at the $\beta$-93 position is responsive to the conformational changes arising from a change in the state of ligation of hemoglobin. In view of the magnitudes of the viscosity-sensitive correlation times, it is likely that they reflect librational motions which extend beyond the probe itself and involve some portion of the polypeptide. It would be unrealistic to assign a size to the effective mobile unit on the basis of the magnitude of the correlation time, in view of the uncertain and complex nature of the librational motion. For a domain rotating freely as a unit, the observed correlation time for CO-hemoglobin would formally correspond to a molecular weight of the order of $6-10 \times 10^3$. However, such a model is unlikely.

Oton et al.[25] have extended these studies to AEDANS conjugates with the $\beta$-93 and $\beta$-112 sulfhydryls of the isolated apo-$\beta$-chain. If measurements were made at sufficiently low concentrations of the $\beta$-93 conjugate so that self-association of the apo-$\beta$-chain was avoided, the time decay of anisotropy could be fitted by equation (66) with the following values of the parameters (for pH 5.4): $\alpha_1, \alpha_2 = 0.06, 0.06$ and $\sigma_1, \sigma_2 = 1.3, 7.3$ ns. In this case, the average fluorescence lifetime was sufficiently long to permit detection of a longer correlation time $(\sigma_2)$, which presumably corresponds to rotation of the entire molecule. The shorter correlation time $(\sigma_1)$ is much smaller than that observed for the hemoglobin tetramer and presumably reflects an enhanced mobility of the probe upon removal of the heme.

In the case of an AEDANS conjugate with the $\beta$-112 sulfhydryl of the apo-$\beta$-chain, only a single rotational mode, of correlation time 3.0 ns, was detected. In this case, the amplitude of the rapid rotational mode appears to be sufficiently large to mask the effects of the rotation of the entire molecule.

In contrast to the behavior of the probes covalently linked to the sulfhydryl groups, the time profile of anisotropy for a complex of 1-anilinonaphthalene-8-sulfonate (1,8-ANS) with the apo-$\beta$-chain showed only a single rotational mode of correlation time 10.4 ns. The 1,8-ANS molecule probably lies within the heme pocket and is effectively immobilized by virtue of its multiple points of contact with the protein.[76] Clearly, it does not sense the localized rotational modes detected by the sulfhydryl specific labels.

A similar result was obtained if the fluorescence of the $\beta$-15 and $\beta$-37 tryptophans was monitored employing static anisotropy measurements. In 40 mM phosphate, pH 5.7, a correlation time of 9.7 ns was computed from a linear Perrin plot obtained by viscosity variation. Under these conditions there was no indication of curvature, which would suggest the presence of internal rotational modes. However, at the lower salt level of 4 mM, there was some indication of significant downward curvature at very low values of $\tau(T/\eta)$, which would be consistent with significant mobility of one or both tryptophans under these conditions.

The mobility of a fluorescent probe which is specifically located within the $\beta$-chain of hemoglobin is thus seen from the preceding results to be very dependent upon the position of the probe. In the case of hemoglobin itself, the mobility of a probe at the $\beta$-93 position is sensitive to the state of ligation. The existence of internal librational modes in hemoglobin may prove to have a significant bearing upon its allosteric properties.

## References

1   G. Weber, *Biochem. J.* **51**, 145, 165 (1952).
2   P. Wahl, in *Biochemical Fluorescence: Concepts* (R. F. Chen and H. Edelhoch, eds.), Marcel Dekker, New York (1975), p. 1.
3.  F. Perrin, *J. Phys.* **7**, 390 (1926).
4.  F. Perrin, *Ann. Phys. (Paris)* **12**, 169 (1929).
5.  F. Perrin, *J. Phys. (Paris)* **5**, 497 (1934).
6.  F. Perrin, *J. Phys. (Paris)* **7**, 1 (1936).
7.  I. Isenberg, R. D. Dyson, and R. Hanson, *Biophys. J.* **13**, 1090 (1973).
8.  A. Grinwald and I. Z. Steinberg, *Anal. Biochem.* **59**, 583 (1974).
9.  A. Grinwald, *Anal. Biochem.* **75**, 260 (1976).

10. A. Gafni, R. L. Modlin, and L. Brand, *Biophys. J.* **15**, 263 (1975).
11. A. Jablonski, *Z. Phys.* **106**, 526 (1936).
12. A. Jablonski, *Acta. Phys. Pol.* **10**, 193 (1950).
13. J. Y. Yguerabide, in *Methods in Enzymology,* Vol. 26, Part C (C. Hirs and S. Timasheff, eds.), Academic Press, New York (1972), p. 98.
14. R. Memming, *Z. Phys. Chem.* **28**, 168 (1961).
15. T. Tao, *Biopolymers* **8**, 609 (1909).
16. G. G. Belford, R. L. Belford, and G. Weber, *Proc. Natl. Acad. Sci. USA* **69**, 1932 (1972).
17. P. Wahl, *C. R. Acad. Sci.* **263**, 1525 (1966).
18. P. Wahl, *Biochim. Biophys. Acta.* **175**, 55 (1969).
19. M. Ehrenberg and R. Rigler, *Chem. Phys. Lett.* **14**, 539 (1972).
20. R. Rigler and M. Ehrenberg, *Q. Rev. Biophys.* **9**, 19 (1976).
21. S. C. Harvey and H. C. Cheung, *Proc. Natl. Acad. Sci. USA* **69**, 3670 (1972).
22. Y. Gottlieb and P. Wahl, *J. Chim. Phys.* **60**, 849 (1963).
23. K. Kinosita, S. Kawato, and A. Ikegami, *Biophys. J.* **20**, 289 (1977).
24. R. D. Dale and J. Eisinger, *Biopolymers* **13**, 1573 (1974).
25. J. Oton, E. Bucci, R. F. Steiner, C. Fronticelli, D. Franchi, J. Montemarano, and A. Martinez, *J. Biol. Chem.* **256**, 7248 (1981).
26. I. Munro, I. Pecht, and L. Stryer, *Proc. Natl. Acad. Sci. USA* **76**, 56 (1979).
27. W. R. Ware, in *Creation and Detection of the Excited State,* Vol. 1, Part A, Marcel Dekker, New York (1971).
28. I. Isenberg, in *Biochemical Fluorescence: Concepts* (R. F. Chen and H. Edelhoch, eds.), Vol. 1, Marcel Dekker, New York (1975), p. 43.
29. R. Schuyler and I. Isenberg, *Rev. Sci. Instrum.* **42**, 813 (1971).
30. J. H. Easter, R. P. DeToma, and L. Brand, *Biophys. J.* **16**, 571 (1976).
31. M. G. Badea and L. Brand, *Methods Enzymol.* **61**, 378 (1979).
32. B. Valeur and G. Weber, *J. Chem. Phys.* **69**, 2393 (1978).
33. G. Weber and M. Shinitzky, *Proc. Natl. Acad. Sci. USA* **65**, 823 (1970).
34. W. C. Galley and R. M. Purkey, *Proc. Natl. Acad. Sci. USA* **67**, 1116 (1970).
35. D. A. Cowburn, E. M. Bradbury, C. Crane-Robinson, and W. B. Gratzer, *Eur. J. Biochem.* **14**, 83 (1970).
36. W. J. Browne, A. C. T. North, and D. C. Phillips, *J. Mol. Biol.* **42**, 65 (1969).
37. C. C. F. Blake, D. F. Koenig, G. A. Mair, A. C. T. North, D. C. Phillips, and V. R. Sarma, *Nature (London)* **206**, 757 (1965).
38. A. L. Barel, J. P. Prisels, E. Maes, Y. Looze, and J. Leonis, *Biochim. Biophys. Acta* **257**, 288 (1972).
39. J. C. Lee and S. N. Timasheff, *Biochemistry* **13**, 257 (1974).
40. E. K. Achter and I. D. A. Swan, *Biochemistry* **10**, 2976 (1971).
41. L. H. Tang, Y. Kubota, and R. F. Steiner, *Biophys. Chem.* **4**, 203 (1976).
42. R. A. Kenner and H. Neurath, *Biochemistry* **10**, 551 (1971).
43. S. Udenfriend, S. Stein, P. Bohlen, W. Dairman, W. Leimgruber, and M. Weigele, *Science* **178**, 871 (1972).
44. D. C. Phillips, *Proc. Natl. Acad. Sci. USA* **57**, 484 (1967).
45. H. B. Bull and K. Breese, *Arch. Biochem. Biophys.* **128**, 488 (1968).
46. J. A. Weltman and G. M. Edelman, *Biochemistry* **6**, 1437 (1967).
47. J. C. Brochon and P. Wahl, *Eur. J. Biochem.* **25**, 20 (1972).
48. P. Wahl, *Biochim. Biophys. Acta* **175**, 55 (1969).
49. J. Yguerabide, H. F. Epstein, and L. Stryer, *J. Mol. Biol.* **51**, 573 (1970).
50. C. Lovejoy, D. A. Holowka, and R. E. Cathou, *Biochemistry* **16**, 3668 (1977).
51. D. A. Holowka and R. E. Cathou, *Biochemistry* **15**, 3373, 3379 (1976).

52. R. C. Siegel and R. E. Cathou, *Biochemistry* **20**, 192 (1981).
53. R. A. Mendelson, M. F. Morales, and J. Botts, *Biochemistry* **12**, 2250 (1973).
54. I. Miller and R. T. Tregear, *J. Mol. Biol.* **70**, 85 (1972).
55. R. A. Mendelson and P. H. C. Cheung, *Biochemistry* **17**, 2140 (1978).
56. M. Miki, P. Wahl, and J. C. Auchet, *Biochemistry* **21**, 3662 (1982).
57. S. Fujime and S. Ishiwata, *J. Mol. Biol.* **62**, 254 (1971).
58. D. D. Thomas, J. C. Seidel, and J. Gergely, *J. Mol. Biol.* **132**, 257 (1979).
59. C. Hall and H. Slayter, *J. Biochem. Biophys. Cytol.* **5**, 11 (1959).
60. R. F. Doolittle, *Adv. Protein Chem.* **27**, 1 (1973).
61. W. E. Fowler, R. R. Hantgan, J. Hermans, and H. P. Erickson, *Proc. Natl. Acad. Sci. USA* **78**, 4872 (1981).
62. P. Johnson and E. Mihalyi, *Biochim. Biophys. Acta* **102**, 476 (1965).
63. R. R. Hantgan, *Biochemistry* **21**, 1822 (1982).
64. R. F. Doolittle, *Horiz. Biochem. Biophys.* **3**, 164 (1977).
65. W. R. Krigbaum and R. S. Hsu, *Biochemistry* **14**, 2542 (1975).
66. J. B. A. Ross, K. W. Rousslang, and L. Brand, *Biochemistry* **20**, 4361 (1981).
67. C. I. Branden, H. Jornvall, H. Eklund, and B. Furugren, *Enzymes,* 3rd edn., Vol. II, Academic Press, New York (1975), p. 103.
68. J. B. A. Ross, C. J. Schmidt, and L. Brand, *Biochemistry* **20**, 4369 (1981).
69. J. R. Lakowicz and G. Weber, Second Biophysical Discussion, Biophysical Society (1980), p. 465.
70. J. R. Lakowicz and G. Weber, *Biochemistry* **12**, 4161, 4171 (1973).
71. E. Bucci, C. Fronticelli, K. Flanigan, J. Perlman, and R. F. Steiner, *Biopolymers* **18**, 1261 (1979).
72. M. Sassaroli, E. Bucci, and R. F. Steiner, *J. Biol. Chem.* **257**, 10136 (1982).
73. J. K. Moffat, *J. Mol. Biol.* **35**, 135 (1971).
74. H. M. McConnell and B. C. McFarland, *Q. Rev. Biophys* **3**, 91 (1970).
75. H. M. McConnell and C. Hamilton, *Proc. Natl. Acad. Sci. USA* **60**, 776 (1968).
76. L. Stryer, *J. Mol. Biol.* **13**, 483 (1965).

# 5

# Plasma Lipoproteins: Fluorescence as a Probe of Structure and Dynamics

## WILLIAM W. MANTULIN AND HENRY J. POWNALL

## 1. Introduction

### 1.1. Native Lipoproteins

The plasma lipoproteins are water-soluble complexes of lipids and proteins that are the primary vehicles for the transport of lipids in blood. Many of their physical and physiologic properties have been reviewed previously.[1-4] This chapter will review human plasma lipoproteins, but animal models will occasionally be incorporated into the discussion. All lipoproteins contain, in varying amounts, free and esterified cholesterol, phospholipids (mainly phosphatidylcholine and sphingomyelin), various acyl glycerides (mainly triglycerides), and proteins (designated as apoproteins). The lipoproteins are operationally defined according to their density as the high-, low-, and very-low-density lipoproteins, HDL, LDL, and VLDL, respectively. Each contains several specialized proteins which are powerful surface-active agents. A given apoprotein may occur in more than one lipoprotein; the compositions of the human plasma lipoproteins are given in Table 1 and some of their physical properties are enumerated in Table 2.

The primary structures of some of the more abundant apoproteins are known and are given elsewhere.[2-4] These include apoA-I, apoA-II, apoC-I, apoC-II, and

WILLIAM W. MANTULIN AND HENRY J. POWNALL • Department of Medicine, Baylor College of Medicine, and The Methodist Hospital, Houston, Texas 77030.

Table 1. Chemical Composition of the Human Serum Lipoproteins[a]

| Particle | Percent protein | Percent phospholipid | Percent triglyceride | Percent cholesterol ester | Percent free cholesterol |
|---|---|---|---|---|---|
| High-density lipoprotein₃ | 55 | 23 | 4.1 | 12 | 2.9 |
| High-density lipoprotein₂ | 41 | 30 | 4.5 | 16 | 5.4 |
| Low-density lipoprotein | 21 | 22 | 11 | 37 | 8.0 |
| Very-low-density lipoprotein | 8 | 18 | 50 | 12 | 7.0 |

[a] From Refs. 1 and 3.

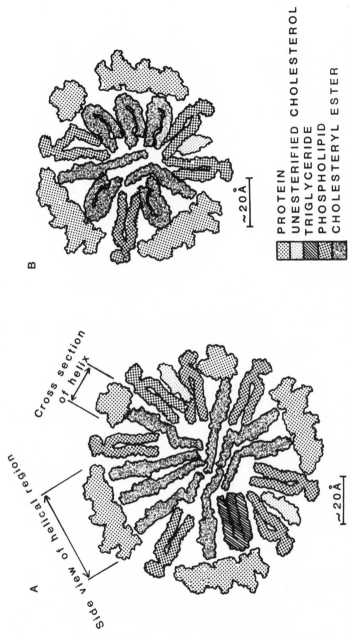

FIGURE 1. Schematic representation of the structure of HDL$_2$ and HDL$_3$. (A) HDL$_2$; (B) HDL$_3$. Models are drawn to scale with the proteins in an α-helical conformation.

*Table 2. Physical Properties of Human Serum Lipoproteins[a]*

| Particle | Particle weight $\times 10^{-6}$ | Hydrated density (g/ml) | Radius (nm) |
|---|---|---|---|
| High-density lipoproteins$_3$ | 0.18 | 1.15 | 3.9 |
| High-density lipoproteins$_2$ | 0.36 | 1.09 | 5.1 |
| Low-density lipoproteins | 2.3 | 1.035 | 9.6 |
| Very-low-density lipoproteins | 19.6 | 0.97 | 20.0 |

[a] From Refs. 1 and 3.

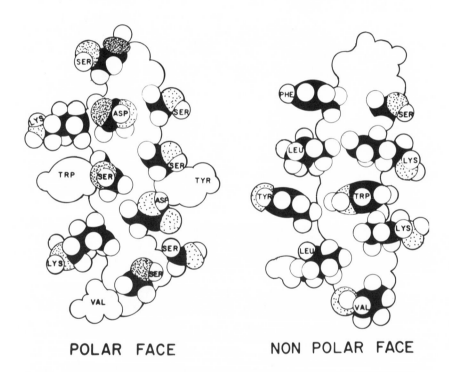

POLAR FACE        NON POLAR FACE

FIGURE 2. Space-filling model of an amphipathic helix according to Segrest *et al.* (Ref. 5) One side of the helix (left) contains the polar residues. The opposite face is nonpolar and, in this particular case, a tyrosine and a tryptophan residue appear in the middle of this segment. Presumably, the blue shift in the intrinsic tryptophan fluorescence that occurs when phospholipid is added to the apolipoproteins is due to the transfer of the hydrophobic face from water to the relatively nonpolar lipid interior. The polar face, presumably, remains coplanar with the polar headgroups of the phospholipids.

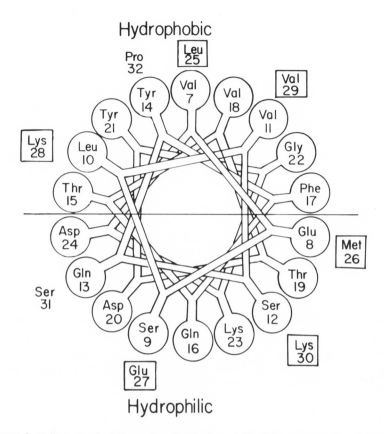

FIGURE 3. Helical wheel of the region 7–32 of apoA-II. This representation shows the segregation of nonpolar and polar amino acid residues that occurs when the protein is α-helical.

apoC-III. ApoB, though easy to isolate in large quantities, has not been sequenced. In spite of significant differences in the composition, size, and physiologic origin of the lipoproteins, all of them can be viewed as having the same general structure as illustrated in Figure 1 for two HDL subfractions $HDL_2$ and $HDL_3$. In this model all of the water-insoluble components, mostly triglyceride and cholesteryl esters, are located in the center of the particle. The polar or amphiphilic components, largely phospholipids, cholesterol, and apoproteins, surround the nonpolar core, thereby providing a barrier between water and the hydrophobic region. The surface components have directionality; the hydroxyl group of cholesterol and the zwitterionic headgroup of the phospholipid form part of the water–lipoprotein interface. The apoproteins on the surface also have directionality, except in a more complex way. In a hypothesis originally proposed by Segrest *et al.*,[5] it was suggested that the apoproteins form an amphipathic helix in lipoproteins. As shown schematically in Figure 2, one face of the helix contains most of the hydrophobic amino acid residues whereas the opposite face contains the charged or hydrophilic amino acid residues. Presumably, the protein lies on the surface of the lipoprotein with the hydrophilic face in contact with the aqueous phase and the hydrophobic side partially penetrating the hydrophobic region of the phospholipids. It is the unusual distribution of the charged and nonpolar amino acid residues of apoproteins which gives rise to their singular behavior, which includes self-association and spontaneous association with phospholipids *in vitro*. The helical wheel in Figure 3 provides a dramatic illustration of the spatial distribution of charged and hydrophobic amino acid residues in one region of a very good lipid-associating apoprotein, apoA-II.

## 1.2. Lipoprotein Function

In addition to their role in lipid transport, the apoproteins are also key components in the regulation of several events related to lipid metabolism. ApoA-I is an activator for human plasma lecithin : cholesterol acyltransferase (LCAT), the enzyme which catalyzes the formation of nearly all plasma cholesteryl esters from lecithin and free cholesterol.[1] ApoC-II is a potent activator of the enzyme lipoprotein lipase (LPL) which catalyzes triglyceride hydrolysis in the fat-rich lipoproteins.[1] ApoB and apoE are important components of the receptor-mediated uptake of plasma lipoproteins by cells. Additional roles for the other apoproteins are assumed to exist but remain to be identified.

## 1.3. Isolation of Lipoproteins and Their Apoproteins

VLDL, LDL, and HDL can be isolated by ultracentrifugation at salt densities of 1.006, 1.063, and 1.21 g/m,[6] respectively, or by a sequential polyanion precipitation.[7] Another procedure combines a single flotation in high salt with molecular sieve chromatography.[8] All three techniques give similar products which are suitable for physical studies. HDL, delipidated by extraction with organic solvents, can be fractionated into its component proteins by gel filtration over Sephadex G-150 in 6 M urea or 3 M guanidinium chloride (Gdm · Cl),[9,10] which gives apoA-I and apoA-II purified to homogeneity and a mixture of apoC. The apoC proteins are not found in large amounts in HDL but rather are isolated from the VLDL of certain subjects with elevated triglycerides. Chromatography of delipidated VLDL in 3 M Gdm · Cl separates apoB from the apoC proteins; the latter, whether from HDL or VLDL, are subfractionated into apoC-I, apoC-II, and apoC-III by ion-exchange chromatography over DEAE

Sephadex in 6 M urea.[11] Delipidated apoB can be obtained in large quantities from LDL by solubilization in sodium dodecyl sulfate (SDS) followed by gel-filtration chromatography in 1.0—2.5 mM SDS.[12]

## 1.4. Reassembled Lipoproteins

Much of our understanding of the structure of apoproteins in lipoproteins has been derived from our ability to reassemble a lipoprotein containing only a few of the naturally occurring components.[13] In practice, this has begun with the assembly of a single pure apoprotein with a single lecithin in which there is no fatty acid heterogeneity. Fluorescence studies, in particular, are best performed with a well-defined lipid—protein particle, thereby eliminating, in part, the complications of multiple emitting sites. For reference purposes, Table 3 gives the number of moles of the aromatic amino acid residues in each of the apoproteins for which this information is available.

Most of the reassembly of lipoproteins to date has been conducted with a single apoprotein and dimyristoylphosphatidylcholine (DMPC), and fluorescence techniques have played an important role in the characterization of the resulting lipid—protein complex. DMPC has been used because this lipid has a gel → liquid crystalline transition, $T_c$, at an experimentally accessible temperature, 24°C. This is important because it has been reported that some apoproteins spontaneously associate with lecithins only in the vicinity of $T_c$.[14-15] This effect is illustrated in Figure 4; the temperature dependence of the fluorescence maximum of apoC-III in the presence of multishelled liposomes of DMPC was recorded at lipid-to-protein ratios extending from 0 to 100. As the data indicate, the free protein (Figure 4e) has a fluorescence maximum of about 350 nm that is characteristic of emission from one or more of its three tryptophans in an

Table 3. Distribution of Aromatic Amino Acids among Human Serum Apolipoproteins

|  | Tryptophan | Tyrosine | Phenylalanine | Molecular weight | Number of amino acid residues |
|---|---|---|---|---|---|
| ApoA-I | 4 | 7 | 6 | 28,331 | 245 |
| ApoA-II[a] | 0 | 8 | 8 | 2 × 8690 | 2 × 77 |
| ApoC-I | 1 | 0 | 3 | 6625 | 57 |
| ApoC-II | 1 | 4 | 2 | 9110 | 79 |
| ApoC-III | 3 | 2 | 4 | 8764 | 79 |

[a] Values given are for the disulfide-linked dimer.

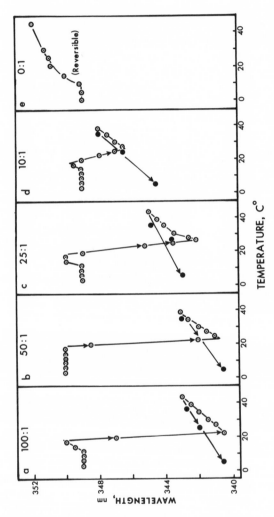

FIGURE 4. Temperature dependence of the intrinsic tryptophan fluorescence of apoC-III. ApoC-III and DMPC were mixed at 5°C in the indicated lipid/protein molar ratios and slowly heated. Spectra were recorded every few degrees. The dramatic change at ~ 24°C is due to the penetration of the protein into the lipid. Independent chromatographic experiments support the view that below 24°C the rate of lipid/protein associations is very slow.

aqueous environment. If various molar ratios of DMPC are mixed with apoC-III, there is no significant shift in the spectral maximum. However, as the sample temperature is increased, a clearly defined blue shift in the fluorescence spectrum occurs at 24°C; additional heating and cooling cycles revealed very little change. Other data have shown that these fluorescence changes are due to the formation of a DMPC/apoC-III complex.

Based upon the fluorescence behavior of tryptophan in water and various organic solvents, the shift has been assigned to the transfer of tryptophan in an aqueous environment to one that is relatively more hydrophobic. However, this environment is probably not as hydrophobic as a pure hydrocarbon, since the fluorescence maximum of tryptophan in hexane is 310 nm. One reason for this difference may be the penetration of water into the lipid bilayer or, perhaps, to the location of the tryptophan in the phospholipid matrix close to the lipid–water interface.

The maximal blue shift in the fluorescence of apoC-III measured as a function of temperature was observed between DMPC–protein molar ratios of 50 and 100, suggesting that the saturation stoichiometry is in this range. This view is supported by the fluorescence titration of apoC-III with DMPC shown in Figure 5. Concomitant with the blue shift, the increase in the fluorescence quantum yield appears to be similar in magnitude to the shift in the wavelength of maximal fluorescence.

A large literature on lipoprotein reassembly has accumulated over the past

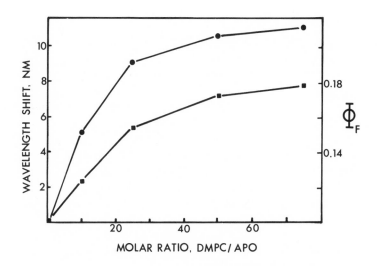

FIGURE 5. Fluorescence titration of apoC-III with DMPC.

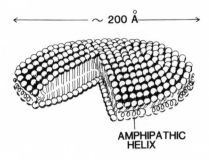

FIGURE 6. Bilayer disc model of a reassembled lipoprotein.

10 years. Although more than one model of an apoprotein–phospholipid complex have been proposed in recent years, evidence has accumulated in favor of a bilayer disc surrounded by an α-helical protein (Figure 6). Most of the fluorescence studies can be understood within the restrictions of this model.

## 2. Structural Studies of Native Lipoproteins, Apoproteins, and Reassembled Lipoproteins Using Intrinsic Protein Fluorescence and Covalently Attached Fluorescence Probes

### 2.1. Native Lipoproteins

Structural studies of native lipoproteins, utilizing intrinsic protein fluorescence, have been hampered by the heterogenous distribution of apoproteins (Tables 1 and 3), and consequently the literature provides greater detail on compositionally simpler systems such as reassembled lipoproteins and apoproteins in solution. Nonetheless the early fluorescence literature on lipoproteins established, on the basis of emission wavelength maxima, that the fluorescence originated from tryptophan residues in HDL[16–18] and LDL, and that, in part, these tryptophan residues were shielded from the aqueous environment by lipid (Table 4).[2,4,6,16–22] After delipidation the apoHDL[16] or apoLDL[20] fluorescence was red-shifted and the quantum yield reduced by the exposure of tryptophan residues to the aqueous environment. In a pH titration experiment it was found that the fluorescence quantum yields of both human[20] and porcine[19] native LDL and apoLDL were essentially invariant between pH 2–10, indicating a lack of any major conformational change in the protein. Above pH 10 fluorescence quenching occurred, presumably due to ionization of

*Table 4. Fluorescence Wavelength Maximum and Quantum Yield for HDL and LDL[a]*

|  | $\lambda_{max}$ (nm) | $\Phi$ | Reference |
|---|---|---|---|
| HDL$_2$ | 333 |  | 18 |
| apoHDL$_2$ | 335 |  | 18 |
| PHDL$_3$ (porcine) | 338 | 0.14 | 17 |
| BHDL (bovine) | 328 | $x^b$ | 16 |
| apoBHDL | 338 | $0.15x^b$ | 16 |
| PLDL$_2$ (porcine) | 336 | 0.09 | 17 |
| PLDL$_1$ (porcine) | 332 |  | 19 |
| PLDL$_2$ | 332 |  | 19 |
| LDL | 330 | $y^c$ | 20 |
| apoLDL | 334 | $0.3y^c$ | 20 |
| LDL | 328 |  | 22 |

[a] Corrected spectra recorded at 25°C.
[b] The quantum yield of apoBHDL is 15% less than that of BHDL.
[c] The quantum yield of apoLDL is 30% less than that of LDL.

carboxyl and tyrosyl residues. In contrast, lipid provides the proteins of HDL protection from denaturation by urea.[18] In an attempt to localize the aromatic residues of the protein in porcine PHDL$_3$ and PLDL$_2$, Badley[17] performed fluorescence quenching experiments with iodide and succinimide. With both quenchers the tryptophan residues of PLDL$_2$ were significantly more shielded than those of PHDL$_3$, and both lipoproteins show curved Stern–Volmer plots. These results support the structural model of protein on the surface of the lipoprotein, since the quenchers are only soluble in the aqueous phase (Fig. 1). Nonetheless, access to the tryptophan residues is hindered by lipid interaction and/or protein secondary and tertiary structure. Molotkovsky *et al.*[23] prepared anthryl derivatives of phospholipid and sphingomyelin to study the differential behavior of these lipids in HDL. Energy transfer experiments between anthryl lipid fluorophors and protein tryptophan show a high efficiency for the sphingomyelin. The kinetics of sphingomyelin incorporation into HDL are biphasic, leading the authors to suggest loosely and strongly bound pools of lipid. From these early, limited, and necessarily complex studies of intrinsic lipoprotein fluorescence, the emphasis has shifted toward understanding the forces governing the structure and stability of simpler related systems; these include apoprotein interactions in solution and apoprotein–lipid reassembly.

## 2.2. Apoproteins

The conformation and behavior of the apoA and apoC proteins are known and have been studied by various physical techniques such as analytical ultracentrifugation and circular dichroic spectroscopy (for a review, see Ref. 24) as well as fluorescence methods. In the lipid-free state the apoproteins are water soluble and exhibit a tendency to self-associate primarily through hydrophobic interactions. Removal of lipid greatly reduces secondary structure in the apoproteins. Excepting apoA-II, which is devoid of tryptophan but contains eight tyrosine residues, the intrinsic fluorescence of the apoA and apoC proteins arises from tryptophan. A number of conformational studies of apoA and apoC have used the intrinsic fluorescence or the fluorescence from a covalently bound probe to delineate apoprotein self-association, unfolding, and hydrodynamic characteristics.

ApoA-I, the most abundant protein of HDL, unfolds in the presence of acid, base, or Gdm · Cl.[25] The unfolding can be monitored by the red shift in tryptophan fluorescence wavelength maximum (333 to 348 nm), loss of fluorescence intensity, or fluorescence depolarization ($p = 0.13-0.05$). Interestingly, apoA-I is less stable to denaturation than typical globular proteins, since it is completely unfolded in 1.5 M Gdm · Cl. By covalently attaching dimethylaminonaphthalene sulfonyl groups (Dns) to bovine apoA-I (apoA-I-Dns) Jonas[26] demonstrated, from the fluorescence depolarization upon dilution, that the self-associated apoA-I (tetramic) began dissociating below $5 \times 10^{-7}$ M. In spite of high sequence homology, a similar experiment with rhesus monkey and human apoA-I did not reveal a concentration-dependent change in fluorescence polarization.[27] The hydrodynamic properties of apoA-I were analyzed from fluorescence polarization data according to the Perrin equation as modified by Weber[28]

$$\frac{1}{p} - \frac{1}{3} = \left(\frac{1}{p_0} - \frac{1}{3}\right)\left(1 + \frac{3\tau}{\rho_\text{h}}\right) \tag{1}$$

where $p$ is the observed polarization, $p_0$ is the limiting polarization in the absence of Brownian rotation, $\tau$ is the fluorescence lifetime, and $\rho_\text{h}$ is the harmonic mean of the rotational relaxation time. For apoA-I-Dns $\tau$ was independent of concentration and therefore $p$ was a function of the rotational relaxation time, which reflects changes in the hydrodynamic volume of the protein. Changes in polarization, induced by temperature variation and isothermally by sucrose addition, gave identical results. The coincidence of these results establishes that the measured $p$ corresponds to a rigidly bound probe measuring total protein rotation. The apparent $\rho_\text{h}$ and fluorescence lifetimes for apoA-I-Dns are presented in Table 5. Fluorescence polarization data, supplemented by sedimentation equilibrium experiments,[26,27] indicate that apoA-I is a moderately asymmetrical particle (prolate ellipsoid, Table 5) and that the mode of self-association involves end-to-end dimerization. Since the apoA-I-Dns

*Table 5. Fluorescence Lifetime, Rotational Relaxation Times, and Axial Ratio of ApoA-I-Dns*

| ApoA-I-Dns | $\tau$ (ns) | $\rho_h$ (ns) | $a/b$ | Reference |
|---|---|---|---|---|
| Bovine (oligomer) | 12.3 | 115 | 3.1 | 25 |
| Rhesus monkey (monomer) | 9.9 | 65 | 5.8 | 26 |
| Human | | | | |
| Monomer | 10.6 | 67 | 6.5 | 26 |
| Oligomer | 10.7 | 142 | 7.4 | 26 |

showed decreased self-association as compared to unmodified apoA-I chemical modification of functional groups must occur at sites retarding oligomerization.

ApoA-II, the second most abundant protein of human HDL, is a disulfide-linked dimer of 77 amino acid residues. ApoA-II self-associates in solution, and these interactions can be disrupted by temperature, alkali, and Gdm · Cl.[28] By monitoring the fluorescence intensity and anisotropy of the tyrosine residues as a function of temperature, a maximum in thermal stability was observed at 25°C. These results were corroborated by difference absorption spectroscopy and circular dichroism.[29] Assuming a dimeric form of self-association, the dissociation at low temperatures (<25°C) is exothermic but endothermic above 25°C. The anisotropy of apoA-II is concentration dependent and, therefore, on the basis of the fluorescence studies it was concluded that the denaturants simultaneously dissociate and unfold apoA-II. Reduced and carboxymethylated apoA-II (Cm apoA-II) exhibits fluorescence properties similar to those of native apoA-II.[30] The kinetics of the reduction reaction can be conveniently followed by the apparent decrease of fluorescence polarization. The Perrin plot has a minimum at roughly 25°C; however, the anisotropy of Cm apoA-II is lower. Cm apoA-II and apoA-II both self-associate as dimers, and much of the secondary structure of these proteins is a consequence of this process.

Since apoA-I and apoA-II are the major protein components of HDL, their mixed protein–protein interactions are important for maintaining HDL organization. Swaney and O'Brien[31] performed cross-linking studies with dimethylsuberimidate between the apoproteins and found a 1:1 mixed oligomer. Cross-linking of oligomeric apoA-I showed no effect on its intrinsic tryptophan emission spectra, suggesting that the tertiary structure of the protein remains unchanged.[32] In a fluorescence study of mixed association between apoA-I and apoA-II, Osborne et al.[33] demonstrated energy transfer from the tyrosine of apoA-II to the tryptophan residues of apoA-I, resulting in a concentration-dependent tryptophanyl emission. Energy transfer between apoA-I and modified nitrotyrosine residues in apoA-II was also demonstrated, and in agreement with the cross-linking studies, a mixed oligomer of equimolar

*Table 6. ApoC-II Fluorescence Properties at 25°C*

| [Gdm·Cl] (M) | $\langle \tau \rangle^a$ (ns) | $\tau_1$ | $\alpha$ | $\tau_2$ (ns) | $P$ (nm) |
|---|---|---|---|---|---|
| 0.0 | 4.40 | 2.53 | 0.56 | 6.77 | 300 |
| 0.4 | 4.31 | 2.09 | 0.41 | 5.78 | 0.252 |
| 1.2 | 3.49 | 1.65 | 0.50 | 5.05 | 0.173 |
| 2.0 | 2.96 | 2.05 | 0.81 | 6.10 | 0.112 |

$a$ $\langle \tau \rangle$ is an arithmetic average of the measured lifetimes.

apoA-I and apoA-II (nitrated) was confirmed by sedimentation velocity experiments.

In a spectroscopic study of the conformational properties of human plasma apoC-II, Mantulin et al.[34] showed that the reversible denaturation by Gdm · Cl proceeded in a sequential fashion. ApoC-II self-association is disrupted at low concentrations of Gdm · Cl, and then at higher Gdm · Cl concentrations (midpoint 1.1 M Gdm · Cl) the monomeric protein is unfolded. Sedimentation equilibrium studies with apoC-II showed that the apoprotein self-associates, but that 0.3 M Gdm · Cl disrupts the aggregation. This disruption of self-association was also monitored by changes in fluorescence polarization and lifetimes. Table 6 lists some of the fluorescence properties of apoC-II at 25°C as a function of Gdm · Cl concentration. The fluorescence lifetime, as measured by the phase/demodulation technique, indicated a multiexponential decay pattern, which was analyzed to yield a long ($\tau_2$) and short ($\tau_1$) lifetime component and the fraction ($\alpha$) of the short component. The multiexponential fluorescence decay of apoC-II is not uncommon, since there are numerous examples of complex decay kinetics for proteins containing a single tryptophan.[35] Figure 7 contains a plot of the fraction ($\alpha$) as a function of increasing Gdm · Cl. There is a transition at 0.3 M Gdm · Cl (indicated by an arrow) and another larger transition with its midpoint at 1.1 M Gdm · Cl which corresponds to the unfolding reaction of the monomeric protein. Fluorescence quenching experiments of apoC-II with acrylamide and iodide ion were analyzed according to the Stern–Volmer law. They indicate that disruption of self-association did not appreciably affect the accessibility of tryptophan to water-soluble quenchers. However, apoC-II unfolded by Gdm · Cl was permeable to quencher diffusion. Even though quenching by acrylamide was apparently more efficient than by iodide, normalization of the data by comparison with quenching of N-acetyl-L-tryptophanamide indicates that the quenching by neutral and charged quenching molecules is roughly equivalent for unfolded apoC-II.

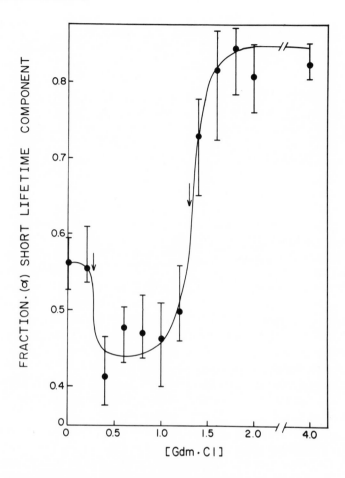

FIGURE 7. The dependence of the fraction ($\alpha$) of the short-lived lifetime component of apoC-II fluorescence on Gdm · Cl concentration.

## 2.3. Reassembled Lipoproteins

Much of the structural information about lipoproteins has come from studies of reassembled lipid–apoprotein complexes (cf. Section 1.4), and a substantial part of this experimental data has concerned changes in apoprotein fluorescence upon interaction with lipid. The use of extrinsic fluorescence probes in (reassembled) lipoproteins will be considered in a following section. ApoA-I, modified apoA-I (by acid, Gdm · Cl, succinylation), and a 25-residue fragment of apoA-I each associate with lysolecithin with an accompanying increase in fluorescence intensity and a small blue shift of the emission wavelength.[36] The binding curves for these associations maximize at a lysolecithin/apoA-I ratio of about 0.9. The calculated binding constant for apoA-I association with lysolecithin micelles at neutral pH was $1.5 \times 10^7 M^{-1}$. Unfortunately, as in many of the early studies of lipoprotein reassembly, the sometimes strong thermal dependence of the association was not considered. Verdery and Nichols[37] also observed the association of apoA-I with lysolecithin micelles by fluorescence intensity and $\lambda_{max}$ measurements. By comparing the Gdm · Cl-induced denaturation profiles of apoA-I with that of the complex, they demonstrated that the transition in the complex occurred at higher Gdm · Cl concentrations, indicating that lysolecithin stabilizes the apoprotein. A very similar stabilization of apoA-I in Gdm · Cl by complex formation with DMPC was observed by a $\lambda_{max}$ red shift.[38] ApoA-II also associates with lysolecithin as well as with hexadecyltrimethylammonium bromide,[39] and these associations can be monitored by increases in fluorescence intensity or polarization. Both lipids stabilized secondary and tertiary structures of the protein in the presence of Gdm · Cl concentrations ($>1$ M) sufficient to denature apoA-II, and in the case of lysolecithin complexes could be formed in the presence of 6 M Gdm · Cl. Association of a variety of lipids with human, bovine, and rhesus monkey apoA-I, covalently labeled with fluorescent dansyl groups (Dns), was quantitated by fluorescence polarization titration.[40,41] Human and bovine apoA-I-Dns associate with myristoyl lysophosphatidylcholine (lyso-MPC) above the critical micelle concentration of the lipid, with an accompanying decrease in fluorescence polarization. The isothermally obtained rotational relaxation time ($\rho_h$), which corresponds to the rotational motion of the whole particle [equation (1)], suggests that the complexes in excess lyso-MPC for both apoproteins are similar ($\rho_h = 60$ ns) (Table 7). The $\rho_h$ values obtained by temperature variation are sensitive to faster motions associated with bonds adjacent to the fluorescence probes. The substantial decrease in these $\rho_h$ values for both human and bovine apoA-I-Dns suggests that the changes in protein secondary structure induced by association with lyso-MPC render the Dns probes more mobile. Isothermally determined $\rho_h$ for complexes of apoA-I with DMPC and egg lecithin indicate that the latter are much larger than those formed with lyso-MPC (Table 7). As in the case of lyso-MPC association, $\rho_h$ values calculated from temperature variation are smaller (Table 7).

The apoC proteins interact with lipid more strongly than do the apoA proteins. The association of apoC-III with egg lecithin vesicles was followed by the $\lambda_{max}$ blue shift (from 350 to 345 nm) at a saturating mole ratio of 100 to 1.[42] This complex formation was coincident with a 30% increase in secondary structure ($\alpha$-helix) as assessed by circular dichroic spectroscopy. Interestingly enough, the shift effected by interaction of apoC-III with egg PC vesicles ($\lambda_{max}$ 348–336 nm) was greater than that observed with multilamellar

Table 7. Fluorescence Properties of ApoA-I-Dns and Lipid Complexes

| Sample | Amphiphile/apoA-I (mol/mol) | $\rho_h$ (ns)[a] | $\rho_h$ (ns)[b] | Reference |
|---|---|---|---|---|
| Human apoA-I | 0 | 74 | 1 | 39 |
| Human apoA-I | 0 | 66 | 54 | 40 |
| + Lyso-MPC | 8 | 76 | 4 | 39 |
| + Lyso-MPC | 23 | 62 | 14 | 39 |
| + DMPC | 38 | 193 | 35 | 39 |
| + DMPC | 633 | 218 | 19 | 39 |
| + egg lecithin | 350 | 301 | 39 | 40 |
| Bovine apoA-I (oligomer) | 0 | 124 | 88 | 39 |
| + Lyso-MPC | 30 | 78 | 39 | 39 |
| + Lyso-MPC | 227 | 60 | 6 | 39 |
| + DMPC | 127 | 254 | 28 | 39 |
| + DMPC | 633 | 314 | 15 | 39 |
| Rhesus apoA-I | 0 | 64 | 51 | 40 |
| + egg lecithin | 350 | 300 | 40 | 40 |

[a] Isothermal at 25°C.
[b] Temperature varied.

liposomes ($\lambda_{max}$ 348–343 nm).[43] Since unsonicated egg lecithin liposomes have a gel → liquid crystalline transition ($T_c$) at −15°C, the effect of lipid structure on apoC-III/lipid complex formation was investigated with lecithins of varying $T_c$.[44,45] ApoC-III mixed with dimyristoyl-(DMPC), dipalmitoyl-(DPPC), or distearoylphosphatidylcholine (DSPC) ($T_c = 23$, 41, and 58°C, respectively) below the corresponding $T_c$ revealed a sharp irreversible decrease in $\lambda_{max}$ (~ 10 nm) as the temperature was increased through $T_c$.[44–46] For DPPC and DSPC liposomes, apoC-III did not appreciably disturb the lipid structure, but DMPC was irreversibly converted to smaller complexes, as determined by light scattering and tryptophan fluorescence.[45] In a macroscopic mixture of DMPC and DPPC (1 : 1) vesicles two phase transitions were detected (22, 39°C), whereas in a microscopic mixture the observed $T_c$ was about 33°C.[46] A comparison of positively and negatively charged quenching probes, pyridinium and iodide ions, was used to assess the effects of egg lecithin association with apoA-I and apoC-III on tryptophan accessibility and net charge near the fluorophor.[47] To correct for steric factors introduced by incorporation of tryptophan into the protein, the quenching constants were normalized by dividing the Stern–Volmer constant for the protein by that of tryptophan, using the same quencher. From these quenching studies, it was determined that (1) apoC-III associates with egg lecithin more efficiently than does apoA-I, (2) roughly a third of the fluorophors are accessible to quencher in the apoprotein, and (3) association of apoC-III with egg PC shields the tryptophan residues from quencher more efficiently than it does in apoA-I. In a similar quenching study with apoC-I and egg lecithin, using potassium iodide and $N$-methylnicotinamide (NMN) as quenchers, it was found that (1) the net charge surrounding the sole tryptophan in apoC-I is negative since NMN quenches very efficiently, and (2) formation of a lipid—apoprotein complex inhibits the quenching efficiency of NMN more than that of iodide, suggesting a charge redistribution.[48] Patterson and Jonas (49) have isolated seven apoC proteins from bovine plasma, and several of these fractions associate with DMPC.[50] Upon association, the apoC degrades the vesicles to smaller particles, which can be isolated by gel permeation. The association results in a blue shift of the apoC tryptophan fluorescence, which is generally saturable at a lipid/protein molar ratio of 100 : 1.[49]

# 3. Extrinsic Fluorescence Probes of Lipoprotein Structure and Function

## 3.1. Native Lipoproteins

The general model of plasma lipoprotein structure contains an apolar core, mainly consisting of triglyceride and cholesteryl esters, surrounded by polar components, mostly phospholipids, cholesterol, and apoproteins (Figure 1). The quantitative interpretation of fluorescence data from extrinsic probes introduced into lipoproteins has suffered from the indefinite localization of the probe. Recent advances in resonance energy transfer theory and fluorescence time resolution experiments have improved our ability to localize the probes. Of necessity, the discussion of many of the early papers, dealing with extrinsic fluorescence probes, will be of a qualitative nature. The binding of 1-anilino-8-naphthalene sulfonate (ANS) to the surface of HDL,[19,51] LDL,[52] and VLDL[32] revealed a saturable increase in fluorescence intensity. High concentrations of ANS (1 mM) altered the structural integrity of the lipoprotein.[53] ANS binding to apoHDL is weak and involves three hydrophobic binding sites per molecule, as compared with ~ 300 per HDL.[19] Analysis of the ANS binding data to lipoproteins yielded linear Scatchard plots, indicative of one type of binding site.[51,52] Comparison with ANS binding to phospholipid dispersions suggests that the binding sites for ANS on the lipoprotein contain phospholipids[19,51,52]; the pH and ionic strength dependence of the ANS binding requires involvement of lipoprotein surface sites.[51,52]

The lipophilic fluorescence probe 1,6-diphenyl-1,3,5-hexatriene (DPH) has been used extensively to assess the fluidity of the hydrophobic core region of lipoproteins. Apparent microviscosity values were determined from the modified

Perrin equation [equation (1)]

$$r_0/r = 1 + C(r)(T\tau/\eta) \qquad (2)$$

where $r_0$ is the limiting anisotropy of DPH, $\tau$ is the fluorescence lifetime, $T$ and $\eta$ are the temperature and microviscosity, respectively, and $C(r)$ is a parameter related to the rotational volume and shape of DPH.[54] The flow activation energy $\Delta E$ can be calculated from the thermal dependence of the microviscosity

$$\eta = A \exp(\Delta E/RT) \qquad (3)$$

where $R$ is the gas constant. Table 8 lists $\eta$ and $\Delta E$ for several lipoprotein systems as determined by equations (2) and (3) from measurements of DPH depolarization. The lipoproteins with low protein content and large volume-to-surface ratios (VLDL and chylomicrons) are more fluid (fluidity is the inverse of viscosity) than the protein-rich lipoproteins. Dispersions prepared from the lipids of HDL and LDL exhibit lower microviscosities than the intact lipoprotein, suggesting that lipid–protein interactions inhibit DPH depolarizing motions[55] (Table 8). Diet-induced changes in lipoprotein composition can also affect the apparent microviscosity of the lipoprotein. Hypercholesterolemia (hc) in rabbits, produced by cholesterol feeding, results in increased cholesteryl ester content and decreased triglyceride content in hc VLDL, as compared to normal VLDL.[56] This compositional change is reflected in a much higher microviscosity in the hc VLDL[57] (Table 8). Comparison of saturated (SAT) and polyunsaturated (PUS) fat diets on human lipoprotein composition and structure indicated that the saturated fat diet uniformly produced slightly more viscous lipoproteins[58] (Table 8). In a qualitative sense the lipoprotein microviscosities summarized in Table 8 are useful, but they have several quantitative deficiencies. First, on the basis of energy transfer between DPH and the tryptophan residues of the proteins in lipoproteins, Jonas[55] located DPH in the lipoprotein core. This location is only approximate, since considerable uncertainty exists concerning the specific location of the donor tryptophans. Furthermore, the thermal dependence of DPH depolarization reveals no thermal transitions in LDL, whereas differential scanning calorimetry detects cholesteryl ester melting in the core.[58] Most likely this melting process encompasses a small fraction of the LDL volume, and thus changes in motion are not detected by DPH, which is distributed into other lipid compartments that do not undergo a phase transition. Thermal transitions in simpler reassembled lipoproteins are detected by DPH. Probably the single most complicating feature of microviscosity studies is that DPH depolarization is hindered in lipid (or lipoprotein) systems and thus contributes both dynamic (time dependent) and orientational

*Table 8. Apparent Fluidity Parameters for Native and Reassembled Lipoproteins*

| Sample | $\eta$ (25°C) (poise) | $\Delta E$ (kcal/mol) | Reference |
|---|---|---|---|
| HDL | 5.0 ± 0.7 | 7.9 ± 1.6 | 52 |
| HDL[a] | | | |
| SAT | 4.3 ± 0.4 | 8.1 ± 1.8 | 55 |
| PUS | 4.0 ± 0.3 | 6.7 ± 1.5 | 55 |
| LDL | 6.1 ± 0.9 | 8.4 ± 1.7 | 52 |
| LDL | | | |
| SAT | 5.8 ± 0.8 | 8.6 ± 1.7 | 55 |
| PUS | 5.4 ± 0.8 | 7.1 ± 1.5 | 55 |
| VLDL | 1.3 ± 0.2 | 7.1 ± 1.4 | 52 |
| VLDL | | | |
| SAT | 0.9 ± 0.1 | 8.0 ± 0.2 | 55 |
| PUS | 0.8 ± 0.1 | 6.5 ± 0.3 | 55 |
| Chylomicrons | 1.0 ± 0.2 | 6.5 ± 1.9 | 52 |
| HDL lipids | 2.0 ± 0.3 | | 52 |
| LDL lipids | 2.4 ± 0.3 | | 52 |
| VLDL lipids | 1.3 ± 0.2 | | 52 |
| Chylomicron lipids | 1.0 ± 0.2 | | 52 |
| HDL$_3$ | 3.4 | | 56 |
| HDL$_3$ + cholesterol | 4.6[b] | | 56 |
| Bovine HDL | 6.1 ± 0.5 | 13 ± 3 | 53 |
| Bovine HDL | 5.9 | | 56 |
| Bovine HDL + cholesterol | 5.8[c] | | 56 |
| Rabbit VLDL | 0.6 ± 0.2[d] | 7.6 ± 1.5 | 54 |
| Rabbit VLDL hc | 4.6 ± 0.3[d] | 7.8 ± 1.5 | 54 |
| DMPC | 2.0 | | 66 |
| DMPC + ApoA-I | 4.2 | | 66 |
| DMPC + bovine apoA-I | 4.2 | | 66 |
| DMPC + bovine apoA-I + cholesterol | 5.5[e] | | 67 |
| DMPC + bovine apoA-I + cholesterol | 6.4[f] | | 67 |

[a] These lipoproteins are isolated from patients on a diet supplemented with saturated (SAT) or polyunsaturated (PUS) fat.
[b] Human HDL$_3$ plus 123 cholesterol molecules.
[c] HDL plus 165 cholesterol molecules.
[d] At 30°C.
[e] DMPC/cholesterol (mol/mol) ratio 4.9.
[f] DMPC/cholesterol (mol/mol) ratio 2.8.

(time independent) components to the calculated microviscosity.[60] The problem is that $C(r)$ in equation (2) is not a simple constant and depends on the reference system used for calibration as well as the time dependence of the anisotropy decay. This concept will be developed more fully in the following section.

Due to recent theoretical advances the resonance energy transfer technique, sometimes referred to as the "spectroscopic ruler," shows great promise in localizing the components of lipoprotein structure. In an early study of lipoprotein structure, Smith and Green[61] examined energy transfer from tryptophan to fluorescent cholesta-5,7,9(11)-trien-3β-ol and its oleate ester in reconstituted HDL and native LDL. Tryptophan quenching efficiency was low, suggesting the lack of specific sterol—protein associations in both HDL and LDL. The unesterified sterol was significantly more efficient at quenching tryptophan fluorescence, especially in LDL. Since energy transfer efficiency is inversely proportional to the sixth power of the distance between donor and acceptor, the weak quenching by the oleate sterol is compatible with a large separation between surface-localized protein and core-associated esterified sterol. To better localize extrinsic probes in the lipoprotein, Schroeder and co-workers combined energy transfer and fluorescence quenching experiments to locate cholestatrienol and *trans*-parinaric acid in the VLDL surface monolayer[62,63] and DPH and *N*-phenyl-1-naphthylamine in a loosely defined VLDL interior core.[64] The probes were localized in the surface by energy transfer from the protein tryptophan residues. The composition of these rat VLDL, varied by perfusing the rat liver with a medium supplemented by palmitic or oleic acids, resulted in an enrichment of the acyl group of triglyceride or cholesteryl ester by the respective fatty acid infused. The fluidity of the core region, as detected by DPH fluorescence polarization, was greater for the VLDL supplemented by the unsaturated fatty acid, in agreement with other diet studies.[58] Phenylnaphthylamine, which is located in a region midway between the core and the surface monolayer, does not show any change in polarization with altered VLDL fatty acid composition.[64]

A new approach to localization of fluorescence probes in lipoproteins by energy transfer provides quantitative results for both polar and nonpolar chromophores.[65] The energy transfer efficiency $T$ is calculated from the reduced fluorescence quantum yield $Q_a$ of the donor at a given acceptor concentration: $T = 1 - Q_a/Q_0$, where $Q_0$ refers to the unquenched emission of the donor. If $R_0$ is defined as the distance between a particular donor—acceptor pair at which energy transfer efficiency is 50%, then the acceptor surface density can be expressed in terms of acceptors per unit surface area, $R_0^2$. Plots of $Q_a/Q_0$ vs. surface density are generally linear and independent of $R_0$. Combinations of this approach with an algorithm for transfer efficiency[66] permits discrimination between probes on the lipoprotein surface and those displaced toward the center or core of the particle. As a fluorescence acceptor, the lipid analogue 5-(*N*-hexadecanoylamino)fluorescein (I, HAF) was shown to bind quantitatively to lipoproteins, and it was localized in the surface by its accessibility to the water-soluble quencher, potassium iodide. Figure 8 presents the localization of fluorescence probes based on the results of energy transfer experiments between

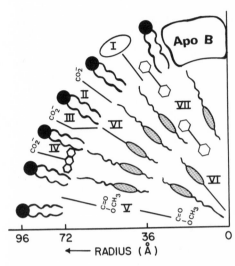

FIGURE 8. Schematic diagram of LDL showing location of several fluorescent probes. LDL is depicted as comprised of a monolayer of phospholipid and protein (cholesterol is omitted) surrounding a core of cholesteryl esters arranged in concentric annuli (triglyceride is omitted). (I) HAF; (II) *trans*-PnA; (III) *cis*-PnA; (IV) AP; (V) *trans*-PnA methyl ester; (VI) *trans*-PnA cholesteryl ester; (VII) DPH.

HAF (I) and the corresponding probe in LDL; the polar donors were the isomeric polyene fatty acids *trans*-parinaric acid (II, *trans*-PnA) and *cis*-parinaric acid (III, *cis*-PnA), and 16-(9-anthroyloxy)palmitic acid (IV, AP). The nonpolar donors consisted of diphenylhexatriene (DPH, VII) and the methyl (V) and cholesteryl esters (VI) of parinaric acid. In LDL the PnA and AP probes are within 15 and 25 Å, respectively, of the surface. The depths of the nonpolar probes DPH (VII) and *trans*-PnA methyl ester (V) are 40–50 and 25–30 Å, respectively, apparently represent partitioning between surface and core regions. The *trans*-PnA cholesteryl ester (VI) is anchored in the core, but, surprisingly, it does not detect thermal melting of LDL cholesteryl esters.[65] A similar distribution of these probes was also found in VLDL.[65]

## 3.2. Reassembled Lipoproteins

An early study of reassembled lipoproteins by extrinsic fluorescence probes examined the interaction of porcine apoHDL with DMPC[67] by comparing the thermal dependence of the fluorescence polarization in DMPC vesicles with that observed in the complex. In general, the two anthroyloxy fatty acid probes and ANS detected a more viscous environment in the complex and a broadening of the lipid melting transition. As chromatographic techniques for separation of apoproteins and isolation of lipid–apoprotein complexes improved, a number of studies reporting the relationship between composition and the physical properties of the complexes appeared. Rosseneu *et al.*[68] measured the DPH-derived apparent microviscosities in complexes of apoA-I, apoA-II, apoC-I, and apoC-III with DMPC at varying molar ratios. In general, the presence of apoprotein increased the lipid melting transition from 24 to about 30°C and decreased the amplitude of the transition. The fact that the lipid melting transition is not completely abolished in the complexes is compatible with the generalized bilayer disc structure shown in Figure 6. Presumably, the middle of the disc contains bulk-phase lipid that can undergo a thermal transition. Both bovine and human apoA-I form micellar complexes with DMPC at lipid–apoprotein molar ratios of 95 : 1,[69] resulting in particles of the general size of HDL. These reassembled lipoproteins can also accommodate up to 33 mol% cholesterol.[70] The thermal dependence of the fluorescence depolarization of DPH in these complexes indicates that the presence of apoprotein and increasing amounts of cholesterol reduces the fluidity of the phospholipid (Table 8).[70] Jonas et al.[71] also report the formation of much larger DMPC/apoA-I (4000 : 1 to 500 : 1) complexes of a vesicular nature, whose apparent microviscosity is midway between that of DMPC vesicles and the micellar complexes. Van Tornout et al.[72] recorded the kinetics of association of apoA-I and apoA-II with DMPC/cholesterol liposomes by observing the time-dependent changes in DPH fluorescence polarization. Optimal association occurs near the lipid phase transition, where maximal disorder of the lipid array enhances association. ApoA-II interacts with the liposomes over a larger temperature and composition range than apoA-I. As mentioned in the previous section, it has been shown that for DPH in lipid systems the time-resolved decay of the anisotropy, $r$, reveals a nonzero value at times long compared with $\tau$.[73] This residual anisotropy $r_\infty$ can be quite large especially at temperatures below the lipid phase transition, and thus apparent microviscosity parameters become artificially elevated. The residual anisotropy reflects the order parameter $S$ of the system ($S^2 = r_\infty/r_0$).[60] To better understand the influence of apoprotein on lipid domains in reassembled lipoproteins, Mantulin et al.[74] used the technique of differential polarized lifetimes[75] to examine the time-resolved motion of DPH in DMPC/apoA-II complexes of the following stoichiometry: 240 : 1, 75 : 1, and 45 : 1.[76] The thermal dependence of the DPH fluorescence lifetime $\tau$, the steady-state anisotropy $r$, and the order parameter $S$ in the DMPC/apoA-II complexes and DMPC control liposomes are presented in Figure 9. From these results it is apparent that increasing protein concentration shifts the observed $T_c$ to higher temperatures and decreases the amplitude of the measured phase transition, as well as increasing the order of the lipid phase. The DMPC adjacent to apoA-II in the complexes ("boundary" lipid) is probably only slightly ordered relative to the bulk lipid, since the $S$ values are similar in all the complexes below $T_c$.

Additional information about the structure of the DMPC/apoA-II complexes was obtained from excimer fluorescence studies of a pyrene-labeled lecithin analog 1-myristoyl-2-[9-(1'-pyrenyl)nonanoylphosphatidylcholine].[74] Because

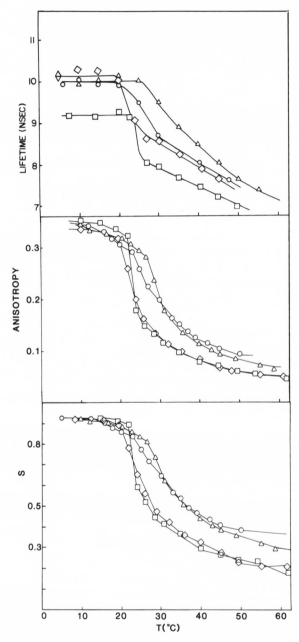

FIGURE 9. Temperature dependence of the fluorescence lifetime (upper panel), steady-state anisotropy (middle panel), and order parameter $S$ (lower panel). DMPC control liposomes (□); 240:1 DMPC/apoA-II complex (◇); 75:1 DMPC/apoA-II complex (○); 45:1 DMPC/apoA-II complex (△).

of its long fluorescence lifetime, the first excited singlet state of pyrene forms excimers (*excited dimers*) with ground-state pyrene. The following equations describe the photophysics of this process.

| | Rate | Process |
|---|---|---|
| $P + h\nu \rightarrow P^*$ | $I_a$ | Absorption, where $I_a$ is the intensity of absorbed light |
| $P^* \rightarrow P + h\nu'$ | $k_t P^*$ | Monomer fluorescence |
| $P^* \rightarrow P$ | $k_t' P^*$ | Monomer radiationless transition |
| $P^* + P \rightarrow P_2^*$ | $k_a(P)(P^*)$ | Excimer formation |
| $P_2^* \rightarrow P^* + P$ | $k_d(P_2^*)$ | Excimer decomposition |
| $P_2^* \rightarrow P_2 + h\nu''$ | $k_1'(P_2^*)$ | Excimer fluorescence |
| $P_2^* \rightarrow P_2$ | $k_{r1}'(P_2^*)$ | Excimer radiationless transition |
| $P_2 \rightarrow 2P$ | $k_d'(P_2)$ | Dimer decomposition |

In addition to the fluorescence of the monomer $P^*$, at short wavelengths a second broad band centered at about 475 nm appears (Figure 10). This has been assigned to excimer emission ($P_2^*$). Since $P_2^*$ is formed from a diffusion-controlled

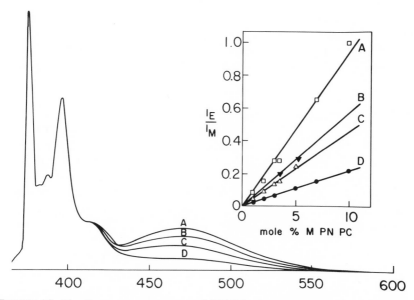

FIGURE 10. The fluorescence spectrum of MPNPC; the structured region at 390 nm is monomer emission and the peak centered at 475 nm corresponds to excimer emission. Inset: The MPNPC concentration dependence of $I_E/I_M$. (A) DMPC control liposomes; (B) 240 : 1 DMPC/apoA-II complex; (C) 75 : 1 DMPC/apoA-II complex; (D) 45 : 1 DMPC/apoA-II complex.

encounter of P with P*, the relative intensity of the excimer band, $I_E$, compared to that of the monomer, $I_M$, is dependent upon the ground state concentrations of pyrene [P] and the viscosity of the medium according to

$$I_E/I_M = [P] \, TK/\eta$$

where $T$ is the absolute temperature, $\eta$ is the viscosity, and $K$ incorporates both instrumental and theoretical constants. The linear dependence of $I_E/I_M$ on the concentration of MPNPC in the DMPC/apoA-II complexes between 0–10% probe doping is a reflection of the random distribution of MPNPC in the complexes (inset, Figure 10). The spectra and inset in Figure 10 were recorded above $T_c$. It is clear from both the spectra and the inset that at any given concentration of MPNPC the $I_E/I_M$ ratio increases inversely with the protein content of the complex. Since excimer formation is a diffusion-controlled process, the presence of apoA-II presumably restricts the lateral diffusion of MPNPC. A similar decrease in $I_E/I_M$ of pyrene was seen in a 50 : 1 DMPC/apoC-III complex as compared to the DMPC alone.[77]

## 3.3. Fluorescence Probes of Lipoprotein Function

The two broad groupings of extrinsic fluorescence probes in the study of lipoprotein function involve studies of enzymatic action on lipoproteins and lipoprotein–cellular receptor interactions. The first group contains studies on the conversion of VLDL to remnant VLDL (LDL-like particles) by the action of lipoprotein lipase (LPL). There are two reports[78,79] that the apparent micro-viscosity (as determined from DPH depolarization measurements) of remnant VLDL is higher than that of native VLDL and indeed is quite similar to that of native LDL. In an innovative application of fluorescence methods, Johnson and co-workers[80,81] demonstrated that the fluorescent phospholipid analogue dansyl phosphatidylethanolamine (DPE) incorporated into VLDL exhibited an increased fluorescence intensity, a 20-nm blue shift of $\lambda_{max}$, and a large increase in polarization during the LPL-induced lipolysis. The lipolysis-associated fluorescence changes are described by a mechanism involving hydrolysis of DPE to lyso-DPE by LPL and subsequent transfer of lyso-DPE to albumin during the reaction.

Utilizing reconstituted LDL in which some of the core cholesteryl esters have been replaced by fluorescent analogues (e.g., a pyrene-labeled cholesteryl oleate), Krieger et al.[82] followed the receptor-mediated endocytosis of LDL in cultured human fibroblasts by fluorescent light microscopy.[82] The fluorescence signal defined the spatial arrangement of the receptors on the cell surface.[83] These reconstituted fluorescent LDL were also used in conjunction with a fluorescence-activated cell sorter to separate receptor-deficient cells, arising from a genetic defect, from normal cells.[84] Matz and Jonas[85] showed that micellar complexes of apoA-I/phosphatidylcholine per cholesterol, prepared from cholate dispersions, acted as good substrates for lecithin : cholesterol acyltransferase (LCAT). After incubation with LCAT, micellar complexes labeled with DNS-apoA-I showed a shorter rotational relaxation time (290 ns) and a smaller Stokes radius (47 Å) than the unreacted complexes (530 ns and 57 Å, respectively).

## 4. Dynamics of Lipid Transfer

One of the fundamental problems concerning plasma lipoproteins has been the identification of the determinants affecting the rate of transfer of lipids, lipid-soluble vitamins, drugs, hormones, and xenobiotics among lipoproteins and between a given lipoprotein and the aqueous phase or cell membrane.[86] Until recently it was believed that many membrane lipids were insoluble and that lipid transfer, if it occurred at all, did so via a mechanism involving the collision of the donor and acceptor lipoproteins. However, before the problem of transfer could be addressed, the chemist needed a reliable method for quantifying the rate of transfer. Perhaps few other areas of lipid dynamics have profited as quickly from a spectral technique as have those that employed pyrene-labeled lipids to measure lipid transfer.[87–91] Some of the many simple analogs that have been studied are shown in Figure 11.

The relationship between the fluorescence spectrum and the distribution of pyrene between lipoproteins is given in Figure 12. If one labels a lipoprotein or lipid vesicle with pyrene ($<$ 10 mol %), both monomer and excimer peaks may be seen in the fluorescence spectrum (Figure 12). If an equal amount of unlabeled lipoprotein is added to the pyrene-labeled lipoprotein, some of the pyrene will transfer to the unlabeled particle, resulting in a decrease in the concentrations (per particle basis) of pyrene. The decrease in local concentration lowers the rate of formation of excimers and thereby a linear decrease in the excimer fluorescence intensity occurs. The decrease of excimer fluorescence as a function of time establishes the rate of transfer of pyrene and its lipid analogues. This is typically done with a large excess of acceptor (unlabeled lipoprotein or phospholipid single bilayer vesicles) to minimize the effects of the reverse transfer reaction and to obtain a large dynamic range in the decrease of $I_E$, which will go to zero at completion of transfer.

Although a considerable range of rate constants may be covered by a given group of compounds, these and other data demonstrate the domination of the rate of transfer by the hydrophobicity of the pyrene-labeled analogue; in one sense[92] this is reduced to the total hydrocarbon surface of the analogue.

In each case, the criteria for transport of the lipid as monomers via the aqueous phase that separates the donor and acceptor compartments is satisfied;

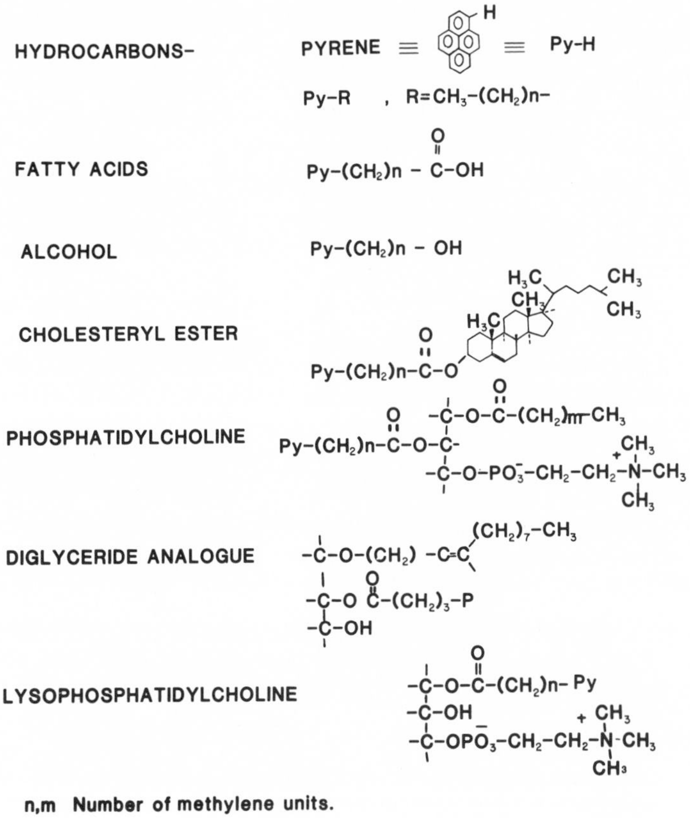

FIGURE 11. Fluorescent probes of lipid transfer.

i.e., the transfer is first order, independent of donor and acceptor concentrations and the identity of the acceptor. Some representative rate data are shown in Figure 13A, B. Figure 13A shows that with a given alkylpyrene, decylpyrene, the kinetics of transfer are first order and the rate of transfer increases with increasing temperature. In contrast, at constant temperature, the rate of transfer decreases with increasing chain length, as shown in Figure 13B. These effects are better illustrated in Figure 14, which shows Arrhenius plots of the transfer of all of the pyrene-labeled alkanes that were studied. In addition to the decrease in the rate that occurs with increasing alkyl chain length, these data show that the activation energy, $E_a$, for this process also increases with hydrocarbon content. A similar behavior was observed for the other analogs and the chain length dependence of $E_a$ for each group of derivatives is given in Figure 15. For the alcohols, acids, and methyl esters the incremental increase in $E_a$ as a function

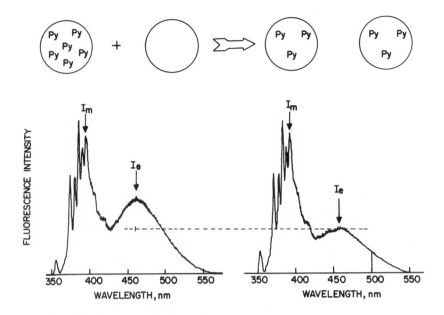

FIGURE 12. Schematic representation of the spectroscopic consequences of pyrene transfer from one compartment to another.

of the number of $CH_2$ groups was nearly constant ($\sim 850$ cal) at both acid and basic pH. In contrast, that observed for the alkylpyrenes was more than twice as high, $\sim 2$ kcal/$CH_2$. We have calculated the values of the free energy ($\Delta G^{\ddagger}$), entropy ($\Delta S^{\ddagger}$), and enthalpy ($\Delta H^{\ddagger}$) of activation from absolute rate theory (Table 9). The chain length dependence of these parameters, illustrated in Figure 16, reveals several regularities that permit one to assign differences in transfer rates and activation parameters to structural determinants contained in each of the transferring species. We note that the incremental activation parameters of the acids, alcohols, and methyl esters are similar; that is, $\Delta H^{\ddagger}/CH_2 \sim 830$ cal; $\Delta S^{\ddagger}/CH_2 = 0$; and $\Delta G^{\ddagger}/CH_2 = 740$ cal. In each of these cases the major contribution of each $CH_2$ to $\Delta G^{\ddagger}$ occurs through an increase in $\Delta H^{\ddagger}$. In contrast, in the pyrene-labeled alkanes, the much larger $\Delta H^{\ddagger}$ for each additional $CH_2$ group is not fully expressed in the $\Delta G^{\ddagger}/CH_2$ because of the large compensatory changes in $\Delta S^{\ddagger}$ that are observed with increasing chain length; that is, $T\Delta S^{\ddagger}/CH_2 = \sim 1200$ cal. As a consequence, the changes in $\Delta G^{\ddagger}/CH_2$ are very similar for all the derivatives irrespective of pH. Comparison of the $\Delta G^{\ddagger}/CH_2$ with the incremental free energy of transfer of a single methylene unit from a hydrocarbon to an aqueous environment, $\Delta G_t$, reveals that these values are quite similar: $\Delta G^{\ddagger}/CH_2 = \Delta G_t/CH_2$. This strong correlation suggests that the

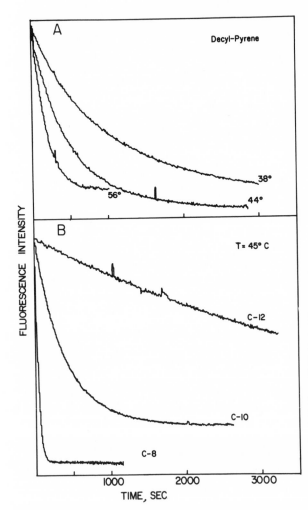

FIGURE 13. Decay of excimer fluorescence as a function of time. (A) 1(3-pyrenyl)decane at the three indicated temperatures; (B) Three different pyrenyl alkanes (number of carbon atoms in sidechain is shown with curve) at one temperature, 45°C. All curves were obtained by mixing labeled single bilayer vesicles of lecithin with those having no label.

transition state for the transfer of these molecules involves a state similar, if not identical, to that of the molecule in pure water. Other correlations between the $\Delta G^{\ddagger}$ and $\Delta G_t$ are consistent with this view. For example, it is known that the $\Delta G_t$ of a given fatty acid is lower at basic than at acid pH. The data of Table 9 and Figure 16 show that the anionic form of fatty acids is transferred faster than the protonated form and that the $\Delta G^{\ddagger}$ for a given fatty acid is lower at pH 9.0. As expected, the pyrenyl alkanes and alcohols, which have no titratable groups between pH 3.0 and 9.0, exhibit no changes in their transfer rates and activation

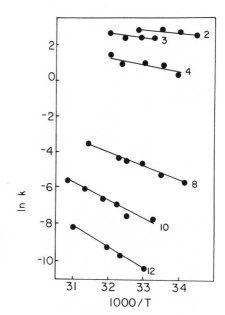

FIGURE 14. Arrhenius plots of the kinetics of transfer of pyrenyl alkanes between single bilayer vesicles of PPOPC. The numbers adjacent to each line are the number of carbon atoms in the aliphatic chain (i.e., 2 = ethyl, etc.).

parameters in this pH range. Consistent with our view that the transition state is similar to that of pure water, $\Delta G_t$ is independent of pH for alkanes and alkanols.

A similar kinetic study has been conducted on the pyrene derivatives of sphingomyelin and phosphatidylcholine (PC), shown in Figure 17A, and some

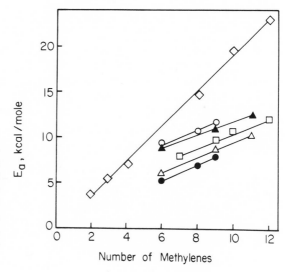

FIGURE 15. Chain length dependence of the activation energy for molecular transfer of pyrenyl alkanes at pH 2.8–7.4 (◇), alcohols (pH 2.8–7.4) (□), methyl esters (pH 7.4) (○), methyl esters (pH 2.8) (●), fatty acids (pH 9.0) (△), and fatty acids (pH 2.8) (▲).

*Table 9. Thermodynamics of the Activated State for the Transfer of Pyrenyl Derivatives between PPOPC Single Bilayer Vesicles at 25°C*

| Derivative | $n^b$ | $\Delta H^\ddagger$ (kcal/mol) | $\Delta\Delta H^\ddagger$ (kcal/mol) | $\Delta S^\ddagger$ (eu/mol) | $\Delta\Delta S^\ddagger$ (eu/mol) | $\Delta G^\ddagger$ (kcal/mol) | $\Delta\Delta G^\ddagger$ (kcal/CH$_2$) |
|---|---|---|---|---|---|---|---|
| Alcohol (7.4)$^a$ | 7 | 7.4 | | 31 | 0 | 16.8 | |
| | 9 | 9.1 | 0.80 | 29 | 0 | 17.7 | 0.76 |
| | 10 | 9.9 | | 30 | | 19.0 | |
| | 12 | 11.4 | | 30 | | 20.5 | |
| Ester (7.4) | 6 | 8.5 | | 30 | | 17.3 | |
| | 8 | 10.1 | 0.83 | 29 | 0 | 18.9 | 0.80 |
| | 9 | 11.0 | | 29 | | 19.7 | |
| Ester (2.8) | 6 | 4.6 | | 40 | | 16.5 | |
| | 8 | 6.4 | 0.90 | 39 | 0 | 18.0 | 0.74 |
| | 9 | 7.3 | | 38 | | 18.7 | |
| Acid (9.0) | 6 | 5.2 | | 32 | | 14.7 | |
| | 8 | 6.8 | 0.84 | 30 | 0 | 15.6 | 0.69 |
| | 9 | 7.6 | | 30 | | 16.5 | |
| | 11 | 9.2 | | 30 | | 18.1 | |
| Acid (2.8) | 7 | 7.9 | | 28 | | 16.2 | |
| | 10 | 10.4 | 0.80 | 27 | 0 | 18.6 | 0.74 |
| | 12 | 11.9 | | 27 | | 19.9 | |
| Alkanes (2.8–9.0) | 3 | 4.9 | | 38 | | 16.1 | |
| | 4 | 6.7 | 1.98 | 35 | −4 | 17.0 | 0.88 |
| | 8 | 14.6 | | 20 | | 20.5 | |
| | 10 | 18.9 | | 11 | | 22.3 | |
| | 12 | 22.7 | | 4 | | 24.0 | |

$^a$ Values in parentheses are the pH at which the experiments were conducted.

$^b$ $n$ is the number of methylene units.

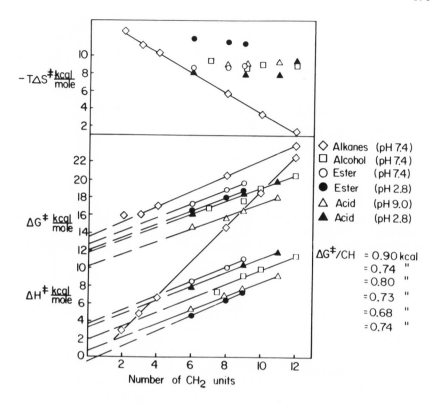

FIGURE 16. Chain length dependence of the transition state parameters for molecular transfer of pyrenyl species between single bilayer vesicles of PPOPC. Alkanes (pH 2.7–7.4), $\Delta G^{\ddagger} = 0.90$ kcal ($\diamond$); alcohols (pH 2.8–7.4), $\Delta G^{\ddagger} = 0.74$ kcal ($\square$); methyl esters (pH 7.4), $\Delta G^{\ddagger} = 0.80$ kcal ($\circ$); methyl esters (pH 2.8), $\Delta G^{\ddagger} = 0.73$ kcal ($\bullet$); fatty acids (pH 9.0), $\Delta G^{\ddagger} = 0.68$ kcal ($\triangle$); fatty acids (pH 2.8), $\Delta G^{\ddagger} = 0.74$ kcal. ($\blacktriangle$).

other phospholipid derivatives in which the polar head has been cleaved or replaced with bases other than choline (Figure 17B). In each instance the rate of transfer from artificial lipoproteins has been measured. Furthermore, the kinetics of transfer of all of these derivatives are consistent with monomer transfer through water as described above. There is a dramatic effect of both chain length and increased unsaturation of the acyl group at the $SN$-2 position on the kinetics of transfer that is shown in Figure 18A, B. The addition of each pair of methylene units to the molecule reduces the transfer rate by a factor of about 10, whereas the addition of each double bond increases the rate by a factor of about 4. This behavior can be understood in terms of the reduced solubility of long-chain alkanes in water when compared to their short-chain and/or unsaturated analogs. If the transition state for transfer is similar or

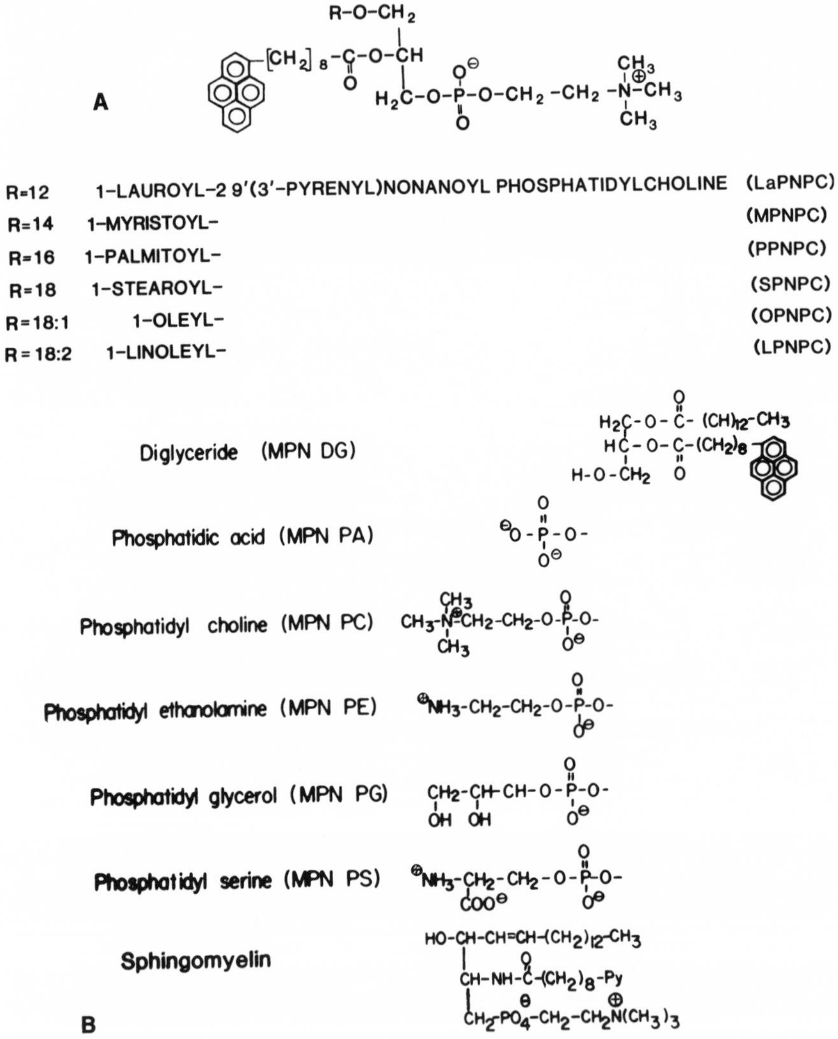

R=12     1-LAUROYL-2 9'(3'-PYRENYL)NONANOYL PHOSPHATIDYLCHOLINE   (LaPNPC)

R=14     1-MYRISTOYL–    (MPNPC)

R=16     1-PALMITOYL–    (PPNPC)

R=18     1-STEAROYL–    (SPNPC)

R=18:1    1-OLEYL–    (OPNPC)

R=18:2    1-LINOLEYL–    (LPNPC)

Diglyceride (MPN DG)

Phosphatidic acid (MPN PA)

Phosphatidyl choline (MPN PC)

Phosphatidyl ethanolamine (MPN PE)

Phosphatidyl glycerol (MPN PG)

Phosphatidyl serine (MPN PS)

Sphingomyelin

FIGURE 17. (A) Structures of pyrene-labeled phosphatidylcholines; (B) structures of pyrene-labeled phospholipids with different headgroups.

identical to that of pure water, then the transfer of the less water-soluble long-chain saturated derivative should require more energy to leave the lipid complex than that required by either shorter-chain or unsaturated lecithins, which are expected to be more water soluble. Sphingomyelin is another important choline-containing lipid that is an important component of membranes and plasma

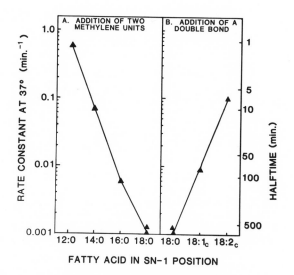

FIGURE 18. Effect of the fatty acid composition of a phospholipid on its rate of transfer between apolipoprotein–phospholipid recombinants. The donor recombinants were DMPC/apoA-II (45/1). Donor recombinants (0.01 mg) with 1 mol% pyrene-labeled PC in STD buffer at 37°C were mixed with identical acceptor complexes (0.25 mg). In Panel A, the fatty acyl composition in the *SN*-1 position of the pyrene-labeled PC contained a different number of methylene units (myristic, palmitic, and stearic acids). In Panel B, the fatty acyl composition is varied in the number of double bonds (stearic, oleic, and linoleic acids).

lipoproteins. The transfer kinetics of a pyrene-labeled sphingomyelin between apoA-II/DMPC complexes is similar to that of the PC analogs except that the rate is faster than some of the PC analogs but slower than that of MPNPC. This observation can be assigned to the effects of the relative hydrocarbon content on the solubility which decreases in the order SPNPC > PPNPC > PySM > MPNPC; this is the same order in which the transfer rate of these four derivatives increases (Table 10).

Finally, we examine the effects of altering the polar headgroup on the rate and activation energy of transfer (Table 10). The transfer rates for PS, PG, PA, and PC are about two to three times faster than that of PE, and the PE transfers between a given lipid compartment occur at a rate that is about five times that of the corresponding DG. Once again, this behavior can be understood in terms of the effects of the headgroup on the relative solubility of these homologs on the transfer rate and the free energy of activation. One would expect the solubility of PS, PG, PC, and PA to be greater than that of the PE by virtue of their zwitterionic or charged headgroup. In contrast, the DG analog would be expected to be the least soluble in water since it has only the hydroxyl group, which is available for association with water. It is, moreover, probable that the

*Table 10. Summary of the Kinetic Data for the Transfer of Pyrene-Labeled Lipids between Model Lipoproteins*

| Donor | Lipid | Rate constant ($min^{-1}$) | Activation energy (kcal/mol) |
|-------|-------|----------------------------|------------------------------|
| POPC/ApoA-I (100 : 1) | MPNPC | 0.09 | 24 |
| | PPNPC | 0.013 | 26 |
| | SPNPC | 0.0026 | 24 |
| | OPNPC | 0.010 | 22 |
| | LPNPC | 0.037 | 24 |
| DMPC/ApoA-II (45 : 1) | PySM | 0.06 | 25 |
| | MPNPE | 0.019 | 25 |
| | MPNPG | 0.047 | 24 |
| | MPNPA | 0.055 | 20 |
| | MPNDG | 0.0038 | 21 |
| | MPNPS | 0.043 | 24 |

DG can diffuse into the middle of the lipid bilayer so that these additional degrees of freedom have the entropic effect of increasing the lipid solubility of DG and thereby increase the difference in the relative solubility of DG in lipid when compared to the other derivatives, which probably have their headgroups located at the lipid surface.

This concept is illustrated in the data in Figures 13–16 in which the rate dependence of a series of pyrene-labeled alkanes, alcohols, fatty acids, and methyl esters is shown as a function of alkyl chain length and temperature.[93]

## 5. Summary and Perspectives

Fluorescence methods have provided new insights into the structure and dynamics of lipoproteins; the methods and the principles derived therefrom are also important to our understanding of membranes. We anticipate additional research activity in two important areas; these are the application of energy transfer techniques to structural studies of lipoproteins and the use of excimer fluorescence to understand lipid dynamics. In both cases the synthesis of new fluorescent analogs of natural lipids will play an important role. We also anticipate that fluorescent lipid analogs will be used to study the mechanism of action of LPL and LCAT as well as the structure of the enzyme—substrate complexes. Finally, we expect that fluorescent lipid analogs, incorporated into native or recombinant lipoprotein, will be used to trace the intracellular metabolism of lipoproteins via video intensification microscopy.

## ACKNOWLEDGMENTS

The authors wish to acknowledge the support of the National Research and Demonstration Center, National Heart, Lung and Blood Institute, The National Institutes of Health, HL-17269 (HJP), HL-26250 (HJP), The American Heart Association, Texas Affiliate, Inc. (WWM). The authors would also like to thank Sarah Myers for editorial assistance, Carol Coreathers for assistance in the preparation of the manuscript, and Susan McNeely-Kelly for providing the line drawings.

# References

1. R. J. Havel, M. S. Brown, and J. L. Goldstein, in: *Metabolic Control and Disease* (P. K. Bondy and L. E. Rosenberg, eds.), Saunders, Philadelphia (1980), p. 393.
2. L. C. Smith, H. J. Pownall, and A. M. Gotto, Jr., *Annu. Rev. Biochem.* **47**, 751–777 (1978).
3. A. Scanu, ed., Symposium on Lipoprotein Structure, Vol. 348, The New York Academy of Sciences (1978).
4. R. L. Jackson, J. D. Morrisett, and A. M. Gotto, Jr., *Physiol. Rev.* **56**, 259–315 (1976).
5. J. P. Segrest, R. L. Jackson, J. D. Morrisett, and A. M. Gotto, *FEBS Lett.* **38**, 247–253 (1974).
6. R. J. Havel, H. A. Eder, and J. M. Brogdan, *J. Clin. Invest.* **34**, 1345 (1955).
7. M. Burstein, H. R. Scholnick, and R. Morfin, *J. Lipid Res.* **11**, 583–595 (1970).
8. L. Rudel, J. A. Lee, M. D. Norris, and J. M. Felts, *Biochem. J.* **139**, 89 (1974).
9. A. M. Scanu, J. Toth, C. Edelstein, S. Koga, and E. Stiller, *Biochemistry* **8**, 3309 (1969).
10. J. A. Reynolds and R. H. Simon, *J. Biol. Chem.* **249**, 3937 (1974).
11. W. V. Brown, R. I. Levy, and D. S. Fredrickson, *J. Biol. Chem.* **244**, 5687–5694 (1969).
12. H. C. H. Steele and J. A. Reynolds, *J. Biol. Chem.* **254**, 1633 (1979).
13. H. J. Pownall, R. L. Jackson, J. D. Morrisett, and A. M. Gotto, in: *Biochemistry of Atherosclerosis* (A. M. Scanu, R. W. Wissler, and G. S. Getz, eds.), Vol. 7, Marcel Dekker, New York (1979), pp. 123–143.
14. H. J. Pownall, J. B. Massey, S. K. Kusserow, and A. M. Gotto, *Biochemistry* **17**, 1183–1188 (1978).
15. J. B. Swaney, *J. Biol. Chem.* **255**, 8791–8797 (1980).
16. A. Jonas, *Biochemistry* **12**, 4503–4507 (1973).
17. R. A. Badley, *Biochim. Biophys. Acta* **379**, 517–528 (1975).
18. C. J. Hart, R. B. Leslie, and A. Scanu, *Chem. Phys. Lipids* **4**, 367–374 (1970).
19. J-I Azume, N. Kashimura, and T. Komano, *J. Biochem.* **83**, 1533–1543 (1978).
20. H. B. Pollard and R. F. Chen, *J. Supramol. Struct.,* 177–184 (1973).
21. J. Gwynne, H. B. Brewer, Jr., and H. Edelhoch, *J. Biol. Chem.* **250**, 2269–2274 (1975).
22. A. Ikai, *J. Biochem.* **79**, 579–688 (1976).
23. J. G. Molotkovsky, Y. M. Manevich, E. N. Gerasimova, I. M. Molotkovskaya, V. A. Poletsky, and L. D. Bergelson, *FEBS Lett..* 573–579, (1982).

24. J. C. Osborne, Jr., and B. Brewer, Jr., *Adv. Protein Chem.* **31**, 253–337 (1977).
25. J. Gwynne, B. Brewer, Jr., and H. Edelhoch, *J. Biol. Chem.* **249**, 2411–2416 (1974).
26. A. Jonas, *Biochim. Biophys. Acta* **393**, 71–82 (1975).
27. D. L. Barbeau, A. Jonas, T.-L. Teng, and A. M. Scanu, *Biochemistry* **18**, 362–369 (1979).
28. G. Weber, *Adv. Protein Chem.* **8**, 415–459 (1953).
29. J. Gwynne, G. Palumbo, J. C. Osborne, Jr., H. B. Brewer, and H. Edelhoch, *Arch. Biochem. Biophys.* **170**, 204–212 (1975).
30. J. C. Osborne, G. Palumbo, H. B. Brewer, and H. Edelhoch, *Biochemistry* **14**, 3741–3746 (1975).
31. J. B. Swaney and K. J. O'Brien, *Biochem. Biophys. Res. Commun.* **71**, 636–642 (1976).
32. J. B. Swaney and K. O'Brien, *J. Biol. Chem.* **253**, 7069–7077 (1978).
33. J. C. Osborne, G. M. Powell, and H. B. Brewer, *Biochim. Biophys. Acta* **619**, 559–571 (1980).
34. W. W. Mantulin, M. F. Rohde, A. M. Gotto, Jr., and H. J. Pownall, *J. Biol. Chem.* **255**, 8185–8191 (1980).
35. A. Grinvald and I. Z. Steinberg, *Biochim. Biophys. Acta* **427**, 663–678 (1976).
36. J. Gwynne, G. Palumbo, H. B. Brewer, and H. Edelhoch, *J. Biol. Chem.* **250**, 7300–7306 (1975).
37. R. B. Verdery and A. V. Nichols, *Biochem. Biophys. Res. Commun.* **57**, 1271–1278 (1974).
38. J. B. Swaney, *J. Biol. Chem.* **255**, 877–881 (1980).
39. G. Palumbo and H. Edelhoch, *J. Biol. Chem.* **252**, 3684–3688 (1977).
40. A. Jonas and D. J. Krajnovich, *J. Biol. Chem.* **252**, 2194–2199 (1977).
41. A. Jonas and S. M. Drengler, *Biochem. Biophys. Res. Commun.* **78**, 1424–1430 (1977).
42. J. D. Morrisett, J. S. K. David, H. J. Pownall, and A. M. Gotto, Jr., *Biochemistry* **12**, 1290–1299 (1973).
43. J. D. Morrisett, H. J. Pownall, and A. M. Gotto, Jr., *Biochim. Biophys. Acta* **486**, 36–46 (1977).
44. H. J. Pownall, J. D. Morrisett, J. T. Sparrow, and A. M. Gotto, *Biochem. Biophys. Res. Commun.* **60**, 779–786 (1974).
45. H. Trauble, G. Middlehoff, and V. W. Brown, *FEBS Lett.* **49**, 269–275 (1974).
46. H. J. Pownall, J. D. Morrisett, and A. M. Gotto, *J. Lipid Res.* **18**, 14–23 (1977).
47. H. J. Pownall and L. C. Smith, *Biochemistry* **13**, 2590–2593 (1974).
48. R. L. Jackson, J. D. Morrisett, J. T. Sparrow, J. P. Segrest, H. J. Pownall, L. C. Smith, H. F. Hoff, and A. M. Gotto, Jr., *J. Biol. Chem.* **249**, 5314–5320 (1974).
49. B. W. Patterson and A. Jonas, *Biochim. Biophys. Acta* **619**, 572–586 (1980).
50. B. W. Patterson and A. Jonas, *Biochim. Biophys. Acta* **619**, 587–603 (1980).
51. M. K. Basu, J. N. Finkelstein, S. Ghosh, and J. S. Schweppe, *Biochim. Biophys. Acta* **398**, 385–393 (1975).
52. S. Ghosh, M. K. Basu, and J. S. Schweppe, *Biochim. Biophys. Acta* **337**, 395–403 (1974).
53. R. A. Muesing and T. Nishida, *Biochemistry* **15**, 2952–2962 (1971).
54. M. Shinitzky and Y. Barenholz, *Biochim. Biophys. Acta* **515**, 367–394 (1978).
55. A. Jonas, *Biochim. Biophys. Acta* **486**, 10–22 (1977).
56. A. Jonas and R. W. Jung, *Biochem. Biophys. Res. Commun.* **66**, 651–657 (1975).

57.  F. J. Castellino, J. K. Thomas, and V. A. Ploplis, *Biochem. Biophys. Res. Commun.* **75,** 857–862 (1977).
58.  H. J. Pownall, J. Shepherd, W. W. Mantulin, L. A. Sklar, and A. M. Gotto, Jr., *Atherosclerosis* **36,** 290–314 (1980).
59.  A. Jonas, L. K. Hesterberg, and S. M. Drengler, *Biochim. Biophys. Acta* **528,** 47–57 (1978).
60.  M. P. Heyn, *FEBS Lett.* **108,** 359–364 (1979).
61.  R. J. M. Smith and C. Green, *Biochem. J.* **137,** 413–415 (1974).
62.  F. Schroeder, E. H. Goh, and M. Heimberg, *J. Biol. Chem.* **254,** 2456–2463 (1979).
63.  R. Schroeder, E. H. Goh, and M. Heimberg, *FEBS Lett.* **97,** 233–236 (1979).
64.  F. Schroeder and E. H. Goh, *J. Biol. Chem.* **254,** 2464–2470 (1979).
65.  L. A. Sklar, M. C. Doody, A. M. Gotto, Jr., and H. J. Pownall, *Biochemistry* **19,** 1294–1301 (1980).
66.  M. C. Doody, L. A. Sklar, H. J. Pownall, J. T. Sparrow, A. M. Gotto, Jr., and L. C. Smith, *Biophys. J.* **25,** 286a (1979).
67.  M. D. Barratt, R. A. Badley, and R. B. Leslie, *Eur. J. Biochem.* **48,** 595–601 (1974).
68.  M. Rosseneu, R. Vercaemst, H. Caster, M. J. Lievens, P. Van Tornout, and P. N. Herbert, *Eur. J. Biochem.* **96,** 357–362 (1979).
69.  A. Jonas, D. J. Krajnovich, and B. W. Patterson, *J. Biol. Chem.* **252,** 2200–2205 (1977).
70.  A. Jonas and D. J. Krajnovich, *J. Biol. Chem.* **253,** 5758–5763 (1978).
71.  A. Jonas, S. M. Drengler, and B. W. Patterson, *J. Biol. Chem.* **255,** 2183–2189 (1980).
72.  P. Van Tornout, R. Vercaemst, M. J. Lievens, H. Caster, M. Rosseneu, and G. Assman, *Biochim. Biophys. Acta* **601,** 509–523 (1980).
73.  L. A. Chen, R. E. Dale, S. Roth, and L. Brand, *J. Biol. Chem.* **252,** 2163–2169 (1977).
74.  W. W. Mantulin, J. B. Massey, A. M. Gotto, Jr., and H. J. Pownall, *Fed. Proc.* **39,** 1766 (1980).
75.  F. G. Prendergast Lakowicz and D. Hogen, *Biochemistry* **18,** 508–519 (1979).
76.  J. B. Massey, A. M. Gotto, Jr., and H. J. Pownall, *J. Biol. Chem.* **255,** 10167–10173 (1980).
77.  Z. Novosad, R. D. Knapp, A. M. Gotto, Jr., H. J. Pownall, and J. D. Morrisett, *Biochemistry* **15,** 3176–3183 (1976).
78.  Y. Barenholz, A. Gafni, and S. Eisenberg, *Chem. Phys. Lipids* **21,** 179–185 (1978).
79.  R. J. Deckelbaum, S. Eisenberg, M. Fainaru, Y. Barenholz, and T. Olivecrona, *J. Biol. Chem.* **254,** 6079–6087 (1979).
80.  J. D. Johnson, M. R. Taskinen, N. Matsuoka, and R. L. Jackson, *J. Biol. Chem.* **255,** 3461–3465 (1980).
81.  J. D. Johnson, M. R. Taskinen, N. Matsuoka, and R. L. Jackson, *J. Biol. Chem.* **255,** 3466–3471 (1980).
82.  M. Krieger, L. C. Smith, R. G. W. Anderson, J. L. Goldstein, Y. J. Kao, H. J. Pownall, A. M. Gotto, Jr., and M. S. Brown, *J. Supramol. Struct.* **10,** 467–478 (1979).
83.  R. G. W. Anderson, J. L. Goldstein, and M. S. Brown, *J. Receptor Res.* **1,** 17–39 (1980).
84.  J. L. Goldstein, M. S. Brown, M. Krieger, R. G. W. Anderson, and B. Mintz, *Proc. Natl. Acad. Sci. USA* **76,** 2843–2847 (1979).
85.  C. E. Matz and A. Jonas, *J. Biol. Chem.* **257,** 4541–4546 (1982).
86.  J. C. Kader, in: *Cell Surface Review* (G. Poste and G. Nicholson, eds.) North-Holland, Amsterdam (1977).
87.  P. Sengupta, E. Sackmann, W. Kuhnle, and H. P. Scholz, *Biochim. Biophys. Acta* **436,** 869–878 (1976).

88. S. C. Charlton, J. S. Olson, K-Y. Hong, H. J. Pownall, D. D. Louis, and L. C. Smith, *J. Biol. Chem.* **251,** 7952–7955 (1976).
89. M. Roseman and T. E. Thompson, *Biochemistry* **19,** 439–444 (1980).
90. M. C. Doody, H. J. Pownall, Y. J. Kao, and L. C. Smith, *Biochemistry* **19,** 108–116 (1980).
91. S. C. Charlton, K-Y. Hong, and L. C. Smith, *Biochemistry* **17,** 3304–3309 (1978).
92. C. Clothia, *Nature (London)* **248,** 338–339 (1974).
93. H. J. Pownall, D. L. M. Hickson, and L. C. Smith, *J. Am. Chem. Soc.* (in press).

# 6

# Fluorescent Dye–Nucleic Acid Complexes

## Robert F. Steiner and Yukio Kubota

## 1. Introduction

During the past 20 years, the interaction of fluorescent dyes with nucleic acids has evolved into one of the most active areas of research in biochemistry and biophysics with ramifications which touch on a wide variety of topics. These extend far beyond the cytological applications which originally drew the attention of biologists and range from structural studies of nucleic acids to molecular genetics and the action of antibiotics.

No effort will be made here to do justice to a major fraction of the vast existing literature. Attention will instead be focused primarily upon the physico-chemical aspects of the interaction of fluorochromes with nucleic acids. Even with this restriction, considerable selectivity is necessary and emphasis will be placed upon those aspects of most intense current interest.

This chapter begins with a survey of the types of systems which have been investigated (Section 2). It then describes a study of the extent of the inter-action between fluorescent dyes and nucleic acids, which yields *binding isotherms* (Section 3), since this is fundamental to all other structural studies. Optical techniques are the most appropriate for a study of the binding inter-action. In particular, fluorescence techniques can provide important information about the nature of binding sites and about the structure and dynamics of both nucleic acids and nucleic acid–dye complexes. The results of the application of these techniques are described in Sections 4–6. This chapter ends with the use of fluorescent dyes to monitor chromosome division (Section 7).

ROBERT F. STEINER • Department of Chemistry, University of Maryland Baltimore County, Catonsville, Maryland 21228. YUKIO KUBOTA • Department of Chemistry, Faculty of Science, Yamaguchi University, Yamaguchi 753, Japan.

## 2. Intercalating and Nonintercalating Dyes

For many purposes it is convenient to group the numerous fluorescent dyes which have been found to interact with nucleic acids into the two classes of *intercalating* dyes,[1-4] which can fit between adjacent base pairs in DNA, and *nonintercalating* dyes, which, because of bulkiness or other factors, must remain external to the bihelical structure.[2]

### 2.1. Intercalating Dyes

In general, the intercalating dyes consist of three coplanar fused aromatic rings. By far the most intensively studied of these are acridine derivatives[1-3] and ethidium bromide,[5-7] of which the most important are depicted in Figure 1. Many of the acridine derivatives have important biological activities, acting as mutagens, carcinogens, and bacteriostatic agents.[1-3] Ethidium bromide also induces several biological effects such as the inhibition of nucleic acid synthesis *in vivo*.[8]

Several of the acridine dyes have found extensive application as cytological stains.[9-11] In particular, the aminoacridines such as proflavine, acridine orange, and acriflavine have the useful property of staining the nuclei of living cells without killing them. RNA and DNA in both fixed and living cells are readily distinguished by virtue of the different colors of the fluorescence which they develop when stained with acridine orange.

The acridine dyes are potent mutagens for viruses. Mutagenic effects have been observed for the T-even bacteriophages[12] and for the polio virus.[13] Amino acid sequence studies upon a protein synthesized under the direction of bacteriophage DNA have indicated that the aminoacridines act by causing a removal or insertion of a single nucleotide.[14] A photodynamic mutagenicity of acridine dyes, especially with bacteria and viruses, has also been reported.[15]

Several acridine dyes, including proflavine and acridine orange, have been shown to accelerate the photoinactivation of viruses.[3,16] The mechanism of this effect is still incompletely understood.

In general, the only aminoacridines which interact strongly with nucleic acids are those which are cationic at the experimental pH. This usually means that a substituted or unsubstituted amino group occurs at the 3-, 6-, or 9-positions of the acridine ring.[17] A loss of the planar aromatic character of one of the three rings, as in the hydrogenation of 9-aminoacridine to form 9-amino-1,2,3,4-tetrahydroacridine, results in a substantial reduction in ability to bind to DNA.[18]

The attachment of a bulky side chain to the 9-amino group of 9-aminoacridine, as in quinacrine (Figure 2), does not hinder binding or intercalation.[18] However, the presence of bulky substituents at the 2- and 7-positions, as in 2,7-di-*t*-butylproflavine (Figure 2), has been shown to prevent intercalation but

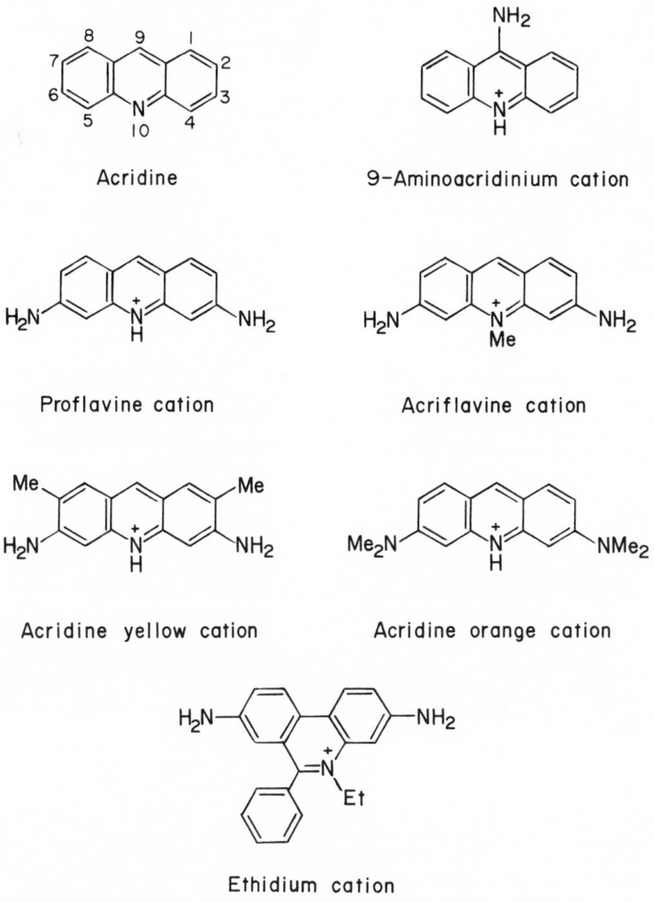

FIGURE 1. The structures of some important intercalating dyes.

to allow the persistence of strong binding.[2,19,20] In this case binding occurs on the exterior of the double helix, perhaps with the insertion of the dye into the major groove of DNA.[20] On the other hand, the introduction of bulky groups to the 10- or 3,6-positions of the acridine ring has been shown to cause a marked decrease in both the apparent binding constant and the apparent maximum number of binding sites per DNA phosphate.[21,22] This suggests that these bulky groups produce some steric hindrance to the interaction.

The presence of an intercalating dye results in characteristic changes in the physical properties of DNA. These include a pronounced increase in intrinsic viscosity and decrease in sedimentation coefficient.[23] In the case of noncircular DNA, the observed effects have been interpreted in terms of an increment in

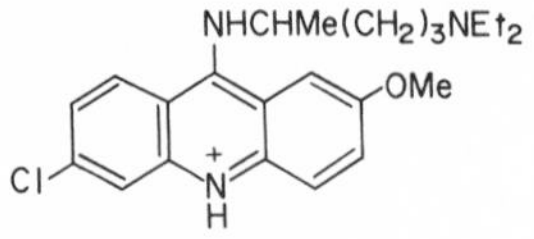

NHCHMe(CH$_2$)$_3$NEt$_2$

Cl

OMe

Quinacrine cation

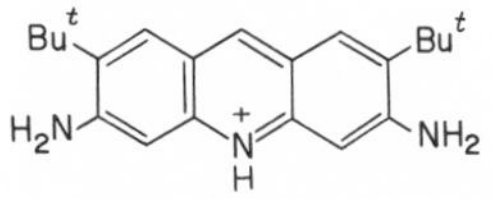

Bu$^t$

Bu$^t$

H$_2$N

NH$_2$

2,7-di-$t$-Butyl-proflavine cation

FIGURE 2. The structures of acridine dyes with bulky substituents.

contour length close to the normal spacing (3.4 Å) between DNA bases for each bound dye molecule.[23] The extension process increases with increasing extents of binding until a limit is approached at about 0.2 mol of dye bound per nucleotide, after which no further change occurs. The increase in contour length has also been observed by autoradiography[24] and light scattering.[25] The available x-ray diffraction evidence is consistent with the occurrence of both unwinding and extension.[4] It is of interest that the viscosity of thermally denatured DNA shows no increase upon interaction with intercalating dyes.[26]

The absorption and fluorescence properties of acridine dyes and ethidium bromide are well known to be altered upon binding to nucleic acids. A characteristic displacement occurs of the electronic absorption spectrum to longer wavelengths when the acridines are bound as single cations.[27,28] The bound aminoacridines acquire optical activity in the region of the displaced absorption bands, exhibiting an extrinsic circular dichroism.[29-31] The quenching of the fluorescence of the aminoacridines such as proflavine[32] and the enhancement of acridine orange[32] and ethidium bromide[7] upon binding are also notable changes. The fluorescence behavior of the bound dye and information obtained from fluorescence measurements will be described in detail in Sections 4–6.

The binding of intercalating dyes with closed circular DNA causes both an unwinding of the helical duplex structure and a simultaneous and equivalent unwinding of the superhelices.[33] The progressive addition of ethidium bromide to the closed circular DNA from SV 40 results in a decrease in sedimentation coefficient, followed by its increase.[33] This has been interpreted as reflecting the conversion of the original negatively supercoiled molecule to a relaxed state and then to a positive supercoil. A circular DNA from the same species which has been converted to a relaxed state by "nicking" one strand shows only a monotonic decrease in sedimentation coefficient with increasing dye level. The bouyant densities of both intact and "nicked" SV 40 DNA decrease in the presence of ethidium bromide, but show a different quantitative dependence.

## 2.2. Nonintercalating Dyes

One class of nonintercalating dyes corresponds to acridine derivatives which have been rendered incapable of intercalation by the introduction of bulky substituents. An example of this category is 2,7-di-*t*-butylproflavine (Figure 2), which has been found to bind to several DNA's of varying base content, as well as to the synthetic polynucleotides poly dA · poly dT and poly d(A-T).[19] However, the changes in hydrodynamic properties characteristic of intercalating dyes have not been observed, leading to the conclusion that this dye is bound to the outside of the double helix.[19] A red shift and intensification of the primary absorption band occurs upon binding. A comparison of binding isotherms obtained by equilibrium dialysis for natural DNAs of varying base content has indicated that binding occurs preferentially to A-T base pairs, while binding measurements with poly d(A-T) have suggested that a single bound dye molecule occupies three base pairs.

The analogous acridine derivatives di-*t*-butylproflavine monoacetate and di-*t*-butylacriflavine show binding properties which are qualitatively similar to the above.[34]

Another category of nonintercalating dyes includes the triphenylmethane derivatives such as *p*-fuchsin, malachite green, methyl green, and crystal violet, as well as the related fluorescent dye auramine. Binding studies have indicated that all of the above, except for *p*-fuchsin, are preferentially bound by A-T base pairs; this appears to be a general characteristic of dyes of this class.[34]

A nonintercalating fluorescent dye of quite different structure (Figure 3) is the bisbenzimidazole derivative Hoechst 33258.[35–39] The prevailing evidence indicates that it is bound within the major groove of DNA with preferential interaction with A-T base pairs. While binding to A-T base pairs results in the development of intense fluorescence, the complexes formed with G-C or with A-BrU base pairs are largely quenched.

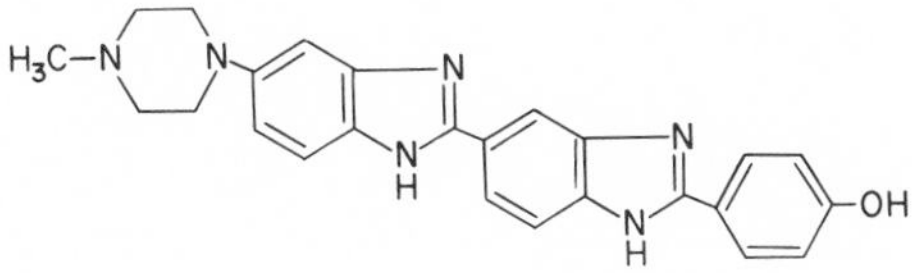

FIGURE 3. The structure of Hoechst 33258.

# 3. Nucleic Acid–Dye Binding Isotherms

## 3.1. Intercalating Dyes

The general equation describing the binding of a ligand by $P$ classes of binding sites in a nucleic acid molecule is[2]

$$r = \sum_{j=1}^{j=P} \frac{n_j k_j c}{1 + k_j c} \tag{1}$$

where $n_j$ and $k_j$ are the number and binding constant, respectively, of sites of type $j$, $r$ is the number of moles of ligand per nucleotide phosphorus, and $c$ is the concentration of free ligand. Equation (1) is simplified to the following form for one class of noninteracting binding sites[2]:

$$r/c = k(n - r) \tag{2}$$

The existing binding data for the aminoacridine dyes with DNA are qualitatively consistent with the presence of two distinct categories of binding modes which correspond to *strong* and to *weak* binding.[2,40] A Scatchard plot of $r/c$ versus $r$ shows two distinct regions. A region of high slope, corresponding to strong binding with a change of free energy of $-(6-9)\,\text{kcal/mol}$, is present for values of $r$ up to about 0.2, and a region of low slope, corresponding to weak binding, extends up to $r = 1$. The latter portion of the curve shows significant cooperativity.

Electrostatic factors are very important in the interaction of the amino-acridines with DNA. This is reflected by the critical dependence of the binding process upon ionic strength. An increase in ionic strength reduces both the strong and weak binding, but the weak binding is considerably more sensitive.[2,41]

### 3.1.1. Strong Binding

In the case of the strong complexes, flow-polarized fluorescence and flow dichroism studies have shown that the acridine rings are bound with their planes approximately parallel (within $\pm 30°$) to the planes of the bases.[42] In the original intercalation model of Lerman the aminoacridine cation was postulated to lie centrally over a base pair with its positive nitrogen atom close to the helical axis.[4,43] In order to accommodate a dye, the double helix was assumed to locally unwind and extend so that the distance between base pairs is increased from 3.4 to approximately 6.8 Å.[4] Lerman[4] estimated the unwinding angle for aminoacridine intercalation to be $-45°$, while Fuller and Waring[5] concluded from x-ray diffraction and model building studies that the unwinding angle is only $-12°$ in the case of ethidium bromide. The ability of dyes to unwind helices has been used to determine superhelical twists in closed circular DNA.[44,45] There are a number of indirect indications supporting intercalation[2,3]; the strongest one is the fact that the contour length of rodlike DNA molecules is increased by the binding of proflavine.[23] The most direct evidence for intercalation stems from x-ray crystallographic structure determinations on complexes between ribodinucleoside monophosphates and aminoacridines.[46–49]

Based upon electric dichroism measurements, Hogan *et al.*[50] calculated that an intercalated dye is not perpendicular to the helix but is tilted from perpendicularity by $21 \pm 7°$ along the long base-pair axis. They proposed that tilting occurs because base pairs neighboring the intercalated dye are flattened by the heterocycle and are therefore incompatible with adjacent propeller-twisted base pairs.[51,52]

A modified intercalation model has also been proposed in which the acridine lies between successive nucleotide bases of the same polynucleotide strand in a plane approximately parallel to the base planes and with the positive ring nitrogen close to a phosphate group.[53] This model can better explain the formation of strong complexes by denatured DNA and by single-stranded RNA.[18] It is also consistent with the observation that intercalation causes an extension of native, but not of denatured, DNA.[2,18] It is of course possible that both mechanisms occur in particular instances. However, it must be pointed out that the structure of the complex would be different in single-stranded and double-stranded polynucleotides.

In the strong binding region at low $r$ values, temperature-jump relaxation studies have shown that proflavine binds to DNA[54,55] and poly rA · poly rU[56] in two kinetically distinguishable steps. In the first step, the dye binds to phosphate groups at the outside by electrostatic attraction; this process is almost diffusion-controlled. The second step, the insertion into the helix, occurs in roughly millisecond time scale, as a first-order process from the outside-bound state. The contribution of outside binding is small when ionic strength is high, but is much more important in low ionic strengths. The hydrodynamic properties of the outside-bound dye are different from those of the intercalated dye, while their optical properties are very similar.[54–56] However, the structure of the outside-bound complex is still obscure. Recent fluorescence temperature-jump studies[57,58] have revealed that there exists different regions in DNA with different kinetics of proflavine binding; one is similar to the binding mechanism of poly d(A-T) in which proflavine binds in a two-step process, while the other is similar to the binding mechanism of poly d(G-C) which is characterized by a single resolvable step. Kinetic studies suggest that the actual mechanism for dye binding in the strong binding region might not be as simple as predicted by the above two models. It is clear that the binding mode depends on the dye structure and the nature of nucleic acids.

In the strong binding region, an anticooperative character of the binding becomes evident. The anticooperativity has been explained in terms of the excluded-site model in which the intercalation cannot occur at sites immediately adjacent to one already occupied.[41,59-61]

There is another evidence that the strong binding process is itself composite in character and involves at least two modes of interaction of dye with DNA. Armstrong *et al.*[41] have examined the binding of proflavine and acridine orange to DNA by a combination of equilibrium dialysis and spectroscopy. While at an ionic strength of 0.2 the binding isotherms could be analyzed in terms of a single complex species, at the lower ionic strength of 0.002 there was indication of two distinct complex species, whose characteristic binding constant differed by an order of magnitude. Armstrong *et al.*[41] attributed the stronger binding mode to an intercalation process and the weaker mode to the binding of a second acridine by the exposed portion of an already intercalated acridine.

There is, moreover, strong evidence that A-T and G-C base pairs are not equivalent with respect to the properties of the intercalated dye molecule. In particular, the fluorescence of many acridine dyes is extensively quenched upon binding to G-C pairs.[62-66] There is also evidence that binding to the nonquenching A-T pairs may be energetically favored in some cases.[2]

## 3.1.2. Weak Binding

Substantial uncertainty still persists with respect to the nature of the weak complexes, whose structure may, moreover, vary for different aminoacridines. It appears likely that the forces stabilizing these complexes are predominantly electrostatic and that the acridine cations are externally bound with their positively charged nitrogens close to the phosphate groups.[2] In the case of dyes such as acridine orange, which have a strong tendency to self-associate,[67] the acridine rings may stack up upon each other in a direction parallel to the helix axis, giving rise to an absorption spectrum which resembles that observed for aggregates of the free dye.[68-70] However, aggregate-type spectra are also observed for dyes such as proflavine, which do not have a pronounced tendency to self-associate in the free state,[71] suggesting that nearest-neighbor interactions of bound dye are important in stabilizing weak complexes formed by these dyes as well.

Weak binding becomes increasingly dominant with decreasing molar ratio of polymer phosphate to dye (P/D). Since it occurs at P/D ratios at which strong binding has attained saturation, it is probably an external binding process, in which the dye molecules are attached to the periphery of the helix. Electrostatic factors are very important in weak binding, and the process is much more sensitive to an increase in electrolyte concentration than in the case of strong binding. The degree of cooperativity depends upon the nucleic acid. The quantum yield of fluorescence is in general greatly reduced over the value observed in the region of strong binding.[2,28] In the case of acridine orange, the green fluorescence ($\lambda_{max} = 505$ nm) which is characteristic of the strongly bound dye species is replaced by a red fluorescence ($\lambda_{max} = 635$ nm) in the region of weak binding.[28,70]

Because of its pronounced tendency to self-associate to form stacked complexes, acridine orange has some distinctive features with respect to its interaction with DNA, synthetic polynucleotides, and polyelectrolytes. For both bihelical and single-stranded polynucleotides, as well as for other linear poly-electrolytes, there is always observed at high P/D ratios a complex characterized by green fluorescence and by an absorption band near 500 nm.[28,70] This has been attributed to "monomeric" bound dye which is devoid of nearest-neighbor contacts.[68,70] With decreasing P/D ratio there is a progressive transition to complexes showing red fluorescence and an absorption band at 470 or 440 nm.[28,68–70] The properties of these complexes presumably reflect the mutual nearest-neighbor interactions of stacked clusters of bound dye molecules.[68–70]

It is possible to account semiquantitatively for the observed dependence of the fluorescence and absorption spectra upon the P/D ratio by means of a simple statistical-mechanical treatment.[68–70] If the polyelectrolyte is represented as a linear array of binding sites, then $Y_0$, the mole fraction of combined dye which is bound at isolated sites and has no nearest-neighbor bound dye molecules, is given by[72]

$$Y_0 = \left(\frac{\theta - \theta_n}{\theta}\right)^2 \tag{3}$$

where $\theta$ is the fraction of occupied sites and $\theta_n = n/B$, where $n$ is the number of first nearest-neighbor pairs of bound dye and $B$ is the total number of sites.

The quantity $\theta_n$ may be computed from the quasichemical expression for the number of first nearest-neighbor pairs of bound dye[73]:

$$\theta_n = \theta - \frac{2\theta(1 - \theta)}{1 + \beta} \tag{4}$$

$$\beta = \{1 - 4\theta(1 - \theta)[1 - \exp(-w/kT)]\}^{1/2} \tag{5}$$

where $w$ is the net free energy of interaction of a nearest-neighbor pair. Upon combining equations (3) and (4), we obtain

$$Y_0 = [2(1 - \theta)/(1 + \beta)]^2 \tag{6}$$

Under conditions of low ionic strength ($10^{-3}$), where binding of acridine orange is close to quantitative, it is possible to analyze the dependence of absorption spectrum or green fluorescence intensity upon the P/D ratio in terms of a distribution of bound dye between the monomeric state and clusters, according to equations (4) and (6). This simplified treatment does not take into account the distribution between different modes of binding and is probably most applicable to the case of single-stranded homopolynucleotides. This approach yields values of $w$ which vary strongly for different polynucleotides; its magnitude increases in the order: DNA < poly rU < poly rA < polyaspartic acid.[68, 70]

## 3.2. Nonintercalating Dyes

While it would probably be premature to generalize about the structurally diverse class of nonintercalating dyes, it is of interest that common features have recurred in several instances. Müller et al.[19] have described the binding to DNA of the nonintercalating proflavine derivative 2,7,-di-t-butyl-proflavine. The presence of bulky substituents renders intercalation unlikely a priori. Müller et al. confirmed this view by sedimentation and viscosity measurements. In both cases the changes characteristic of intercalating dyes were not observed.[19]

Despite the nonintercalating character of this dye, there was strong indication of two distinct modes of binding, which corresponded to binding affinities which differed by an order of magnitude.[19, 20] It was possible to distinguish the two binding modes clearly by kinetic measurements, using a temperature-jump technique. In both cases the second-order rate for the forward rate constant was 10–100 times faster than the corresponding quantity for the intercalation of proflavine into DNA.[19]

Equilibrium dialysis studies likewise indicated the presence of at least two binding modes: strong and weak binding.[19] The intercept on the $r/c$ axis of a plot of $r/c$ versus $r$ was found to be dependent upon the base composition of DNA, increasing progressively with increasing AT content. This intercept is equal to $\Sigma k_i n_i$, where $n_i$ is the number of binding sites of type $i$ and $k_i$ is its intrinsic binding constant. A detailed analysis of the binding isotherms suggested a model for which a bound dye molecule occupies three base pairs, at least two of which must be A-T pairs.

In parallel to the behavior of other acridine dyes, the intensity of fluorescence was substantially increased for DNA complexes at high P/D ratios.[19, 20]

In order to interpret optical and hydrodynamic properties, Bontemps et al.[20] have proposed that the strong complex results from the external binding with the insertion of the dye into the major groove of DNA. However, the nature of the weak complex is still not completely elucidated.

The bisbenzimidazole dye Hoechst 33258 provides a second example of a nonintercalating fluorescent dye (Figure 3), which has an altogether different structure. At pH 7 this dye shows a strong primary absorption band at 338 nm.[35-37] The progressive addition of DNA or poly d(A-T) causes a biphasic change in absorption spectrum, an initial decrease in absorbance at low P/D ratios being followed at larger P/D ratios by a gradual increase and shift to longer wavelengths.[39] While the complex formed at low P/D ratios is almost nonfluorescent, that arising at higher ratios has an intense green fluorescence ($\lambda_{max}$ = 460 nm).

The binding isotherms of Hoechst 33258 with different polydeoxyribonucleotides were found to be very dependent upon the base composition and upon the ionic strength.[35,36] While the dye is bound more strongly by poly d(A-T) or poly (dA) · poly (dT) than by poly (dG) · poly (dC) at all ionic strengths, binding by the latter polymer is much more sensitive to an increase in electrolyte concentration, so that in 0.4 M NaCl, binding by poly d(A-T) is still quite strong while that by poly (dG) · poly (dC) is almost abolished. The implication is that a preferential binding to A-T base pairs exists, which is accentuated at higher salt concentrations.

The quantum yield of fluorescence is also dependent upon the base composition, being higher by orders of magnitude for the poly d(A-T) complexes than for the complexes with poly (dG) · poly (dC).[35,36] Parallel studies with polyribonucleotides have indicated that the bihelical species poly (rA) · poly (rU) also forms an intensely fluorescent complex at high P/D ratios, while the single-stranded polymers poly rA and poly rU do not.[39]

It follows from the above results that Hoechst 33258 is bound preferentially by A-T (or A-U) base pairs in bihelical polynucleotides to form fluorescent complexes at high P/D ratios. While the structure of the complexes is still uncertain, it has been proposed that the dye lies within the wide helical groove.[34-39]

In view of the sensitivity to ionic strength, it is probable that electrostatic factors make a significant contribution to the free energy of binding. The quenching of fluorescence observed at low P/D ratios can probably be attributed to nearest-neighbor interactions of bound dye molecules.

The fluorescence of Hoechst 33258 is strongly quenched by BrU. The incorporation of BrU into DNA is reflected by a pronounced drop of fluorescence intensity even at high P/D ratios.[35,36] Since the bromine atom of BrU projects into the major groove of the DNA double helix, quenching of the dye fluorescence by BrU might indicate interaction of the dye with the bromine atom in this groove.[74]

## 4. Fluorescence Lifetimes and Quantum Yields

Fluorescence spectroscopy is one of the most sensitive techniques available for studying the binding interaction of small molecules with macromolecules and for studying the structure and dynamics of macromolecules.[75,76] A combination of steady-state fluorescence techniques and time-resolved emission spectroscopy in the nanosecond or picosecond time scale can provide much valuable information concerning (1) the microenvironments at binding sites, (2) the specific interaction between dyes and binding sites, (3) the intramolecular orientation and the local motion of bound dye, (4) the size and the shape of nucleic acids, and (5) energy transfer processes from nucleic acid bases to dye molecules and between bound dye molecules.

In this section, we first describe fluorescence lifetimes and quantum yields of acridine dyes and ethidium bromide, the most extensively studied fluorescent dyes, upon binding to nucleic acids, and then describe the fluorescence behavior of mononucleotide—dye complexes. Fluorescence decay measurements are particularly useful for distinguishing different types of binding sites. In the sections that follow, particular attention will be paid to two fluorescence techniques of most intense current interest, anisotropy decay (Section 5) and radiationless energy transfer (Section 6). These techniques allow us to detect the torsional mobility of intercalated dye and to determine the conformational structure of nucleic acids and nucleic acid—dye complexes.

## 4.1. DNA—Acridine Dye Complexes

### 4.1.1. Heterogeneity of Binding Sites: Emitting and Quenching Sites

The interaction of acridine dyes with nucleic acids is of special interest because these dyes induce a wide variety of biological effects[1-3]; these effects may be attributed to the specific interaction between dyes and nucleic acids.

The optical properties of dyes are well known to be altered upon binding to DNA, depending on various factors such as P/D, ionic strength, and the base composition of DNA. Here our attention is focused on the results obtained at high P/D ratios where the contribution of free dye and energy transfer between bound dye molecules are negligible. In the cases of acridine orange and proflavine, the most widely studied acridine dyes, the absorption spectra shift to longer wavelengths while the fluorescence spectra shift to shorter wavelengths.[32] These spectral properties are very similar to those observed when the solvent changes from water to ethanol or when the dyes are solubilized into the micelle of sodium dodecyl sulfate. Such behavior, however, is easily interpreted in terms of the polarizability of the medium, by changing the hydrophilic environment into the hydrophobic one. It therefore appears that the binding interaction of dyes with nucleic acids is hydrophobic in nature.[77]

Weill and Calvin[32] were the first to show that the fluorescence quantum yield of the dye is modified by its intercalation between DNA base pairs, increasing by a factor of about 3 in the case of acridine orange, but decreasing by a factor of about 3 in the case of proflavine. This difference is surprising in view of the fact that the quantum yields of both dyes are increased by changing the solvent from water to glycerol or when the dyes are solubilized into the micelle. While acridine orange, upon intercalation, shows a threefold increase in fluorescence lifetime corresponding to its increase in quantum yield, proflavine shows no variation.[78] Tubbs *et al.*[79] have found that the fluorescence quantum yield of bound acriflavine strongly depends on the base composition of DNA, decreasing almost linearly with increasing GC content. Similar behavior has also been observed for proflavine, quinacrine, and 9-aminoacridine.[62–66,80–83] The quantum yields of these dyes upon binding to poly (dG) · poly (dC) are nearly zero or very small as compared to the corresponding values of free dyes, whereas those upon binding to poly d(A-T) remain practically unchanged. In view of these results, it seems reasonable to conclude that there are at least two classes of strong binding sites. One class (G-C base pair) strongly quenches the fluorescence of bound dye, while the other (A-T base pair) does not alter its fluorescence quantum yield. The first is called "quenching sites" and the second is called "emitting sites."[62] In the case of acridine orange, both G-C and A-T pairs seem to act as emitting sites.

There is, moreover, strong evidence that G-C pairs act as quenching sites. Ramstein and Leng[84] have indicated that proflavine also interacts with chemically methylated DNA, in which the methylation occurs at the $N^7$ of the guanine base. The similarity of the optical properties for the complexes with DNA and methylated DNA shows that proflavine is also intercalated in the methylated DNA. However, the fluorescence of proflavine upon binding to methylated DNA is enhanced as compared to that of free dye and can be quenched by iodide ions. This result implies that G-C pairs play a major role in the quenching of the dye fluorescence and the dye bound to methylated DNA may be exposed partially to the solvent.

Kubota[65] has examined the lifetime and quantum yield characteristics of various acridine derivatives upon binding to DNA. Summarizing the results obtained by Kubota[65] and other workers,[63,64,66] we find that acridine dyes are grouped into three classes with respect to the fluorescence behavior.

(1) Class I: 3-aminoacridine, proflavine, acriflavine, acridine yellow, quinacrine, and 9-aminoacridine. A preferential quenching of the dye fluorescence by G-C pairs is observed.

(2) Class II: 3-dimethylaminoacridine, acridine orange, 3,6-bis-methylamino-acridine, 3,6-bis-ethylaminoacridine, and 3,6-bis-diethylaminoacridine. Fluorescence properties show no dependence on the kind of binding sites.

(3) Class III: 10-methylacridinium cation. All binding sites almost completely quench the dye fluorescence.

From comparison of Classes I and II, it can be noted that the substitution by alkyl groups of one or both hydrogens of the amino groups leads to a change of the dye from Class I to II. It seems likely that, in Class I, the quenching of fluorescence arises from a specific interaction between the G-C pair and the amino groups of the acridine ring. Experiments with mononucleotide–dye systems provide potential evidence for the specificity of the DNA base–dye interaction; this will be discussed later.

### 4.1.2. Heterogeneity of Emitting Sites as Revealed by Fluorescence Decay Measurements

Recent studies by nanosecond pulse fluorometry have shown that the fluorescence decay behavior of the DNA–dye complexes is not so simple, but rather complicated.[81,85–89] This is, of course, expected from the content of DNA bases and the existence of different binding modes.

Latt et al.[90,91] have indicated that the fluorescence decay of quinacrine bound to different DNAs is nonexponential, but the range of the lifetime is much smaller than the variations of the corresponding quantum yield. This anomaly was attributed to prerelaxation phenomena and/or to the heterogeneity of the binding sites. However, the binding interaction of quinacrine seems to be much more complex because the fluorescence decay of the dye itself deviates from exponentiality and depends on the pH of the solution.[87] This nonexponentiality was ascribed to the coexistence of several ionic and thus fluorescent species.[87,92]

Georghiou[85] has indicated that the fluorescence decay of proflavine bound to DNA is nonexponential and that the deviation from exponentiality becomes more pronounced as the GC content of DNA increases. For M. lysodeikticus DNA, which has a 72% GC content, a very fast component corresponding to a lifetime much smaller than 1 ns was detected and ascribed to the dye bound in the vicinity of G-C base pairs. On the other hand, a long-lived component was ascribed to the dye bound near A-T base pairs.

Furthermore, Duportail et al.[86] have reported that the fluorescence decays of several acridine dyes which belong to Class I can be resolved into two exponential components corresponding to a short and a long lifetime and that the relative proportions of these two components depend only slightly on the DNA base composition but do not depend on the structure of acridine dyes. They proposed that lifetime heterogeneity corresponds to two discrete steps in the complex formation disclosed by kinetic studies;[54,55] the short lifetime arises from a semiintercalated or an externally bound dye, and the long lifetime arises from a totally intercalated dye. However, their results should be reexamined because they did not take into account artifacts, anisotropic contributions to the observed decay,[93,94] and energy-dependent effects of the photomultiplier tube,[95] which may introduce significant errors in determinations of the decay parameters.

Kubota and co-workers[81,88,89] have systematically investigated the binding

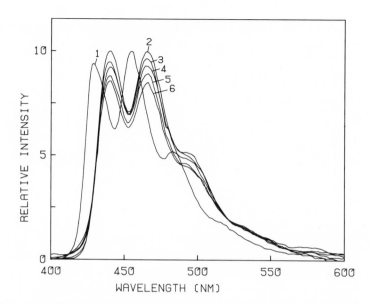

FIGURE 4. Fluorescence quantum spectra of 9-aminoacridine ($8 \times 10^{-6}$ to $1.3 \times 10^{-5}$ M) in the presence of various DNAs in 0.005 M phosphate buffer (pH 6.9) at 25°C: (1) free; (2) poly d(A-T) (P/D = 136); (3) *Cl. perfringens* DNA (P/D = 400); (4) T2 DNA (P/D = 101); (5) calf thymus DNA (P/D = 205); (6) *E. coli* DNA (P/D = 417). The excitation wavelength was 390 nm. The units of the ordinate are arbitrary; the maximum of each spectrum is properly reduced to avoid overlapping. (From Kubota and Motoda Ref. 89.)

interaction of acridine dyes with DNAs of various base composition by a combination of steady-state quantum yield and transient decay measurements. It has been found that the fluorescence decay of acridine orange, which belongs to Class II, is exponential upon binding to DNA and is very similar to that of the dye upon binding to poly d(A-T) or poly (dG) · poly (dC), indicating similar environments for all fluorescing species.[81] In contrast, the decay behavior of dyes which belong to Class I is found to be much more complex. We describe here in some detail the results obtained with 9-aminoacridine[88,89] and acridine yellow,[96] both of which have strong mutagenic activity.[1,2]

The fluorescence quantum yield ($\phi_F$) of the dye bound to DNA strongly depends on its base composition. The value of $\phi_F$ decreases with increasing GC content and is nearly zero for the poly (dG) · poly (dC)–9-aminoacridine complex (Table 1). Figures 4 and 5 depict the fluorescence spectra of 9-amino-acridine and acridine yellow, respectively, in the presence of various DNAs at high P/D ratios. The shape or the maximum of the fluorescence spectrum of 9-aminoacridine bound to DNA can be seen to be almost identical with that of the dye bound to poly d(A-T), regardless of the GC content. On the other hand,

Table 1. Fluorescence Decay Parameters and Quantum Yields for 9-Aminoacridine, Acridine Yellow, and Their Complexes with DNAs of Various Base Composition[a]

| DNA | GC (%) | P/D | $\tau_1$ | $\alpha_1$ | $\tau_2$ | $\alpha_2$ | $\tau_3$ | $\alpha_3$ | $\phi_F$ |
|---|---|---|---|---|---|---|---|---|---|
| **9-Aminoacridine[b]** | | | | | | | | | |
| Free | | | 15.8 | 1.00 | | | | | 0.96 |
| Poly d(A-T) | 0 | 136 | 31.3 | 1.00 | | | | | 0.73 |
| Cl. perfringens DNA | 30 | 400 | 28.9 | 0.40 | 12.3 | 0.21 | 2.1 | 0.39 | 0.129 |
| T2 DNA | 34 | 404 | 28.6 | 0.32 | 12.8 | 0.17 | 2.2 | 0.51 | 0.090 |
| Calf thymus DNA | 42 | 395 | 27.8 | 0.29 | 11.2 | 0.15 | 1.8 | 0.56 | 0.042 |
| | | 328[c] | | | | | | | 0.026 |
| E. coli DNA | 50 | 417 | 26.8 | 0.06 | 12.6 | 0.29 | 1.7 | 0.65 | 0.028 |
| M. lysodeikticus DNA | 72 | 400[d] | 27.6 | 0.20 | 12.9 | 0.17 | 1.7 | 0.63 | <0.002 |
| Poly (dG) · poly (dC) | 100[e] | 130 | | | | | | | ≈0 |
| **Acridine yellow[f]** | | | | | | | | | |
| Free | | | 5.2 | 1.00 | | | | | 0.42 |
| Poly d(A-T) | 0 | 108 | 6.9 | 1.00 | | | | | 0.53 |
| Cl. perfringens DNA | 30 | 407 | 6.9 | 0.49 | 1.6 | 0.51 | | | 0.25 |
| Calf thymus DNA | 42 | 400 | 6.7 | 0.27 | 1.5 | 0.73 | | | 0.14 |
| E. coli DNA | 50 | 399 | 6.8 | 0.20 | 1.3 | 0.80 | | | 0.11 |
| M. lysodeikticus DNA | 72 | 400 | 6.9 | 0.05 | 1.2 | 0.95 | | | 0.05 |
| Poly (dG) · poly (dC) | 100[g] | 101 | 4.8 | 0.39 | 1.3 | 0.61 | | | 0.07 |

[a] The solvent was 0.005 M phosphate buffer (pH 6.0) at 25°C. The decay was observed at 455 and 495 nm, respectively, for 9-aminoacridine and acridine yellow; the results obtained here are not dependent on the emission wavelength. $\chi^2$ values ranged from 1.1 to 1.9. $\tau$ is given in nanoseconds, and the amplitudes ($\alpha$) are normalized to unity.
[b] From Kubota and Motoda (Ref. 89).
[c] Denatured DNA.
[d] The fluorescence intensity was too weak to observe decay curves.
[e] According to the manufacturer (Miles), G = 49% and C = 51%.
[f] Kubota and Motoda, unpublished results.
[g] G = 47% and C = 53%.

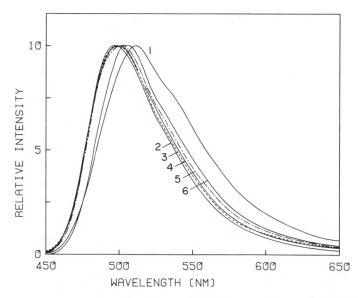

FIGURE 5. Fluorescence quantum spectra of acridine yellow ($5.0 \times 10^{-6}$ M) in the presence of various DNAs in 0.005 M phosphate buffer (pH 6.9) at 25°C; (1) free; (2) poly d(A-T) (P/D = 108); (3) *Cl. perfringens* DNA (P/D = 200); (4) calf thymus DNA (P/D = 799); (5) *M. lysodeikticus* DNA (P/D = 400); (6) poly (dG) · poly (dC) (P/D = 101). The excitation wavelength was 400 nm. The maximum of each spectrum is normalized.

the shape of the fluorescence spectrum of the bound acridine yellow as well as its maximum are dependent on the GC content. A blue shift and a narrowing of the fluorescence band are observed upon binding to DNA as compared to the spectrum of the free dye. The maximum of the fluorescence band can be seen to shift progressively toward longer wavelengths with increasing GC content. This change indicates that the fluorescence spectrum may be assigned to the superposition of emissions of the dye bound near A-T and G-C base pairs. In view of these results, it can be concluded that G-C base pairs almost completely quench the fluorescence of the bound 9-aminoacridine, while acridine yellow bound in the neighborhood of G-C pairs is still fluorescent although its fluorescence is substantially quenched.

Nanosecond pulse fluorometry revealed that the decay kinetics of 9-aminoacridine and acridine yellow upon binding to poly d(A-T) obeys a single-exponential decay law, whereas that of both dyes upon binding to DNA does not.[88,89,96] Figures 6 and 7 display typical decay curves obtained with 9-aminoacridine and acridine yellow, respectively, upon binding to *Cl. perfringens* DNA. The analysis of the observed decay curves was performed by the method

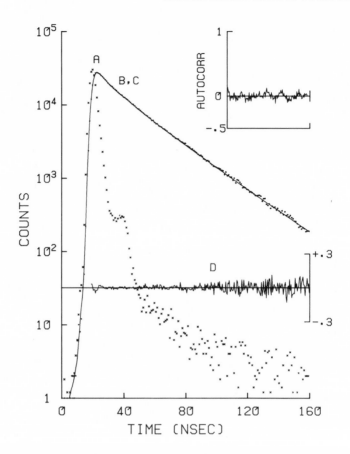

FIGURE 6. Three-component analysis of the fluorescence decay of the *Cl. perfringens* DNA–9-aminoacridine complex (P/D = 201) in 0.005 M phosphate buffer (pH 6.9) at 25°C. The concentration of 9-aminoacridine was $1.3 \times 10^{-5}$ M, and the decay was observed at 455 nm. Curve A is the apparatus response function. Curve B is the observed decay curve. The smooth curve C shows the computed decay curve. Curve D is the weighed residuals. The inset is the autocorrelation function of the residuals. Parameters obtained: $\tau_1 = 29.20$ ns, $\tau_2 = 12.23$ ns, $\tau_3 = 2.17$ ns, $\alpha_1 = 0.123$, $\alpha_2 = 0.070$, $\alpha_3 = 0.130$, and $\chi^2 = 1.82$. The amplitudes normalized to unity are $\alpha_1 = 0.38$, $\alpha_2 = 0.22$, and $\alpha_3 = 0.40$. (From Kubota and Motoda Ref 89.)

of nonlinear least squares,[97] assuming that the fluorescence decay $I(t)$ is a sum of exponentials:

$$I(t) = \sum_{i=1}^{n} \alpha_i \exp\left(-t/\tau_i\right) \tag{7}$$

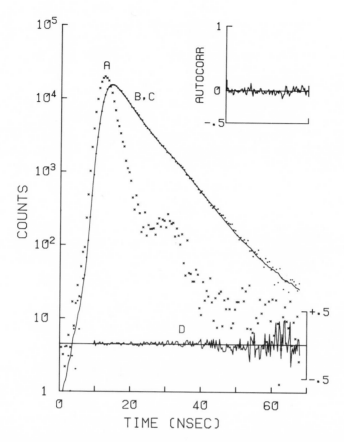

FIGURE 7. Two-component analysis of the fluorescence decay of the *Cl. perfringens* DNA–acridine yellow complex (P/D = 407) in 0.005 M phosphate buffer (pH 6.9) at 25°C. The concentration of acridine yellow was $5.0 \times 10^{-6}$ M, and the decay was observed at 495 nm. Legends to each curve are as described for Figure 6. The computed decay curve is based on the following parameters: $\tau_1 = 6.87$ ns, $\tau_2 = 1.60$ ns, $\alpha_1 = 0.193$, $\alpha_2 = 0.201$, and $\chi^2 = 1.33$. The amplitudes normalized to unity are $\alpha_1 = 0.49$ and $\alpha_2 = 0.51$.

where $\alpha_i$ and $\tau_i$ are the amplitude and lifetime, respectively, of the $i$th component. The reduced $\chi^2$, the weighed residuals, and the autocorrelation function of the residuals[97] indicate that the decay kinetics of 9-aminoacridine upon binding to DNA is consistent with a three-exponential decay law, while that of acridine yellow is compatible with a double-exponential decay law (Figures 6 and 7). Typical sets of the decay parameters obtained for both dyes complexed with DNA at high P/D ratios are summarized in Table 1.

It can be seen in Table 1 that each lifetime is almost independent of the GC content of DNA, while the proportion of the amplitudes ($\alpha$'s) is dependent on it. In view of the observation that the decay parameters do not show any significant dependence on the emission wavelength, it is concluded that the fluorescence decay characteristic of the DNA–dye complex is a result of the heterogeneity of the emitting sites.[89]

Because of the strong quenching ability of guanine residues or G-C base pairs, one could expect that the greater contact between the bound dye and the G-C pair causes the more efficient quenching of the dye fluorescence.[98–100] Kubota and Motoda[89] proposed that the decay behavior of the DNA–9-aminoacridine complex has its origin in different binding sites, depending on the distance between the bound dye and the G-C pair. The component $\tau_1$, having almost the same lifetime as the poly d(A-T)–9-aminoacridine complex, is attributed to the dye intercalated in A-T : A-T sites far from G-C pairs. This is consistent with the finding that there is a continuous increase of $\alpha_1$ with decreasing GC content and its striking decrease in the case of denatured DNA (Table 1). The intermediate ($\tau_2$) and short lifetimes ($\tau_3$) may arise from the partly intercalated dye in A-T : A-T sites adjacent to G-C pairs and in A-T : G-C sites, respectively. This interpretation is also in harmony with the dependence of $\alpha_2$ and $\alpha_3$ upon the base composition of DNA.

It is interesting that the decay behavior of the nonmutagenic dye 9-amino-10-methylacridinium cation upon binding to DNA is very similar to that of the mutagenic dye 9-aminoacridine.[89]

The decay and fluorescence spectral features of the bound acridine yellow are quite different from those of the bound 9-aminoacridine. A red shift of the fluorescence band with increasing GC content indicates that acridine yellow bound in the vicinity of G-C pairs is, to a considerable extent, exposed to the surrounding solvent. In addition, there is a relatively high quantum yield of the dye upon binding to poly (dG) · poly (dC) compared to that of 9-aminoacridine. These observations may result from some differences between quenching interactions for acridine yellow and for 9-aminoacridine. It is clear that the G-C pair has stronger quenching ability for 9-aminoacridine (Table 1).

In view of the findings that $\alpha_1$ increases with decreasing GC content and $\tau_1$ is the same as the lifetime of the poly d(A-T)–acridine yellow complex (Table 1), the component $\tau_1$ (long lifetime) is attributed to the dye bound to A-T : A-T sites. On the other hand, the component $\tau_2$ (short lifetime) is ascribed to the partly intercalated or externally bound dye in G-C : G-C or G-C : A-T sites. This is consistent with the fact that $\alpha_2$ increases with increasing GC content and $\tau_2$ is equal to the major component (1.3 ns) of the dye upon binding to poly (dG) · poly (dC) (Table 1). Surprisingly, there is considerable contribution of the component $\tau_1$ in the case of poly (dG) · poly (dC), suggesting the presence of two different forms of the complex. This may be due, in part, to a slight excess of C (47% G, 53% C) in the sample of poly (dG) · poly (dC) studied.

Similar decay characteristics have also been observed for proflavine and acriflavine.[96] However, the contribution of the component $\tau_2$ (short lifetime) was found to increase pronouncedly in the order proflavine < acriflavine < acridine yellow. Of the three dyes, acridine yellow may have the strongest tendency to externally bind or partly intercalate because of the presence of the methyl groups at the 2,7-positions of the acridine ring (Figure 1).

The results presented here suggest that even a small change in the dye structure may cause a large effect on the binding mode. Further study is still necessary for elucidating the nature of the emitting sites.

## 4.2. Nucleic Acid–Ethidium Bromide Complexes

Ethidium bromide (Figure 1) is a trypanocidal dye which elicits several other biological effects.[8] This dye is widely used in spectrofluorometric studies because of its striking fluorescence enhancement upon binding to double-helical RNA's and DNA's. Fuller and Waring[5] have suggested from x-ray diffraction and model building studies that ethidium bromide intercalates between adjacent base pairs as in the case of acridine dyes such as proflavine.[4] Direct evidence for the intercalation has been obtained from optical studies of the solution complexes of ethidium with dinucleoside monophosphates,[101,102] as well as x-ray crystallographic studies of ethidium complexes with dinucleoside monophosphates in the solid state.[46,103,104]

It is generally recognized that strong fluorescence enhancement arises from intercalation of the dye into the double-helical regions of nucleic acids, but there is also evidence for additional nonintercalative, less fluorescent sites which result from the external binding by electrostatic interactions.[7] In contrast to acridine dyes such as proflavine, ethidium bromide exhibits no difference in its fluorescence intensity upon binding between G-C or A-T base pairs.[7,105]

Le Pecq and Paoletti[7] have suggested that the fluorescence enhancement is due to the immersion of the ethidium bromide molecule in the hydrophobic region of nucleic acids where it is protected against quenching by the collisions with the surrounding solvent. Burns[106] has observed that both fluorescence lifetimes and quantum yields of ethidium bromide upon binding to nucleic acids are increased when compared to those of the free dye and attributed the enhancement to a conformational change of the dye such that a forbidden transition becomes allowable. Olmsted and Kearns[107] have carried out experiments which include measurements of solvent and deuterium isotope effects on fluorescence lifetimes and quantum yields. They observed the following: there is an about 3.5-fold increase in the lifetime of free ethidium in going from $H_2O$ to $D_2O$; addition of small amounts of $H_2O$ to nonaqueous solvents decreases the fluorescence whereas addition of small amounts of $D_2O$ enhances it; and the ethidium bromide triplet yield is enhanced upon intercalation or upon deuteration. In order to account for these observations, Olmsted and Kearns[107] proposed that the major pathway for deactivation of free ethidium in aqueous solution involves excited-state proton transfer from the ethidium amino groups to water. The enhancement of the dye fluorescence observed upon binding to nucleic acids is attributed to a reduction in the rate of excited-state proton transfer. Other mechanisms[7,106] are shown to be inconsistent with the above observations.

The mechanism proposed by Olmsted and Kearns[107] suggests that inter-calation sites on different nucleic acids may yield different fluorescence properties, depending on the degree of exposure to solvent of the intercalated ethidium molecule. There is evidence that the fluorescence quantum yield and/or lifetime of ethidium bromide is different when bound to tRNA or when bound to DNA.[7,106,108,109] There is also evidence that the fluorescence behavior depends upon the base sequence.[102,110,111] Garland et al.[110] have studied the fluorescence lifetimes of ethidium bromide complexed with dinucleoside monophosphates. The 21-ns lifetime observed for the $C_pG$–ethidium complex is characteristic of the intercalation complex formed by this pyrimidine $(3',5')$–purine dimer with the ligand.[102] The relatively short lifetime (7.7 ns) observed for the $A_pU$ complex suggested that the bound ethidium is more exposed to solvent than in the intercalation complex formed with $C_pG$. The average lifetime (17.4 ns) of the $U_pA$ complex was resolvable into two components, suggesting the presence of two different forms of the $U_pA$–ethidium complex in solution. These results are in harmony with the obser-vations of Reinhardt and Krugh[102] and Kastrup et al.[111] which demonstrated a clear preference for ethidium to bind more strongly to the pyrimidine–purine sequence when compared to the purine–pyrimidine sequence.

## 4.3. Bifunctional Intercalating Dyes

New intercalating dyes which can bisintercalate in DNA have recently been developed for further elucidating the dye–nucleic acid interaction.[112-115]

Le Pecq et al.[112] have reported that diacridines, of which the two acridine rings are linked by chains of various lengths, are very useful as fluorescence probes for detecting the specific base sequences of DNA. The fluorescence of the monomeric unit is quenched when the acridine ring is in contact with a G-C pair. The fluorescence intensity of the acridine dimer varies as the fourth power of the AT content, suggesting that four consecutive A-T pairs are required for the fluorescing sites. Owing to efficient resonance energy transfer, when only one of the two acridine rings is in contact with a G-C pair, the fluorescence of the dimer is quenched. This fluorescence characteristic might make it possible to probe the specific DNA sequences on chromosomes.

Another interesting category of bifunctional intercalating dyes includes an ethidium homodimer and an acridine ethidium heterodimer.[113,117] These dimers were found to intercalate only one of their chromophores.[117] The fluorescence quantum yield for emission from the phenanthridinium chromophore is greatly increased when either the ethidium dimer or the acridine ethidium dimer is bound to DNA. The efficient energy transfer from the acridine to the phenanthridinium ring is observed in the acridine ethidium dimer upon binding to DNA, depending on the square of the AT content. Because of the strong enhancement of fluorescence upon binding to DNA, the acridine ethidium dimer could be a useful probe for the study of chromosomes.

The binding affinity of these dyes to DNA and the geometry of the inter-calated complex are expected to be different from those observed for ethidium bromide and acridine dyes.[112,114–116]

## 4.4. Mononucleotide–Dye Complexes

As mentioned earlier, binding interactions of dyes with nucleic acids are considerably more complex because of the content and sequence of nucleic acid bases as well as their stereochemical arrangement. One of the approaches for understanding the nucleic acid–dye interaction is to study the complex formation between dyes and mononucleotides.[85,98–100,117–119]

An indication that a specific complex is formed between acridine dyes and nucleosides or mononucleotides has been provided by absorption spectral shifts and changes in the fluorescence intensity of dyes in the presence of nucleosides or mononucleotides.[120–122] Recent systematic investigations by steady-state and transient fluorescence decay measurements have confirmed that all four mononucleotides, AMP, GMP, TMP, and CMP, form molecular complexes with various acridine derivatives in aqueous solution.[85,98–100,117–119]

In general, the presence of nucleotides markedly affects the fluorescence properties of dyes. Table 2 summarizes the relative fluorescence quantum yields ($\phi/\phi_0$) of dyes in the presence of an excess of nucleotides, where $\phi_0$ and $\phi$ are the fluorescence quantum yields of dyes in the absence and the presence of nucleotides, respectively. As can be seen in Table 2, acridine dyes are classified in four groups with respect to the fluorescence behavior. This behavior is in strong parallelism with the fluorescence behavior of dyes upon binding to DNA. While all nucleotides enhance the fluorescence of acridine orange (Group 1), they quench the fluorescence of 10-methylacridinium cation (Group 4). The fact that GMP quenches the fluorescence of acridine dyes which belong to Class I indicates that guanine residues are responsible for the quenching of the dye fluorescence upon binding to DNA (Groups 2 and 3). It is somewhat surprising that AMP in addition to GMP also quenches the fluorescence of 9-aminoacridine and 9-amino-10-methylacridinium cation (Group 3). From the fluorescence properties of these dyes upon binding to poly d(A-T)[89] or double-stranded

Table 2. Relative Fluorescence Quantum Yields ($\phi/\phi_0$) of Acridine Dyes in the Presence of an Excess of Mononucleotides[a]

| | Dyes | AMP | GMP | TMP | CMP | DNA[b] |
|---|---|---|---|---|---|---|
| Group 1 | Acridine orange[c] | 2.50 | 1.65 | 1.64 | 1.60 | 3.00 |
| | Ethidium bromide[d] | 1.57 | 1.60 | 1.10 | 1.10 | 8.7 |
| Group 2 | Proflavine[c] | 1.25 | 0.04 | 1.10 | 1.05 | 0.39 |
| | Acriflavine[d] | 1.24 | 0.046 | 1.13 | 1.33 | 0.30 |
| | Acridine yellow[d] | 1.06 | 0.072 | 1.00 | 1.00 | 0.33 |
| | 9-Methylaminoacridine[d] | 1.90 | 0.15 | 1.40 | 1.30 | — |
| Group 3 | 9-Aminoacridine[c] | 0.074 | 0.030 | 1.10 | 1.10 | 0.047 |
| | 9-Amino-10-methylacridinium[d] | 0.060 | 0.010 | 1.15 | 1.15 | 0.031 |
| Group 4 | 10-Methylacridinium[c] | 0.011 | 0.013 | 0.070 | 0.076 | <0.005 |

[a] The solvent was 0.1 M phosphate buffer (pH 7.0) at 25°C. The concentration of mononucleotides was 0.05 M.
[b] Calf thymus DNA (P/D = 200—400). The solvent was 0.005 M phosphate buffer (pH 6.9) at 25°C.
[c] From Kubota et al. (Ref. 119).
[d] Kubota and Motoda, unpublished results.

poly rA,[123] it is likely that the 6-amino group of the adenine ring free from hydrogen bonding plays a role in the quenching of the dye fluorescence.

We present here a brief delineation of the methodology for obtaining information about the dynamics of the dye–mononucleotide interaction from quenching data.[100] The data can be analyzed according to the following kinetic scheme which takes into account complex formation in the ground state as well as in the excited singlet state[124,125]:

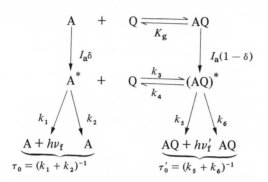

In this scheme, A and Q refer to the dye and quencher molecules, respectively, and $I_a$ is the total light quanta absorbed by the solution; $\delta = \{1 + (\epsilon'/\epsilon) K_g[Q]\}^{-1}$ is the fraction of light absorbed by A, where $\epsilon$ and $\epsilon'$ are the molar extinction coefficients of A and AQ at the excitation wavelength; $K_g = [AQ]/[A][Q]$ is the equilibrium constant for the ground-state equilibrium.

From the proposed kinetic scheme the total fluorescence intensity at time $t$, $I_F(t)$, is given by[100]:

$$I_F(t) = k_1[A^*] + k_5[(AQ)^*] = A_1 \exp(-\lambda_1 t) + A_2 \exp(-\lambda_2 t) \tag{8}$$

where

$$\lambda_1, \lambda_2 = \tau_1^{-1}, \tau_2^{-1} = \tfrac{1}{2}[(k_1 + k_2 + k_3[Q] + k_4 + k_5 + k_6)$$
$$\mp \{(k_4 + k_5 + k_6 - k_1 - k_2 - k_3[Q])^2 + 4k_3 k_4[Q]\}^{1/2}] \tag{9}$$

Equation (9) yields the following relationship for the $\lambda$:

$$\lambda_1 + \lambda_2 = k_1 + k_2 + k_3[Q] + k_4 + k_5 + k_6 \tag{10}$$

$$\lambda_1 \lambda_2 = (k_1 + k_2)(k_4 + k_5 + k_6) + k_3(k_5 + k_6)[Q] \tag{11}$$

Once one determines $\lambda_1, \lambda_2$ and the unquenched lifetime $\tau_0$, one can obtain the rate constants $k_3, k_4$, and $k_5 + k_6$ from plots of $\lambda_1 + \lambda_2$ and $\lambda_1 \lambda_2$ against $[Q]$.

These results are in qualitative agreement with the well-known excimer or exciplex mechanism.[126-128]

The following steady-state equation can also be derived:[98,125]

$$\frac{\phi}{\phi_0} = \frac{\delta + k_4 \tau_0' + R(1 - \delta + k_3 \tau_0 [Q])}{1 + k_4 \tau_0' + k_3 \tau_0 [Q]} \tag{12}$$

where $\phi$ is the apparent fluorescence quantum yield of the dye in the presence of quencher and $R = \phi_0'/\phi_0$ is the ratio of the quantum yield of the complex AQ to that of free A.

If the contribution of the complex AQ to the fluorescence emission is negligible in all circumstances, we have

$$\phi_0' \ll \phi_0 \qquad k_4 \tau_0' \ll 1 \tag{13}$$

and thus, equation (12) reduces to equation (14)

$$\phi/\phi_0 = \delta/(1 + k_3 \tau_0 [Q]) \tag{14}$$

In cases where equation (13) holds, the decay kinetics follows a single-exponential decay law and the change in the dye fluorescence lifetime $\tau$ due to increasing quencher concentration is given by

$$\tau_0/\tau = 1 + k_3 \tau_0 [Q] \tag{15}$$

A combination of steady-state and transient fluorescence experiments now allows us to determine the ground and excited-state equilibrium constants as well as the rate constants of complex formation ($k_3$) and dissociation ($k_4$). The ground-state equilibrium constant $K_g$ is also determined from the observed absorption change.[98-100,117-119] On the other hand, the equilibrium constant $K_e$ for the excited-state equilibrium can also be estimated from $K_g$ and from the spectral shift due to the complex formation, using the approximate equation[124,129]

$$\log K_e = \log K_g + (0.625/T)\Delta\tilde{\nu} \tag{16}$$

where $\Delta\tilde{\nu}$ is the 0–0 band shift ($cm^{-1}$) of the free and complexed dye for the $S_0 \rightarrow S_1$ transition.

The quantitative analysis just described has been successfully applied to the GMP–proflavine,[98] mononucleotide–9-aminoacridine,[99,100,117] and mononucleotide–10-methylacridinium systems.[119] The fluorescence decays of the AMP–9-aminoacridine systems were found to obey a double-exponential

decay law, reflecting the contribution of the complex AQ to the total fluorescence.[99,100] The other systems showed a single-exponential decay unless
the nucleotide concentration was extremely high.[99,100,119]

A red spectral shift upon formation of the complex indicates that there is a
substantial increase in the strength of interaction in the excited singlet state of
the dye. Typical values obtained for $K_g$ and $K_e$ are as follows: $K_g = 310\,M^{-1}$
and $K_e = 1550\,M^{-1}$ for the GMP–proflavine system,[98] $K_g = 125\,M^{-1}$ and
$K_e = 230\,M^{-1}$ for the GMP–9-aminoacridine system,[100] and $K_g = 175\,M^{-1}$ and
$K_e = 1210\,M^{-1}$ for the AMP–9-aminoacridine system.[100] The enthalpy and
entropy changes for the excited-state complex formation are somewhat larger
than those for the ground-state complex formation. This means that the
excitation of the dye may result in increased stability of the complex. These
results are important for elucidating the dynamics of the excited electronic
states of the photoinactivating dyes, since the inactivation involves the preferential destruction of guanine residues.[3,16]

The magnitudes obtained for the rate constant $k_3$ ($3 \times 10^9$ to $8 \times 10^9\,M^{-1}$
$s^{-1}$) are as expected for a diffusion-controlled forward reaction.[99,100,119] It is,
on the other hand, found that the backward rate constant $k_4$ is much smaller
than $k_3$, in agreement with the finding $K_e \gg K_g$.[99,100]

The thermodynamic properties upon formation of the complex are as
expected for charge transfer and for hydrogen-bonding interactions.[98,100,117]
There would be two possibilities for nonradiative relaxation in $(AQ)^*$–inter-
system crossing in $(AQ)^*$ and charge transfer between nucleic acid bases and
acridine dyes.[98,100] However, the exact mechanism is still not completely
understood. The study of transient intermediates by laser photolysis will answer
the problem.

Schreiber and Daune[66] have found that the fluorescence of several acridine
dyes having mutagenicity is quenched by G-C pairs upon binding to DNA and
attributed frame-shift mutations to the electronic modification of G-C pairs or
guanine residues. On the contrary, there is strong evidence that GMP or G-C
pairs also quench the fluorescence of the nonmutagenic dyes 10-methyl-
acridinium cation[119] and 9-amino-10-methylacridinium cation.[89,99,100]
Therefore, the quenching behavior may not be directly related to the
mutagenicity. It is expected that even a small change in the dye structure may
cause a large effect on the binding interaction of dyes with DNA.[2] Perhaps the
geometry of the intercalated dye may play a significant role in the biological
activity of the dye.

The study of the interaction of fluorescent dyes with dinucleotides or oligo-
nucleotides is also helpful for further understanding the nucleic acid–dye
interaction.[101,102,111,130] It has been demonstrated that there is a base-sequence
specificity of the strong binding of ethidium bromide to dinucleotides or
tetranucleotides.[101,102,111] Ethidium bromide exhibits a clear preferential

binding to pyrimidine–purine sequences as compared to purine–pyrimidine sequences.

## 5. Decay of Fluorescence Anisotropy

### 5.1. General Considerations

Fluorescence anisotropy is another major observable phenomenon in emission spectroscopy. If the fluorescence sample is excited by a plane-polarized and infinitely short flash, the fluorescence anisotropy becomes time dependent, its decay being defined as[131]

$$r(t) = \frac{I_{\parallel}(t) - I_{\perp}(t)}{I_{\parallel}(t) + 2I_{\perp}(t)} = \frac{d(t)}{s(t)} \tag{17}$$

where $I_{\parallel}(t)$ and $I_{\perp}(t)$ are the fluorescence intensities observed through a polarizer whose transmission axis is aligned parallel and perpendicular, respectively, to the direction of polarization of the exciting light. The function $s(t)$ is proportional to the decay of total fluorescence intensity and, in the simplest case, decays as a single exponential:

$$s(t) = s_0 \exp(-t/\tau) \tag{18}$$

where $\tau$ is the lifetime of the excited singlet state. The anisotropy decay $r(t)$ depends on the rotational motion of fluorescent molecules. For a rigid spherical molecule, the anisotropy decays as a single exponential[132]:

$$r(t) = r_0 \exp(-t/\phi) \tag{19}$$

where $r_0$ is the limiting anisotropy ($-0.2 \leqslant r_0 \leqslant +0.4$) and $\phi$ is the rotational correlation time. The quantity $r_0$ is given by[131]

$$r_0 = 0.6 \cos^2 \alpha - 0.2 \tag{20}$$

where $\alpha$ is the angle between the absorption and emission transition moments of the fluorescent molecule. If both moments are in the same direction ($\alpha = 0°$), we obtain $r_0 = 0.4$. On the other hand, the rotational correlation time $\phi$ is related to the size of the rotating molecule by the equation

$$\phi = V\eta/kT = 1/6D \tag{21}$$

where $V$ is the hydrated volume of the molecule, $\eta$ is the viscosity of the medium, $k$ is the Boltzmann constant, $T$ is the absolute temperature, and $D$ is the rotational diffusion coefficient.

The anisotropy decay for molecules other than spheres is complex. When the molecule has no symmetry properties, five exponentials appear in $r(t)$.[133-135] For prolate and oblate ellipsoids, the function $r(t)$ decays as a sum of three exponentials if the absorption and emission transition moments are in the same direction[136]:

$$r(t) = r_0 \sum_{i=1}^{3} r_i \exp\left(-t/\phi_i\right) \tag{22}$$

where the rotational correlation times $\phi_i$ are related to the rotational diffusion coefficients characterizing a Brownian rotation about the symmetry axis and about an axis that is perpendicular to the symmetry axis of the rotating molecule, and $r_i$ is dependent on the orientation angle between the transition moment of the fluorescent label and the symmetry axis of the ellipsoid. The reader is referred to other articles for details of the theoretical treatment of anisotropy decay.[136,137]

The steady-state anisotropy $\langle r \rangle$ can be obtained from the time-dependent anisotropy by taking the time average of $r(t)$ since a continuous light source may be considered as a sum of an infinite number of flashes. If $s(t)$ and $r(t)$ are single exponentials like equations (18) and (19), $\langle r \rangle$ is given as follows[137]:

$$\langle r \rangle = \frac{\int_0^\infty d(t)\,dt}{\int_0^\infty s(t)\,dt} = \frac{1}{\tau} \int_0^\infty r(t) \exp\left(-t/\tau\right) = r_0/(1 + \tau/\phi) \tag{23}$$

This form is equivalent to the well-known Perrin's equation.[138]

Using an appropriate fluorescent probe, we can obtain important information on the size, conformation, and flexibility of a macromolecule from both the steady-state and transient anisotropy measurements. It should be pointed out, however, that the transient measurements are much superior to the steady-state ones; the latter give only the time-average anisotropy, while the former is more direct and more reliable.

There are several requirements for choice of a fluorescent probe: (1) the fluorescence quantum yield of the probe is high, (2) the probe has a relatively long and single lifetime, and (3) the probe is rigidly bound to macromolecules within the excited-state lifetime. The trypanocidal dye, ethidium bromide, has become a standard probe for analyzing nucleic acid structure because of its relatively long lifetime and high quantum yield upon binding to nucleic acids.

We next describe some typical experimental results obtained with nucleic acid–dye complexes and then describe the anisotropy decay due to energy transfer whose data provide information on the unwinding angle of the DNA helix caused by intercalation of the dye.

## 5.2. tRNA–Ethidium Bromide Complexes

It is well established that ethidium bromide strongly binds to double-stranded RNA as well as DNA by intercalating between adjacent base pairs in the double helix.[5–7] Tao et al.[108] have shown that there is a single, strong ethidium binding site on yeast tRNA$^{Phe}$ and on unfractionated tRNA. NMR studies have revealed that this site is located between base pairs 6 and 7 on the amino acid acceptor stem.[139] In such location, it is thought that the bound dye must be rigidly held with respect to the macromolecule within the lifetime of the excited singlet state.

The first attempt to measure the anisotropy decay of the tRNA–ethidium bromide complex has been made by Tao et al.[108] who found that the anisotropy decay is magnesium dependent. For unfractionated yeast tRNA in the presence of 0.003 M $Mg^{2+}$, the anisotropy decays with a single exponential of $\phi = 24.8$ ns. This value corresponds to a molecular volume $V = 114{,}000$ Å$^3$ on the assumption that the tRNA is spherical in shape. By taking the degree of hydration to be 1.2 g $H_2O$/g tRNA, the molecular volume of the tRNA can be calculated to be 75,000 Å$^3$. The discrepancy between the observed and calculated molecular volumes was attributed to deviation from spherical geometry.[108] If the tRNA is assumed to be a prolate ellipsoid, the axial ratio ranges from 2.0 to 3.0, depending on the degree of hydration and the orientation angle of the transition moment of ethidium bromide with respect to the long axis of the ellipsoid. This result is not in conflict with the tertiary structure of yeast tRNA$^{Phe}$ disclosed by subsequent x-ray crystallographic studies.[140,141]

For the unfractionated yeast tRNA in the absence of $Mg^{2+}$, the fact that the anisotropy decay obeys a double-exponential decay law like $r(t) = 2.1 \exp(-t/16.8 \text{ ns}) + 1.7 \exp(-t/39.4 \text{ ns})$ rules out a spherical geometry. If the tRNA is assumed to be a prolate ellipsoid, the values of the two correlation times, together with the ratio of the two amplitudes, lead to an axial ratio of 4.6. This result means that there is a magnesium-dependent conformational change and that the tRNA becomes more elongated as $Mg^{2+}$ is removed.[108]

Anisotropy decay measurements have also been carried out with the fluorescence from the Y base of tRNA$^{Phe}$ in the presence of $Mg^{2+}$.[142] The fluorescence of the Y base yields a correlation time of 10 ns, whereas the fluorescence of the intercalated ethidium gives $\phi = 24.5$ ns. This discrepancy suggests that some degree of flexibility exists for either the Y base itself or the anticodon loop of the tRNA.[142]

## 5.3. DNA–Ethidium Bromide and DNA–Acridine Dye Complexes

Ellerton and Isenberg[143] have studied the binding of proflavine to DNA by measuring the steady-state fluorescence polarization at high P/D ratios. They found that the degree of polarization ($p$) for the DNA–proflavine system is about 0.375 at 4°C, significantly lower than that observed for the glycerol–proflavine system ($p = 0.468$) and that the slope of a Perrin plot is much larger than that expected for the motion of the whole DNA molecule. They inferred that the depolarization of the DNA–proflavine complex may arise from some form of local flexibility of the DNA helix.

Kubota[144] has determined the mean rotational relaxation times ($\rho_h$) for various acridine dyes complexed with calf thymus DNA from Perrin plots. The $\rho_h$ values were of almost the same order of magnitude ($\rho_h \simeq 50$ ns) for proflavine, acridine orange, 3,6-bis-methylaminoacridine, and 3,6-bis-ethylamino-acridine. However, these values are much too small for the rotation of the entire DNA molecule. Furthermore, the $\rho_h$ value for 3,6-bis-diethylaminoacridine (32 ns) was significantly lower than those for the other acridines. This result implies that the torsional mobility of the intercalated dye may depend on the dye structure.

In order to obtain more direct information, the decays of fluorescence anisotropy for the DNA–ethidium bromide and DNA–acridine dye complexes have been measured by using nanosecond pulse fluorometry.

Wahl and collaborators[109,145,146] have found that the anisotropy decay of the calf thymus DNA–ethidium bromide complex can be fitted well with two rotational correlation times; one of these was 24 ns and the other was infinite. They attributed the 24-ns correlation time to the torsional motion of the inter-calated dye which results from the local deformation of the DNA helix and the infinite one to a global DNA molecular motion. Genest and Wahl[147] have reexamined the anisotropy decay of ethidium bromide upon binding to calf thymus DNA (42% GC) and M. lysodeikticus DNA (72% GC). It has been shown that the base composition of DNA does not influence the anisotropy decay and that the anisotropy decay can be described by either of the following empirical formulas[147]:

$$r(t) = 0.32 \, (0.5 \exp \, (-t/23 \text{ ns}) + 0.5) \tag{24}$$

$$r(t) = 0.32 \, (0.35 \exp \, (-t/15 \text{ ns}) + 0.65) \tag{25}$$

The difference between the two curves is presumably a measure of the precision of the original data.

It has also been shown that the anisotropy decay for various acridine dyes upon binding to calf thymus DNA is very similar to that of the bound ethidium bromide.[148] For example, the following empirical equations were obtained[148]:

$$r(t) = 0.34 \, (0.62 \exp \, (-t/10.6 \, \text{ns}) + 0.38) \qquad \text{for proflavine} \qquad (26)$$

$$r(t) = 0.33 \, (0.71 \exp \, (-t/10.2 \, \text{ns}) + 0.29) \qquad \text{for acridine orange} \qquad (27)$$

In general, the anisotropy decay characterizing a limited local motion of a chromophore bound to a macromolecule or a membrane is given by the following formula[137]:

$$r(t) = r_0[\alpha \exp \, (-t/\phi) + (1 - \alpha)] \qquad (28)$$

where $\alpha$ is a constant expressed by

$$r_\infty = r_0(1 - \alpha) = r_0 \frac{(3 \, \langle \cos^2 \, \omega \rangle - 1)}{2} \qquad (29)$$

where $r_\infty$ is the anisotropy at $t \to \infty$, $\omega$ is the rotation angle between the absorption moment at $t = 0$ and the emission moment at $t \to \infty$, and $\langle \cos^2 \, \omega \rangle$ is the average of $\cos^2 \, \omega$.

Combining equations (24)–(27) with equation (29) yields the following values for $\omega$: $\omega = 35°$ or $29°$ for ethidium bromide, $\omega = 40°$ for proflavine, and $\omega = 44°$ for acridine orange. The result means that bound dye molecules behave as if they could wobble within the restricted angles in their binding sites. The magnitude of oscillation for ethidium bromide is a little smaller than those for acridine dyes. This limited oscillation in the case of ethidium bromide is presumably due to the presence of a bulky phenyl group attached to the phenanthridinium ring (Figure 1).

It has also been demonstrated that the anisotropy decay of ethidium bromide upon binding to synthetic polynucleotides, poly d(A-T) and poly (rA–rU), is very similar to that of the dye upon binding to DNA.[149,150]

In recent papers two theoretical models have been presented to interpret the depolarization caused by internal rotatory Brownian motion in the DNA helix: an elastic model of semiflexible chain macromolecules[151] and a model which consists of a series of identical rigid rods connected by torsion springs.[152] Both models predict that the decay of the fluorescence anisotropy of an intercalated dye should be complex. If DNA behaves as a uniformly elastic rod, the anistropy decay $r(t)$ is expected to obey the following decay law[151]:

$$r(t) = r_0 \left\{ \frac{1}{4} + \frac{3}{4} \exp \left[ -\frac{2kT}{\pi} \, (t/b^2 \eta C)^{1/2} \right] \right\} \qquad (30)$$

where $C$ is the torsional rigidity, $\eta$ the solvent viscosity, $k$ the Boltzmann constant, and $b$ the helix radius. The rigid-rod and torsion spring model of Allison and Schurr[152] also predicts a nonexponential, $\exp(-\alpha t^{1/2})$, where $\alpha$ is a constant, torsional relaxation at intermediate times, but predicts an initial exponential decay characteristic of uncoupled rod motion at sufficiently short times.

Very recently picosecond time-dependent anisotropy techniques with subnanosecond time resolution have been used to monitor the reorientation of ethidium bromide intercalated in DNA and to test the theoretical models.[153–155] Millar et al.[153,154] analyzed their data in terms of the elastic model of Barkley and Zimm[151] and demonstrated that the observed anisotropy decay follows the decay kinetics represented by equation (30) (Figure 8). Thomas et al.[155] also found that only the intermediate decay zone formula [$\exp(-t^{1/2})$ decay law] of the torsion spring model[152] fits the data. Both research groups were able to provide an accurate value for the torsional rigidity of DNA, $C = 1.3 \times 10^{-19}$ erg cm, which is in excellent agreement with the values estimated from supercoiling data ($1.1 \times 10^{-19}$ erg cm) and persistence length data ($1.75 \times 10^{-19}$ erg cm).[153–155]

Furthermore, Millar et al.[156] extended picosecond techniques to synthetic

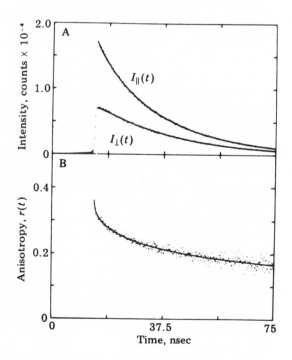

FIGURE 8. (A) Experimental $I_\parallel(t)$ and $I_\perp(t)$ curves for the ethidium–DNA complex in 0.1 M Tris-HCl, pH 7.7, 0.15 M NaCl. (B) Fluorescence anisotropy decay $r(t)$ constructed from the experimental data in (A). The solid lines in (A) and (B) are the best fit of the data to equation (30), with $\tau = 22.6$ ns, $r_0 = 0.36$, and $b^2 \eta C = 2.6 \times 10^{-35}$ erg² s. (Reproduced from Millar et al., Ref. 153.)

polynucleotide–ethidium complexes. They showed that an intact double helix has a torsional rigidity, $C = 1.3 \times 10^{-19}$ erg cm, independent of the base sequence or helical conformation, and that a triple helix is more rigid ($C = 3.1 \times 10^{-19}$ erg cm). It has also been shown that denatured DNA is considerably more flexible than the intact double helix, thus suggesting the influence of secondary structure on internal motions.[153,156] These picosecond techniques now open up the possibility for studying conformational dynamics in a wide range of macromolecules.

## 5.4. Anisotropy Decay Due to Energy Transfer

It has long been known that fluorescence depolarization occurs by singlet–singlet energy transfer between like molecules.[144] Wahl and collaborators[100,137,138,145] have found that the initial slope of the anisotropy decay of ethidium bromide when bound to DNA is increased with decreasing P/D ratio. They attributed this phenomenon to the existence of energy transfer between dye molecules bound to the same DNA molecule. They developed elegant techniques for estimating the unwinding angle of the DNA helix from anisotropy decay data.

It is generally assumed that the anisotropy decay of the fluorochrome upon binding to a macromolecule such as DNA can be described by[137]

$$r(t) = r_B(t) r_{ET}(t) \tag{31}$$

where $r_B(t)$ is the Brownian anisotropy decay measured in the absence of energy transfer and $r_{ET}(t)$ is the anisotropy decay due to energy transfer; $r_B(t)$ is given by equation (28). If $\theta$ is the angle between the transition moment of the initially excited dye, that of the dye excited at time $t$ via energy transfer, $r_{ET}(t)$, is given by[137]

$$r_{ET}(t) = \tfrac{3}{2}(\langle \cos^2 \theta(t) \rangle - 1) \tag{32}$$

where $\langle \cos^2 \theta(t) \rangle$ is the average of $\cos^2 \theta(t)$.

It is well established that the intercalation of a dye molecule into the DNA helix results in the extension (3.4 Å per bound dye) and local unwinding of the helix.[4,5,42] Since the efficiency of energy transfer is closely dependent on the orientation angle between two chromophores on a macromolecule, it is expected that anisotropy decay data in the presence of energy transfer permit the determination of the change in the unwinding angle of the DNA helix induced by intercalation.

A quantitative interpretation of the anisotropy decay[146,147] has been made on the basis of Förster resonance transfer.[159-162] According to the Monte Carlo method described by Paoletti and Le Pecq,[163] the effect of energy transfer on the anisotropy decay has been calculated.[146] The intercalated dye molecules are assumed to distribute randomly among equivalent sites under the conditions of the excluded-site model in which two adjacent sites cannot be simultaneously occupied.[59,60]

The time course of the energy migration is simulated for statistically generated configurations of the DNA–ethidium bromide complex. The average value of the anisotropy $r_{ET}(t)$ is then calculated for a great number of

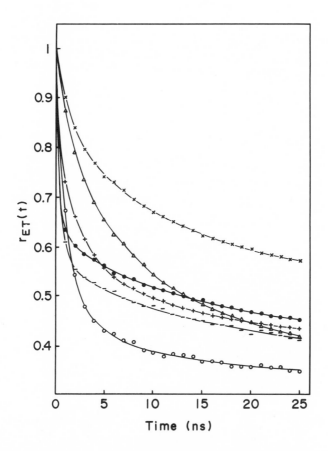

FIGURE 9. $r_{ET}(t)$ curves for the DNA–ethidium bromide complex (P/D = 7.35), computed by the Monte Carlo method with different values of δ: (×) δ = + 20°; (△) δ = + 14°; (●) δ = − 20°; (−) δ = − 12°; (○) δ = 0°. (Reproduced from Genest et al., Ref. 146.)

simulations; $r_{ET}(t)$ is computed with different values of the unwinding angle $\delta$, and the decay function $d(t)$ is determined according to equations (31) and (33),

$$d(t) = r(t)s(t) \tag{33}$$

then convolved with the apparatus response function. The resultant $d(t)$ is then compared with the experimental one by inspecting the weighted residuals and the deviation function.[146]

Figure 9 shows several $r_{ET}(t)$ curves at $P/D = 7.35$ computed by the Monte Carlo method, whereas Figure 10 shows a comparison between the calculated and experimental $d(t)$ curves. A detailed analysis led to the conclusion that the intercalated ethidium unwinds the DNA helix by an angle $\delta = -(16 \pm 4)°$.[146]

Genest and Wahl[147] have reexamined the anisotropy decay by studying the influence of the direction of the dye transition moment as well as the influence of the dye distribution along the DNA helix. A reasonable value of the unwinding angle ($\delta = -18°$) was obtained by assuming that the transition moment direction lies along the long axis of the ethidium molecule.[147]

The results obtained from the anisotropy decay data are in fair agreement with the unwinding angle as proposed by Fuller and Waring[5] ($\delta = -12°$) and with the result of Bauer and Vinograd.[33] On the other hand, somewhat higher values of the unwinding angle ranging from $\delta = -23$ to $\delta = -33°$ have been found with closed circular DNAs.[164-166] In circular DNAs, the intercalation may be subjected to some strains which are not present in linear DNA's.

Paoletti and Le Pecq,[163] on the contrary, have proposed a winding angle of $14°$ on the basis of static depolarization data. It should be noted, however, that the static anisotropy measures the time average of $r(t)$ as expressed by equation (23). If the anisotropy decay is computed with $\delta = 14°$, it is first higher and then lower than the experimental decay (Figure 10).[146] This could explain the results of Paoletti and Le Pecq.

The same technique has been applied to ethidium bromide complexed with double-stranded polynucleotides, poly d(A-T) and poly (rA-rU).[149,150] The unwinding angle for the poly d(A-T)—ethidium complex was found to be $-17 \pm 2°$.[149] This value is practically identical with the value obtained with calf thymus DNA,[146,147] suggesting that there is not much difference between the conformation of poly d(A-T) and the conformation of calf thymus DNA in solution. On the other hand, the unwinding angle for the poly (rA-rU)—ethidium complex was found to be $-38°$ by assuming that poly (rA-rU) possesses the A structure of RNA.[150] This result is consistent with the model building studies which predict that an ethidium molecule induces a larger unwinding angle in the case of the A structure than in the case of the B structure.[167]

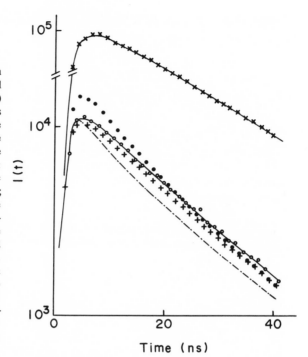

FIGURE 10. Comparison between the experimental (○) and computed $d(t)$ curves for the calf thymus DNA–ethidium bromide complex (P/D = 7.35). The computation was made with different values of δ: (●) δ = +14°; (−) δ = −20°; (+) δ = −12°; (−·−·−) δ = 0°. The upper curve shows the comparison between the experimental (×) and computed (−) $s(t)$ curves. The best-fit $s(t)$ function was found to be 0.12 exp $(-t/10\,\text{ns}) + 0.88\exp(-t/23\,\text{ns})$, where the amplitudes are normalized to unity. (Reproduced from Genest *et al.*, Ref. 146.)

# 6. Radiationless Energy Transfer

## 6.1. Singlet–Singlet Energy Transfer

### 6.1.1. Theoretical Background

Radiationless transfer of electronic excitation energy, which should be differentiated from the trivial process of direct reabsorption of emitted radiation, may occur between donor D and acceptor A molecules as much as 50–100 Å or more apart, provided that there is a reasonable overlap between the donor emission spectrum and the acceptor absorption spectrum. The theory of dipole–dipole energy transfer has been developed by Förster[159–162] and successfully applied to studies of conformational properties of biological macro-molecules, such as distances between two chromophores.[168–173] Stryer and Haugland[168,172] have developed energy transfer techniques as a spectroscopic ruler.

According to Förster's theory, the rate of energy transfer $k_{tr}$ and the quantum efficiency of transfer $E$ are given by[159-162]

$$k_{tr} = \frac{9000 \, (\ln 10)\kappa^2 \phi_D}{128\pi^5 n^4 N R^6 \tau_D} \qquad J(\tilde{\nu}) = \frac{1}{\tau_D}\left(\frac{R_0^6}{R}\right) \qquad (34)$$

$$E = R_0^6/(R_0^6 + R^6) \qquad (35)$$

where $n$ is the refractive index of the medium, $N$ is Avogadro's number, $\phi_D$ and $\tau_D$ are the fluorescence quantum yield and lifetime of the donor in the absence of the acceptor, respectively, $R$ is the distance between donor and acceptor chromophores, and $J(\tilde{\nu})$ is the spectral overlap integral of donor emission and acceptor absorption,

$$J(\tilde{\nu}) = \int_0^\infty F_D(\tilde{\nu})\epsilon_A(\tilde{\nu})\tilde{\nu}^{-4}\,d\tilde{\nu} \qquad (36)$$

where $F_D(\tilde{\nu})\,d\tilde{\nu}$ is the normalized fluorescence quantum intensity of the donor in the wave number range $\tilde{\nu}$ to $\tilde{\nu} + d\tilde{\nu}$, so that $\int_0^\infty F_D(\tilde{\nu})\,d\tilde{\nu} = 1$, and $\epsilon_A(\tilde{\nu})$ is the molar extinction coefficient of the acceptor at the wave number $\tilde{\nu}$. The quantity $\kappa^2$, the dipolar orientation factor, is defined by

$$\kappa^2 = (\cos\theta_{DA} - 3\cos\theta_D\cos\theta_A)^2 \qquad (37)$$

where $\theta_{DA}$ is the angle between the transition moments of the donor and acceptor, and $\theta_D$ and $\theta_A$ are the angles between these moments and the direction joining them, respectively. The quantity $R_0$ is the critical transfer distance at which the transfer efficiency is 50%:

$$R_0(\text{Å}) = (9.79 \times 10^3)[J(\tilde{\nu})\kappa^2\phi_D n^{-4}] \qquad (38)$$

The diffusion motion between donor and acceptor molecules affects the transfer rate. For a fluid solution of low viscosity under the condition $R_0 \ll (2D\tau_D)^{1/2}$, where $D$ is the sum of the diffusion coefficients of the donor and acceptor, the transfer rate is time independent. From the reaction scheme below,

$$\mathrm{D}^* \;+\; \mathrm{A} \;\xrightarrow{\;k_T\;}\; \mathrm{D} \qquad + \qquad \mathrm{A}^*$$

$$k_D \swarrow \quad \searrow k'_D \qquad\qquad\qquad k_A \swarrow \quad \searrow k'_A$$

$$\mathrm{D} + h\nu_D \quad \mathrm{D} \qquad\qquad\qquad \mathrm{A} + h\nu_A \quad \mathrm{A}$$

we have the following equations for the time dependence of the fluorescence intensity of D and the quantum efficiency of transfer E, respectively[174]:

$$I_D(t) = I_D(0) \exp\{-(1/\tau_D + k_T[A])t\} \tag{39}$$

$$E = k_T[A]/(k_D + k'_D + k_T[A]) = 1 - \tau/\tau_D \tag{40}$$

where $\tau_D = (k_D + k'_D)^{-1}$ and $\tau = (k_D + k'_D + k_T[A])^{-1}$ are the fluorescence lifetimes of the donor in the absence and presence of the acceptor, respectively.

For a rigid solution of high viscosity under the condition $R_0 \gg (2D\tau_D)^{1/2}$, where D and A molecules remain effectively stationary during the energy transfer, the transfer rate becomes time dependent. We obtain the following equations for the fluorescence response function of the donor and E, respectively[174]:

$$I_D(t) = I_D(0) \exp[-t/\tau_D - 2\gamma(t/\tau_D)^{1/2}] \tag{41}$$

$$E = \pi^{1/2}\gamma \exp(\gamma^2)(1 - \text{erf}\,\gamma) \tag{42}$$

The parameter $\gamma = [A]/[A]_0$ is the molar concentration expressed relative to the critical molar concentration $[A]_0$ of the acceptor, defined by

$$[A]_0 = 3000/2\pi^3 N R_0^3 \tag{43}$$

Irrespective of the kinetics, the quantum yield of the donor in the presence of A ($\phi$) expressed relative to that in the absence of A ($\phi_D$) is given by the following simple expression:

$$\phi/\phi_D = 1 - E \tag{44}$$

The predicted dependence of the transfer rate on $R^{-6}$ has been experimentally demonstrated,[168,175,176] and energy transfer techniques have been established as one of the powerful tools for estimating the distance between chromophores attached to macromolecules.[168,169,171,173] Since $\kappa^2$ can theoretically have any value between 0 and 4, the error introduced by the uncertainty in the orientation factor may introduce large errors in $R$. It is customary to assume the average value for $\kappa^2$, that is, $\kappa^2 = 2/3$. This is justified only if the orientations of donor and acceptor chromophores are random and completely averaged within the lifetime of the transfer process.[177–179] Therefore the question arises as to how to treat $\kappa^2$ if this condition is not met. Dale and Eisinger[177–179] have suggested a method for setting limits on the value of $\kappa^2$.

## 6.1.2. Applications of Energy Transfer

*6.1.2.a. tRNA Studies.* Beardsley and Cantor[169] were the first to apply energy transfer techniques to measure the distance between two chromophores attached to yeast $tRNA^{Phe}$. In their experiments, the energy donor was the naturally occurring fluorescent Y base and the energy acceptors were acridine dyes covalently attached to the 3'-end of the tRNA. From quantum yield measurements, they determined the transfer efficiency using equation (44), and showed that the distance between the Y base and a dye bound to the 3'-end of the tRNA is somewhere in the range 40–60 Å. The uncertainty for the value of $\kappa^2$ may cause some error in calculating $R_0$. The rotational correlation time of the Y base is shown to be considerably shorter than that of the entire tRNA molecule, suggesting that the orientation of the Y base is not rigidly fixed to the macromolecule.[142] Therefore Beardsley and Cantor[169] assumed an average value of 2/3 for $\kappa^2$.

Blumberg *et al.*[178] have examined critically the reorientation freedom of the Y base by analyzing both the steady-state and time-dependent anisotropy of the Y base fluorescence. From limiting values of $\kappa^2$, the separation between the Y base and acriflavine bound at the 3' terminus of yeast $tRNA^{Phe}$ is calculated to be between 34 and 61 Å, this range being independent of the particular model chosen to describe the reorientation freedom of the donor and acceptor. The upper limit is reasonably consistent with about 75 Å separation between the 3' terminus and the anticodon triplet determined by x-ray crystallographic studies.[140,141]

Previous work has shown that there is a single, strong ethidium binding site on yeast $tRNA^{Phe}$.[108,139] Singlet–singlet energy transfer has been used to measure the distance between this site and dansyl hydrazine attached to the 3'-end of *E. coli* unfractionated tRNA and $tRNA^{Phe}$.[180] The distance is found to be between 33 and 40 Å for both tRNA's. This is consistent with the results of NMR studies which placed the ethidium binding site of yeast $tRNA^{Phe}$ between base pairs 6 and 7 on the aminoacyl stem.[139]

The tertiary structure of tRNA has been deeply explored by Yang and Söll[181] by determining several distances between fluorescent labels attached to specific places in the tRNA molecule. They showed that tRNA in solution indeed has a cloverleaf-like conformation similar to the one in the crystal.[140,141] Based on energy transfer techniques, it may now be possible to detect conformational transitions in tRNA when it interacts with aminoacyl-tRNA synthetases or with the ribosome.

*6.1.2.b. Unwinding Angle of the DNA Helix.* Wahl and co-workers[109,146,147,149,150,158] have studied singlet–singlet energy transfer between intercalated dye molecules by measuring anisotropy decay. They showed that their anisotropy data are well explained by Förster's theory, and they estimated the unwinding angle of the DNA helix when ethidium bromide intercalated between DNA base pairs. The results have already been described in Section 5.4.

*6.1.2.c. Fluorescence Decay Kinetics Due to Dye–Dye Energy Transfer.* As found in the DNA–acridine orange[32, 81] and DNA–ethidium bromide com-

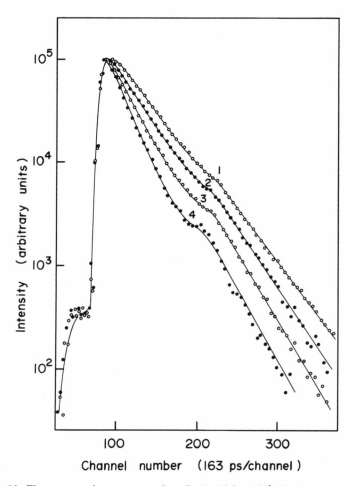

FIGURE 11. Fluorescence decay curves of proflavine ($4.8 \times 10^{-6}$ M) when bound to calf thymus DNA. The solvent was 0.005 M phosphate buffer (pH 6.9) at 23°C. The excitation and emission wavelengths were 430 and 500 nm, respectively. Solid lines are attempts to fit to $\exp(-t/\tau_D)$ for (1) P/D = 214 and to fit to $\exp[-t/\tau_D - 2\gamma(t/\tau_D)^{1/2}]$ for (2) P/D = 51, (3) P/D = 20, and (4) P/D = 11. The best fits were obtained when $\tau_D$ and $\gamma$ were the following: (1) $\tau_D = 6.40$ ns, (2) $\tau_D = 6.40$ ns and $\gamma = 0.07$, (3) $\tau_D = 6.40$ ns and $\gamma = 0.40$, and (4) $\tau_D = 6.40$ ns and $\gamma = 0.75$. (From Kubota and Steiner, Ref. 81.)

plexes,[7] energy transfer between bound dye molecules causes a decrease in both polarization and quantum yield. Here we describe another phenomenon of dye–dye energy transfer.

Kubota and Steiner[81] have shown that the fluorescence decay of proflavine bound to calf thymus DNA was exponential above P/D = 100 but markedly deviated from exponentiality with decreasing P/D ratio (Figure 11). The fluorescence quantum yield and mean lifetime are decreased with decreasing P/D ratio, whereas the fluorescence polarization remains almost unchanged above P/D = 15. In order to interpret this result, it has been assumed that excitation energy is efficiently transferred from the dye bound to the fluorescing site (A-T : A-T site) to the dye bound to the quenching site (G-C : G-C or G-C : A-T site). With the intercalation model,[4,42] the movement of the dye is restricted between base pairs during the excited lifetime. Then the fluorescence decay of proflavine in the presence of energy transfer is predicted to have the form expressed by equation (41). By assuming a set of values for $\tau_0$ and $\gamma$ in equation (41), several $I_D(t)$ functions are tested so that the numerically computed convolution with the response function of the pulse may fit the experimental decay curve. Figure 11 displays the calculated best-fit curves. If intercalated dye molecules are assumed to be uniformly distributed between base pairs, the decay data in Figure 11 leads to the following $R_0$ values: $R_0 = 38$ Å at P/D = 51, $R_0 = 29$ Å at P/D = 20, and $R_0 = 21$ Å at P/D = 11. The $R_0$ values obtained at P/D = 15–20, ranging from 27 to 29 Å, are in good agreement with the value ($R_0 = 26.4$ Å) calculated by using equation (38). This implies that the dye binding, on the average, may result in an isotropic distribution.

## 6.2. Nucleic Acid Base–Dye Energy Transfer: Sensitized Fluorescence

When dyes such as proflavine, acridine orange, and ethidium bromide complexed with nucleic acids are excited by ultraviolet light below 310 nm, sensitized fluorescence of the dyes can be observed.[7,32,182–184] This phenomenon has been attributed to energy transfer from the first excited singlet state of purine and pyrimidine bases to intercalated dye molecules.[32]

Two mechanisms for energy transfer have been proposed:[170] one takes place from a direct transfer from the nucleic acid base to the dye (Förster resonance transfer), and the other arises from the base-to-base energy migration along the nucleic acid chain (exciton transfer) which precedes the base-to-dye transfer.

In terms of the Förster theory for resonance excitation transfer between weakly coupled molecules,[159–162] it has been shown that bases which are located two to five base pairs away from dye molecules are capable of efficient singlet energy transfer[7,183] and that the measured effective transfer distances correspond to transfer times of approximately 20 ps.[183] Therefore transfer from the nucleic acid bases to the dye would be efficient if the fluorescence lifetime of DNA were comparable to the transfer time.

The rise time of sensitized fluorescence of acridine orange complexed with DNA has been directly measured by laser excitation to be less than 20 ps, the limit of the instrumentation, and possibly as short as a few picoseconds.[185] The result can be interpreted either in terms of exciton migration originating from base–base coupling or resonance excitation transfer.[184–186] Although there is no direct spectroscopic evidence for strong coupling between bases leading to an exciton band wider than the vibrational width of the spectra of the nucleic acid bases,[187] weak coupling leading to singlet energy migration along the DNA chain might not be excluded.

Rayner et al.[188] have studied the efficiency of singlet energy transfer from nucleic acid bases to quinacrine by fluorescence sensitization techniques. The quantum yield of energy transfer, $\phi_{ET}$, is the sum of the transfer efficiencies for each base, $q_{ET_i}$, multiplied by the probability that an $i$th base would be excited by an incident photon. Since the probability varies inversely with the P/D ratio, one has[188]

$$(P/D)\phi_{ET} = 4 \sum_{i=1}^{\infty} q_{ET_i} \qquad (45)$$

Exclusive contribution from the nearest-neighbor bases ($i = 1$) yields a maximum value of 4 for $(P/D)\phi_{ET}$, since $q_{ET_i} \leqslant 1$. It was found that the value of $(P/D)\phi_{ET}$ is constant over a P/D ratio of 5 to 43 and its average value is $4.8 \pm 0.7$.[188] In harmony with the calculated value for the critical transfer distance ($R_0 \leqslant 5.6$ Å), this result shows that sensitized dye fluorescence is mainly due to the short-range interactions between the dye and nearest-neighbor bases. It therefore appears that fluorescence sensitization due to exciton migration along the DNA chain is not important in the case of quinacrine.

Exchange and charge transfer interactions, in addition to resonance dipole–dipole interaction, are expected to contribute to the overall interaction since there is a nonnegligible overlap between the orbitals of the dye and nearest-neighbor bases.[188]

The relative orientations between the transition moments of the nucleic acid bases and of the dye affect the efficiency of energy transfer. Ander[189] has studied sensitized fluorescence in DNA-dye complexes after excitation by a tunable dye laser and found that energy transfer depends on the excitation wavelength and the base composition of DNA. Energy transfer models and some molecular structures for the intercalation complex have been proposed.[189]

The energy transfer between the nucleic acid bases and the dye is helpful in understanding photochemical reactions of nucleic acids[170] or photobiological actions of dyes.[3] For example, the decrease of quantum yield of thymine dimerization in DNA–dye complexes is related to the trapping of energy by the dye.[183] Laser photolysis in the picosecond or subpicosecond time scale may give deeper insight into the mechanism of energy transfer.

## 6.3. Triplet–Singlet and Triplet–Triplet Energy Transfers

All of the energy transfer experiments described above have been between singlets: $D^*(singlet) + A(singlet) \rightarrow D(singlet) + A^*(singlet)$. Provided that the acceptor transition is allowed and the absorption of the acceptor overlaps the donor luminescence, the following types of energy transfer also occur by dipole–dipole interaction:[174] $D^*(singlet) + A(triplet) \rightarrow D(singlet) + A^*(triplet)$, $D^*(triplet) + A(singlet) \rightarrow D(singlet) + A^*(singlet)$, and $D^*(triplet) + A(triplet) \rightarrow D(singlet) + A^*(triplet)$. Measurements of these types of energy transfer are usually possible in rigid matrices.

### 6.3.1. Triplet–Singlet Energy Transfer

Isenberg et al.[190] have observed sensitized delayed fluorescence of acridine orange and other acridine dyes bound to DNA in aqueous glasses at 77°K. The mechanism for the delayed fluorescence was attributed to triplet–singlet resonance transfer between the DNA triplet and the dye singlet.[190] The existence of sensitized delayed fluorescence has been confirmed for ethidium bromide complexed with DNA[7] and acridine dyes complexed with mononucleotides.[191]

Pearlstein et al.[192] have quantitatively interpreted the nonexponential phosphorescence decay of poly rA upon binding of ethidium bromide and proflavine in terms of triplet–singlet energy transfer based on Förster's mechanism. It was found that the decay data fit well the form predicted by the theory [equation (41)]. From decay analysis, the mean packing density of nucleotides in folded poly rA was estimated to be about $1\ \text{nm}^{-3}$.

### 6.3.2. Triplet–Triplet Energy Transfer

As reported for the DNA-9–aminoacridine complexes,[193] excitation energy transfer between DNA triplet and dye triplet induces a delayed phosphorescence. Another type of delayed fluorescence has been observed when the DNA–acridine dye complexes were excited by visible light.[194] The delayed fluorescence has been interpreted as the result of the triplet–triplet annihilation which arises from two dye triplets placed very close together.[194] The occurrence of these delayed emissions implies that there must be a close spatial proximity of the donor and acceptor, in agreement with the intercalation model.

# 7. Cytological Applications

## 7.1. Acridine Dyes

The aminoacridines, including proflavine, acridine orange, and acriflavine, have been extensively used as cytological stains because of their ability to differentiate between RNA and DNA and to stain the nuclei of living cells without killing them.[9,195,196] These properties have made them very useful as vital stains with a high degree of affinity for nucleic acids.

In the case of acridine orange the differentiation between bihelical and single-stranded RNA can be related to the characteristic features of the interaction of this dye with nucleic acids in solution.[196] For the range of nucleic acid/dye ratios typical of stained cells, intercalation is strongly favored over external binding for DNA, resulting in an intense green fluorescence for acridine orange complexes with cellular DNA. In the case of single-stranded nucleic acids, complexes in which the dye is externally bound and stacked are more favored, leading to red fluorescence.

Thermal treatment of chromosomes produces characteristic changes in their staining properties with acridine orange. Human lymphocyte chromosomes, when stained at temperatures below 60°C, exhibit only a uniform green fluorescence. Heating above 85°C produces chromosomes with red fluorescence. This presumably reflects the thermal denaturation of the DNA and the disruption of the bihelical structure to form single strands, which bind acridine orange as externally stacked complexes.[196] Intermediate temperatures produce a banded pattern, probably reflecting the preferential thermal denaturation of particular regions. Formaldehyde treatment, which blocks DNA reannealing, enhances chromosome banding and displaces its occurrence to lower temperatures.

If acridine orange staining indeed detects chromosomal regions which are susceptible to thermal denaturation, it would appear likely that these correspond to regions which are rich in A-T base pairs.[196] A confirmation of this view is provided by quinacrine staining. Under conditions which produced banding, quinacrine-positive regions displayed a red fluorescence when stained with acridine orange, while quinacrine-negative regions showed a green fluorescence. Since quinacrine fluorescence is quenched by G-C pairs,[63,64] these findings are consistent with the above model.

Another category of fluorescent acridine derivatives, which have been widely used as cytological stains, includes quinacrine and quinacrine mustard.[197–199] The characteristic fluorescent banding patterns of chromosomes stained with quinacrine mustard have been useful for the longitudinal differentiation of metaphase chromosomes. There is substantial evidence that the brilliantly fluorescent regions of quinacrine-stained chromosomes consist of regions with a high AT content, including clusters relatively free of G-C base pairs.[63,64,87,90]

## 7.2. Bisbenzimidazole Dyes

The bisbenzimidazole dye Hoechst 33258 has been found to possess distinctive staining properties for mammalian chromosomes.[200] The staining of metaphase chromosomes of the mouse, other than the Y chromosome, results in intense fluorescence of the centromere regions, with less intense fluorescence of the bulk of the chromatids. A faint banding pattern is sometimes observed. The Y chromosome exhibits diffuse bright fluorescence without differentiation of the centromere region.

In the metaphase chromosomes of the Algerian hedgehog, staining with Hoechst 33258 results in a qualitatively different pattern. While most of the chromosomal material exhibits medium or bright fluorescence, several individual chromosomes show extensive regions of weak fluorescence.[200] The weakly fluorescing regions have been attributed to autosomal constitutive heterochromatin. Since this is in contrast to the behavior observed with mouse chromosomes, this fluorochrome appears to be capable of discriminating between different forms of constitutive heterochromatin.

The specific quenching of the fluorescence of Hoechst 33258 by BrdU has been utilized by Latt and co-workers to analyze DNA replication in cell nuclei by a microfluorometric technique.[201-203] Since BrdU may replace thymine in newly synthesized DNA, this approach provides a means of detecting an uneven distribution of thymine. In the case of human leukocyte DNA into which BrdU has been incorporated for an entire replication cycle, a pronounced difference in the fluorescence intensity of parts of sister chromatids of the Y chromosome has been observed.[201]

This asymmetry of fluorescence of Y chromosomes from cells incorporating BrdU has been attributed to an uneven distribution of thymidine between the two DNA strands.[201] The strand with the greater amount of thymidine would act as a template for the intercorporation of less BrdU than its partner. The fluorescence of Hoechst 33258 combined with the chromatid containing the greater proportion of BrdU would be quenched to a greater extent than that of the corresponding region of its sister chromatid. In order to be detected microscopically by this means, the differences in base composition must persist over millions of base pairs.

Latt and co-workers[201-203] have developed a number of applications of the detection of DNA synthesis by fluorescence microscopy, using Hoechst 33258. The incorporation of BrdU has proven particularly useful for the differentiation of sister chromatids and the detection of sister chromatid exchange. When cells have undergone replication twice in a BrdU-containing medium, one sister chromatid will contain BrdU in both DNA strands, while the other chromatid will contain BrdU in only one DNA strand. Such chromatid pairs can be differentiated by Hoechst 33258 fluorescence. A single replication in a BrdU-containing medium followed by a replication in the absence of BrdU results in sister chromatids in which only one chromatid contains BrdU. In both the above cases the two sister chromatids will exhibit unequal fluorescence intensities when stained with Hoechst 33258. This property may be used to monitor the occurrence of sister chromatid exchange.[201–203]

Sister chromatid exchanges are very responsive to chromosome damage by many agents, such as mitomycin C. The techniques described above can thus be used to monitor the toxic effects of drugs upon chromosomes.

## References

1. A. Albert, *The Acridines*, Arnold, London (1966).
2. A. R. Peacocke, in: *Heterocyclic Compounds: Acridines* (R. M. Acheson, ed.), Vol. 9, Interscience, New York (1973), pp. 723–757.
3. E. R. Lochmann and A. Micheler, in: *Physico-Chemical Properties of Nucleic Acids* (J. Duchesne, ed.), Vol. 1, Academic Press, New York (1973), pp. 223–267.
4. L. S. Lerman, *J. Mol. Biol.* **3**, 18–30 (1961).
5. W. Fuller and M. J. Waring, *Ber. Bunsenges. Phys. Chem.* **68**, 805–808 (1964).
6. M. J. Waring, *J. Mol. Biol.* **13**, 269–282 (1965).
7. J. B. Le Pecq and C. Paoletti, *J. Mol. Biol.* **27**, 87–106 (1967).
8. B. A. Newton, in: *Metabolic Inhibitors* (R. M. Hochster and J. H. Quastel, eds.), Vol. 2, Academic Press, New York (1963), pp. 285–310.
9. F. Bukatsch and M. Haitinger, *Protoplasma* **34**, 515–523 (1940).
10. A. Krieg, *Experientia* **10**, 172–173 (1954).
11. J. A. Armstrong, *Exp. Cell Res.* **11**, 640–643 (1956).
12. R. I. DeMars, *Nature (London)* **172**, 964 (1953).
13. R. Dulbecco and M. Vogt, *Virology* **5**, 236–243 (1958).
14. E. Terzaghi, Y. Okada, G. Streisinger, J. Emrick, M. Inouye, and A. Tsugita, *Proc. Natl. Acad. Sci. USA* **56**, 500–507 (1966).
15. R. B. Webb and H. E. Kubitschek, *Biochem. Biophys. Res. Commun.* **13**, 90–94 (1963).
16. J. D. Spikes and R. Livingstone, in: *Advances in Radiation Biology* (L. Augenstine, R. Mason, and M. Zelle, eds.), Vol. 3, Academic Press, New York (1969), pp. 29–121.
17. F. W. Morthland, D. P. H. De Bruyn, and N. H. Smith, *Exp. Cell Res.* **7**, 201–214 (1954).
18. D. S. Drummond, V. F. W. Simpson-Gildemeister, and A. R. Peacocke, *Biopolymers* **3**, 135–153 (1965).
19. W. Müller, D. M. Crothers, and M. J. Waring, *Eur. J. Biochem.* **39**, 223–234 (1973).
20. J. Bontemps, C. Houssier, and E. Fredericq, *Biophys. Chem.* **2**, 301–315 (1974).
21. G. Löber and G. Achtert, *Biopolymers* **8**, 595–608 (1969).
22. Y. Kubota, Y. Eguchi, K. Hashimoto, M. Wakita, Y. Honda, and Y. Fujisaki, *Bull. Chem. Soc. Jpn.* **49**, 2424–2426 (1976).
23. G. Cohen and H. Eisenberg, *Biopolymers* **8**, 45–55 (1969).
24. J. Cairns, *Cold Spring Harbor Symp. Quant. Biol.* **27**, 311–318 (1962).
25. Y. Mauss, J. Chambron, M. Daune, and H. Benoit, *J. Mol. Biol.* **27**, 579–589 (1967).
26. D. S. Drummond, N. S. Pritchard, V. F. W. Simpson-Gildemeister, and A. R. Peacocke, *Biopolymers* **4**, 971–987 (1966).
27. L. Michaelis, *Cold Spring Harbor Symp. Quant. Biol.* **12**, 131–142 (1947).

28. R. F. Steiner and R. F. Beers, *Arch. Biochem. Biophys.* **81**, 75–92 (1959).
29. D. M. Neville, Jr., and D. F. Bradley, *Biochim. Biophys. Acta* **50**, 397–399 (1961).
30. K. Yamaoka and R. A. Resnik, *J. Phys. Chem.* **70**, 4051–4066 (1966).
31. A. Blake and A. R. Peacocke, *Biopolymers* **4**, 1091–1104 (1966).
32. G. Weill and M. Calvin, *Biopolymers* **1**, 401–417 (1963).
33. W. Bauer and J. Vinograd, *J. Mol. Biol.* **33**, 141–171 (1968).
34. W. Müller and F. Gautier, *Eur. J. Biochem.* **54**, 385–394 (1975).
35. S. A. Latt, G. Stetten, L. A. Juergens, H. F. Willard, and C. D. Scher, *J. Histochem. Cytochem.* **23**, 493–505 (1975).
36. S. A. Latt and J. C. Wohlleb, *Chromosoma (Berl.)* **52**, 297–316 (1975).
37. D. E. Comings, *Chromosoma (Berl.)* **52**, 229–243 (1975).
38. I. Hilwig and A. Gropp, *Exp. Cell Res.* **81**, 474–482 (1973).
39. R. F. Steiner and H. Sternberg, *Arch. Biochem. Biophys.* **197**, 580–588 (1979).
40. A. R. Peacocke and J. B. H. Skerrett, *Trans. Faraday Soc.* **52**, 261–279 (1956).
41. R. W. Armstrong, T. Kuruczec, and U. P. Strauss, *J. Am. Chem. Soc.* **92**, 3174–3181 (1970).
42. L. S. Lerman, *Proc. Natl. Acad. Sci. USA* **49**, 94–102 (1963).
43. L. S. Lerman, *J. Cell. Comp. Physiol. Suppl. 1* **64**, 1–18 (1964).
44. H. Bujard, *J. Mol. Biol.* **33**, 503–505 (1968).
45. J. C. Wang, *J. Mol. Biol.* **43**, 25–39 (1969).
46. C. C. Tsai, S. C. Jain, and H. M. Sobell, *Proc. Natl. Acad. Sci. USA* **72**, 628–632 (1975).
47. N. C. Seeman, R. O. Day, and A. Rich, *Nature (London)* **253**, 324–326 (1975).
48. T. D. Sakore, S. C. Jain, C. C. Tsai, and H. M. Sobell, *Proc. Natl. Acad. Sci. USA* **74**, 188–192 (1977).
49. H. S. Shieh, H. M. Berman, M. Dabrow, and S. Neidle, *Nucleic Acids Res.* **8**, 85–97 (1980).
50. M. Hogan, N. Dattagupta, and D. M. Crothers, *Biochemistry* **18**, 280–288 (1979).
51. M. Hogan, N. Dattagupta, and D. M. Crothers, *Proc. Natl. Acad. Sci. USA* **75**, 195–199 (1978).
52. M. Levitt, *Proc. Natl. Acad. Sci. USA* **75**, 640–644 (1978).
53. N. J. Pritchard, A. Blake, and A. R. Peacocke, *Nature (London)* **212**, 1360–1361 (1966).
54. H. J. Li and D. M. Crothers, *J. Mol. Biol.* **39**, 461–477 (1969).
55. J. Ramstein and M. Leng, *Biophys. Chem.* **3**, 234–240 (1975).
56. D. E. V. Schmenchel and D. M. Crothers, *Biopolymers* **10**, 465–480 (1971).
57. J. Ramstein, M. Ehrenberg, and R. Rigler, *Stud. Biophys.* **81**, 73–74 (1980).
58. J. Ramstein, M. Ehrenberg, and R. Rigler, *Biochemistry* **19**, 3938–3948 (1980).
59. D. M. Crothers, *Biopolymers* **6**, 575–584 (1968).
60. W. Bauer and J. Vinograd, *J. Mol. Biol.* **47**, 419–435 (1970).
61. J. D. McGhee and P. H. Hippel, *J. Mol. Biol.* **86**, 469–489 (1974).
62. J. C. Thomes, G. Weill, and M. Daune, *Biopolymers* **8**, 647–659 (1969).
63. U. Pachmann and R. Rigler, *Exp. Cell Res.* **72**, 602–608 (1972).
64. B. Weisblum and P. L. de Haseth, *Proc. Natl. Acad. Sci. USA* **69**, 629–632 (1972).
65. Y. Kubota, *Chem. Lett.,* 299–304 (1973).
66. J. P. Schreiber and M. P. Daune, *J. Mol. Biol.* **83**, 487–501 (1974).
67. V. Zanker, *Z. Phys. Chem.* **199**, 225–258 (1952).
68. D. F. Bradley and M. K. Wolf, *Proc. Natl. Acad. Sci. USA* **45**, 944–952 (1959).
69. A. L. Stone and D. F. Bradley, *J. Am. Chem. Soc.* **83**, 3627–3634 (1961).
70. R. F. Steiner, I. Weinryb, and R. Kolinski, *Biochim. Biophys. Acta* **209**, 306–319 (1970).

71. G. R. Haugen and W. H. Melhuish, *Trans. Faraday Soc.* **60**, 386–394 (1964).
72. J. Botts and M. Morales, *Trans. Faraday Soc.* **49**, 696–707 (1959).
73. R. Fowler and E. Guggenheim, *Statistical Thermodynamics,* Cambridge University Press, Cambridge (1939).
74. W. C. Galley and R. M. Purckey, *Proc. Natl. Acad. Sci. USA* **69**, 2198–2202 (1972).
75. C. R. Cantor and T. Tao, in: *Procedures in Nucleic Acid Research* (G. L. Cantoni and D. R. Davies, eds.), Vol. 2, Harper and Row, New York (1971), pp. 31–93.
76. R. F. Chen and H. Edelhoch (eds.), *Biochemical Fluorescence: Concepts,* Vol. 1, Marcel Dekker, New York (1975); Vol. 2, Marcel Dekker, New York (1976).
77. G. Löber, H. Schütz, and V. Kleinwächter, *Biopolymers* **11**, 2439–2459 (1972).
78. G. Weill, *Biopolymers* **3**, 567–572 (1965).
79. R. K. Tubbs, W. E. Ditmars, Jr., and Q. Van Winkle, *J. Mol. Biol.* **9**, 545–557 (1964).
80. L. M. Chan and J. A. McCarter, *Biochim. Biophys. Acta* **204**, 252–254 (1970).
81. Y. Kubota and R. F. Steiner, *Biophys. Chem.* **6**, 279–289 (1977).
82. Y. Kubota, K. Hirano, and Y. Motoda, *Chem. Lett.,* 123–126 (1978).
83. G. Baldini, S. Doglia, G. Sassi, and G. Lucchini, *Int. J. Biol. Macromol.* **3**, 248–252 (1981).
84. J. Ramstein and M. Leng, *Biochim. Biophys. Acta* **281**, 18–32 (1972).
85. S. Georghiou, *Photochem. Photobiol.* **22**, 103–109 (1975).
86. G. Duportail, Y. Mauss, and J. Chambron, *Biopolymers* **16**, 1397–1413 (1977).
87. D. J. Arndt-Jovin, S. A. Latt, G. Strinkler, and T. M. Jovin, *J. Histochem. Cytochem.* **27**, 87–95 (1979).
88. Y. Kubota, Y. Motoda, and Y. Fujisaki, *Chem. Lett.,* 237–240 (1979).
89. Y. Kubota and Y. Motoda, *Biochemistry* **19**, 4189–4197 (1980).
90. S. A. Latt, S. Brodie, and S. H. Munroe, *Chromosoma (Berl.)* **49**, 17–40 (1974).
91. S. A. Latt and S. Brodie, in: *Excited States of Biological Molecules* (J. B. Birks, ed.), Wiley, New York (1976), pp. 178–189.
92. A. Andreoni, R. Cubeddu, S. De Silvestri, and P. Laporta, *Opt. Commun.* **33**, 277–280 (1980).
93. R. D. Spencer and G. Weber, *J. Chem. Phys.* **52**, 1654–1663 (1970).
94. M. Shinitzky, *J. Chem. Phys.* **56**, 229–235 (1972).
95. P. Wahl, J. C. Auchet, and B. Donzel, *Rev. Sci. Instrum.* **45**, 28–32 (1974).
96. Y. Kubota and Y. Motoda, unpublished results.
97. A. Grinvald and I. Z. Steinberg, *Anal. Biochem.* **59**, 583–598 (1974).
98. M. G. Badea and S. Georghiou, *Photochem. Photobiol.* **24**, 417–423 (1976).
99. Y. Kubota and Y. Motoda, *Chem. Lett.,* 1375–1378 (1979).
100. Y. Kubota and Y. Motoda, *J. Phys. Chem.* **84**, 2855–2861 (1980).
101. T. R. Krugh, F. N. Wittlin, and S. P. Cramer, *Biopolymers* **14**, 197–210 (1975).
102. C. G. Reinhardt and T. R. Krugh, *Biochemistry* **17**, 4845–4854 (1978).
103. C. C. Tsai, S. C. Jain, and H. M. Sobell, *J. Mol. Biol.* **114**, 301–315 (1977).
104. S. C. Jain, C. C. Tsai, and H. M. Sobell, *J. Mol. Biol.* **114**, 317–331 (1977).
105. F. M. Pohl, T. M. Jovin, W. Baehr, and J. J. Holbrook, *Proc. Natl. Acad. Sci. USA* **69**, 3805–3809 (1972).
106. V. W. F. Burns, *Arch. Biochem. Biophys.* **133**, 420–424 (1969).
107. J. Olmsted, III, and D. R. Kearns, *Biochemistry* **16**, 3647–3654 (1977).
108. T. Tao, J. H. Nelson, and C. R. Cantor, *Biochemistry* **9**, 3514–3524 (1970).
109. D. Genest and P. Wahl, *Biochim. Biophys. Acta* **259**, 175–188 (1972).
110. F. Garland, D. E. Graves, L. W. Yielding, and H. C. Cheung, *Biochemistry* **19**, 3221–3226 (1980).
111. R. V. Kastrup, M. A. Young, and T. R. Krugh, *Biochemistry* **17**, 4855–4865 (1978).
112. J. B. Le Pecq, M. Le Bret, J. Barbet, and B. Roques, *Proc. Natl. Acad. Sci. USA* **72**, 2915–2919 (1975).

113. B. Gaugain, J. Barbet, R. Oberlin, B. P. Roques, and J. B. Le Pecq, *Biochemistry* **17**, 5071–5078 (1978).
114. B. Gaugain, J. Barbet, N. Cappelle, B. P. Roques, and J. B. Le Pecq, *Biochemistry* **17**, 5078–5088 (1978).
115. L. P. G. Wakelin, M. Romanos, T. K. Chen, D. Glaubiger, E. S. Canellakis, and M. J. Waring, *Biochemistry* **17**, 5057–5063 (1978).
116. J. Barbet, B. P. Roques, S. Combrisson, and J. B. Le Pecq, *Biochemistry* **15**, 2642–2650 (1976).
117. Y. Kubota, *Chem. Lett.*, 311–316 (1977).
118. Y. Kubota, H. Nakamura, M. Morishita, and Y. Fujisaki, *Photochem. Photobiol.* **27**, 479–481 (1978).
119. Y. Kubota, Y. Motoda, Y. Shigemune, and Y. Fujisaki, *Photochem. Photobiol.* **29**, 1099–1106 (1979).
120. G. Tomita, *Biophysik* **4**, 118–128 (1967).
121. G. Tomita, *Z. Naturforsch. Teil B* **23**, 922–925 (1968).
122. S. Yamabe, *Arch. Biochem. Biophys.* **130**, 148–155 (1969).
123. Y. Kubota and Y. Motoda, *Bull. Chem. Soc. Jpn.* **53**, 3468–3473 (1980).
124. A. Weller, *Prog. React. Kinet.* **1**, 189–214 (1961).
125. W. M. Vaughan and G. Weber, *Biochemistry* **9**, 464–473 (1970).
126. J. B. Birks, *Photophysics of Aromatic Molecules,* Wiley-Interscience, New York (1970), Chapter 7.
127. W. R. Ware, D. Watt, and J. D. Holmes, *J. Am. Chem. Soc.* **96**, 7853–7860 (1974).
128. M. H. Hui and W. R. Ware, *J. Am. Chem. Soc.* **98**, 4718–4727 (1976).
129. A. Weller, *Z. Elektrochem.* **61**, 956–961 (1957).
130. J. Reuben, B. M. Baker, and N. R. Kallenbach, *Biochemistry* **17**, 2915–2919 (1978).
131. A. Jablonski, *Bull. Acad. Pol. Sci. Ser. Sci. Math. Astr. Phys.* **8**, 259–264 (1960).
132. A. Jablonski, *Z. Naturforsch. Teil A* **16**, 1–4 (1961).
133. G. G. Belford, R. L. Belford, and G. Weber, *Proc. Natl. Acad. Sci. USA* **69**, 1392–1393 (1972).
134. T. J. Chuang and K. B. Eisenthal, *J. Chem. Phys.* **57**, 5094–5097 (1972).
135. M. Ehrenberg and R. Rigler, *Chem. Phys. Lett.* **14**, 539–544 (1972).
136. T. Tao, *Biopolymers* **8**, 609–632 (1969).
137. P. Wahl, in: *Biochemical Fluorescence: Concepts* (R. F. Chen and H. Edelhoch, eds.), Vol. 1, Marcel Dekker, New York (1975), pp. 1–41.
138. F. Perrin, *J. Phys. Radium* **5**, 497–511 (1934).
139. C. R. Jones and D. R. Kearns, *Biochemistry* **14**, 2660–2665 (1975).
140. F. L. Suddath, G. J. Quigley, A. McPherson, D. Sneden, J. J. Kim, S. H. Kim, and a. Rich, *Nature (London)* **248**, 20–24 (1974).
141. J. D. Robertus, J. E. Ladner, J. T. Finch, D. Rhodes, R. S. Brown, B. F. C. Clark, and A. Klug, *Nature (London)* **250**, 546–551 (1974).
142. K. Beardsley, T. Tao, and C. R. Cantor, *Biochemistry* **9**, 3524–3532 (1970).
143. N. F. Ellerton and I. Isenberg, *Biopolymers* **8**, 767–786 (1969).
144. Y. Kubota, *Bull. Chem. Soc. Jpn.* **46**, 2630–2633 (1973).
145. P. Wahl, J. Paoletti, and J. B. Le Pecq, *Proc. Natl. Acad. Sci. USA* **65**, 417–421 (1970).
146. D. Genest, P. Wahl, and J. C. Auchet, *Biophys. Chem.* **1**, 266–278 (1974).
147. D. Genest and P. Wahl, *Biophys. Chem.* **7**, 317–323 (1978).
148. Y. Kubota and R. F. Steiner, *Bull. Chem. Soc. Jpn.* **50**, 1502–1505 (1977).
149. J. L. Tichadou, D. Genest, P. Wahl, and G. Aubel-Sadron, *buophys. Chem.* **3**, 142–146 (1975).
150. P. Wahl, D. Genest, and J. L. Tichadou, *Biophys. Chem.* **6**, 311–319 (1977).
151. M. D. Barkley and B. H. Zimm, *J. Chem. Phys.* **70**, 2991–3007 (1979).

152. S. A. Allison and J. M. Schurr, *Chem. Phys.* **41**, 35–59 (1979).
153. D. P. Millar, R. J. Robbins, and A. H. Zewail, *Proc. Natl. Acad. Sci. USA* **77**, 5593–5597 (1980).
154. D. P. Millar, R. J. Robbins, and A. H. Zewail, in: *Picosecond Phenomena II* (R. Hochstrasser, W. Kaiser, and C. V. Shank, eds.), Springer Series in Chemical Physics Vol. 14, Springer, Berlin (1980), pp. 331–335.
155. J. C. Thomas, S. A. Allison, C. J. Appellof, and J. M. Schurr, *Biophys. Chem.* **12**, 177–188 (1980).
156. D. P. Millar, R. J. Robbins, and A. H. Zewail, *J. Chem. Phys.* **74**, 4200–4201 (1981).
157. T. Förster, *Fluoreszenz Organischer Verbindungen,* Vandenhoek and Ruprecht, Göttingen (1951), pp. 160–180.
158. D. Genest and P. Wahl, in: *Dynamical Aspects of Conformation Changes in Biological Macromolecules* (C. Sadron, ed.), Reidel, Dordrecht (1973), pp. 367–379.
159. T. Förster, *Ann. Phys.* **2**, 55–75 (1948).
160. T. Förster, *Z. Naturforsch. Teil A* **4**, 321–327 (1949).
161. T. Förster, *Discuss. Faraday Soc.* **27**, 7–17 (1959).
162. T. Förster, in: *Modern Quantum Chemistry* (O. Sinanoglu, ed.), Academic Press, New York (1965), pp. 93–137.
163. J. Paoletti and J. B. Le Pecq, *J. Mol. Biol.* **59**, 43–62 (1971).
164. J. C. Wang, *J. Mol. Biol.* **89**, 783–801 (1974).
165. D. C. Pulleyblank and A. R. Morgan, *J. Mol. Biol.* **91**, 1–13 (1975).
166. L. F. Liu and J. C. Wang, *Biochim. Biophys. Acta* **395**, 405–412 (1975).
167. W. J. Pigram, W. Fuller, and L. D. Hamilton, *Nature New Biol.* **235**, 17–19 (1972).
168. L. Stryer and R. P. Haugland, *Proc. Natl. Acad. Sci. USA* **58**, 719–726 (1967).
169. K. Beardsley and C. R. Cantor, *Proc. Natl. Acad. Sci. USA* **65**, 39–46 (1970).
170. M. Gueron and R. G. Shulman, *Annu. Rev. Biochem.* **37**, 571–596 (1968).
171. I. Z. Steinberg, *Annu. Rev. Biochem.* **40**, 83–114 (1971).
172. L. Stryer, *Annu. Rev. Biochem.* **47**, 819–846 (1978).
173. H. Fairclough and C. R. Cantor, *Methods Enzymol.* **48F**, 347–379 (1978).
174. J. B. Birks, *Photophysics of Aromatic Molecules,* Wiley-Interscience, New York (1970), Chapter 11.
175. S. A. Latt, H. T. Cheung, and E. R. Blout, *J. Am. Chem. Soc.* **87**, 995–1003 (1965).
176. H. Bucher, K. H. Drexhage, M. Fleck, H. Kuhn, D. Mobius, F. P. Schafer, J. Sondermann, W. Sperling, P. Tillmann, and J. Wiegand, *Mol. Cryst.* **2**, 199–230 (1967).
177. R. E. Dale and J. Eisinger, *Biopolymers* **13**, 1573–1605 (1974).
178. W. E. Blumberg, R. E. Dale, J. Eisinger, and D. M. Zukerman, *Biopolymers* **13**, 1607–1620 (1974).
179. R. E. Dale and J. Eisinger, in: *Biochemical Fluorescence: Concepts* (R. F. Chen and H. Edelhoch, eds.), Vol. 1, Marcel Dekker, New York (1975), pp. 115–284.
180. B. D. Wells and C. R. Cantor, *Nucleic Acids Res.* **4**, 1667–1680 (1977).
181. C. H. Yang and D. Söll, *Proc. Natl. Acad. Sci. USA* **71**, 2838–2842 (1974).
182. J. C. Sutherland and B. M. Sutherland, *Biopolymers* **9**, 639–653 (1970).
183. M. Kaufmann and G. Weill, *Biopolymers* **10**, 1983–1987 (1971).
184. F. Van Nostrand and R. M. Pearlstein, *Chem. Phys. Lett.* **39**, 269–272 (1976).
185. S. L. Shapiro, A. J. Campillo, V. H. Kollman, and W. B. Goad, *Opt. Commun.* **15**, 308–310 (1975).
186. A. Anders, *Opt. Commun.* **26**, 339–342 (1978).
187. M. Gueron, J. Eisinger, and R. G. Shulman, *J. Chem. Phys.* **47**, 4077–4091 (1967).
188. D. M. Rayner, A. G. Szabo, R. O. Loutfy, and R. W. Yip, *J. Phys. Chem.* **84**, 289–293 (1980).

189. A. Anders, *Appl. Phys.* **18**, 333–338 (1979).
190. I. Isenberg, R. B. Leslie, S. L. Baird, Jr., R. Rosenbluth, and R. Bersohn, *Proc. Natl. Acad. Sci. USA* **52**, 379–387 (1964).
191. Y. Kubota, *Bull. Chem. Soc. Jpn.* **43**, 3126–3130 (1970).
192. R. M. Pearlstein, F. Van Nostrand, and J. A. Nairn, *Biophys. J.* **26**, 61–72 (1979).
193. W. C. Galley, *Biopolymers* **6**, 1279–1296 (1968).
194. Y. Kubota, *Bull. Chem. Soc. Jpn.* **43**, 3121–3125 (1970).
195. P. P. H. De Bruyn, R. C. Robertson, and R. S. Farr, *Anat. Rec.* **108**, 279–307 (1950).
196. M. Bobrow, *Cold Spring Harbor Symp. Quant. Biol.* **38**, 435–440 (1973).
197. T. Caspersson, J. Lindsten, G. Lomakka, A. Moller, and L. Zech, *Int. Rev. Exp. Pathol.* **11**, 1–72 (1972).
198. B. Dutrillaux, in: *Molecular Structure of Human Chromosomes* (J. J. Yunis, ed.), Academic Press, New York (1977), pp. 233–266.
199. S. A. Latt, *Annu. Rev. Biophys. Bioeng.* **5**, 1–37 (1976).
200. I. Hilwig and A. Gropp, *Exp. Cell Res.* **75**, 122–126 (1972).
201. S. A. Latt, R. L. Davidson, M. S. Lin, and P. S. Gerald, *Exp. Cell Res.* **87**, 425–429 (1974).
202. M. S. Lin, S. A. Latt, and R. L. Davidson, *Exp. Cell Res.* **87**, 429–433 (1974).
203. S. A. Latt, *Proc. Natl. Acad. Sci. USA* **71**, 3162–3166 (1974).

# Index